Electrical
Double Layers
in
Biology

Electrical Double Layers in Biology

Edited by
Martin Blank
Department of Physiology
Columbia University
New York, New York

PLENUM PRESS • NEW YORK AND LONDON

Library of Congress Cataloging in Publication Data

Electrochemical Society Symposium on Electrical Double Layers in Biology (1985:
Toronto, Canada)
  Electrical double layers in biology.

  "Proceedings of the Electrochemical Society Symposium on Electrical Double Layers
in Biology, held May 14–15, 1985, in Toronto, Canada"—T.p. verso.
  Includes bibliographies and index.
  1. Biochemistry—Congresses. 2. Bilayer lipid membranes—Congresses. I. Blank,
Martin, date. II. Title.
QP517.B53S27   1985   574.19′283   85-31708
ISBN 0-306-42218-2

Proceedings of the Electrochemical Society Symposium on Electrical Double Layers in Biology,
held May 14–15, 1985, in Toronto, Canada

© 1986 Plenum Press, New York
A Division of Plenum Publishing Corporation
233 Spring Street, New York, N.Y. 10013

PREFACE

    A number of apparently unrelated phenomena in biological systems (e.g., biopolymer aggregation, cell-cell interactions, ion transport across membranes) arise from the special properties of charged surfaces. A symposium entitled "Electrical Double Layers in Biology", which took place at the Toronto meeting of the Electrochemical Society, 12-17 May 1985, focused on the common features of these phenomena. The papers presented at that symposium are collected here and they illustrate ways in which an understanding of electrical double layers can elucidate a problem in Biology. An example of this approach can be seen from the paper I presented on ion transport and excitation, where the "unusual" ion flows during nerve excitation are actually expected if one includes the effects of electrical double layers at membrane surfaces. Furthermore, the selectivity of the ion channels in these membranes can be better understood on this basis. Other presentations account for such observations as the changes in spacing between muscle proteins during contraction, the interactions of red cells to form rouleaux, the electrical properties of algal cell membranes, electrokinetic potentials during blood flow in arteries, etc.

    I trust that these papers will indicate the value of electrochemistry in the study of biological systems, an area of research usually called Bioelectrochemistry, and will encourage biologists to use these ideas when approaching related problems.

Martin Blank
Biological Sciences Division
Office of Naval Research

CONTENTS

DONNAN POTENTIAL AND SURFACE POTENTIAL OF A CHARGED MEMBRANE AND EFFECT

OF ION BINDING ON THE POTENTIAL PROFILE

Shinpei Ohki and Hiroyuki Ohshima

Department of Biophysical Sciences
School of Medicine
State University of New York at Buffalo
Buffalo, New York 14214 U.S.A.

ABSTRACT

A model is presented for the electrical potential distribution across a charged biological membrane which is in equilibrium with an electrolyte solution. We assume that a membrane has charged surface layers of thickness d on both surfaces of the membrane, where the fixed charges are distributed at a uniform density N within the layers, and that these charged layers are permeable to electrolyte ions. It is demonstrated that this model smoothly unites two different concepts, the Donnan potential and the surface potential (or the Gouy-Chapman double layer potential). Namely, the present model leads to the Donnan potential when $d \gg 1/\kappa'$ (κ' is the Debye-Hückel parameter of the surface charge layer) and to the surface potential as $d \to 0$, keeping the product Nd constant. It is also shown that the potential distribution depends significantly on the thickness d of the surface charge layer when $d \lesssim 1/\kappa'$, as well as on the types of ion binding to the charge sites of the surface charge layer. The conventional method to estimate surface potential and surface charge density from cell electrophoretic mobility by assuming the zero thickness of the surface charge layer may result in largely underestimated values of those for the membrane having the non-zero thickness of the surface charge layer.

INTRODUCTION

When a membrane possessing fixed charges is in equilibrium with an electrolyte solution, an electric potential difference is generally established between the membrane and the solution.

There have been two entirely different approaches to describe this potential difference (Davies and Rideal, 1961). One approach considers the potential difference to be the Donnan equilibrium potential. This approach has been used particularly for studies of ion transport processes through membranes (Toerell, 1953). In the other approach it is regarded as the surface potential, i.e., the Gouy-Chapman diffused double layer

1

potential, which is familiar in Colloid Sciences (Verwey and Overbeek, 1948).

When a membrane which is permeable to electrolyte ions and contains fixed negative charge groups at a uniform density N, is in equilibrium with a symmetrical electrolyte solution of concentration n and valence Z, the Donnan potential ψ_D relative to the external bulk solution is expressed as (Davies and Rideal, 1961; Ohki, 1965):

$$\psi_D = -\frac{kT}{Ze} \ln \left[\frac{N}{2Zn} + \left\{ \left(\frac{N}{2Zn}\right)^2 + 1 \right\}^{1/2}\right] = -\frac{kT}{Ze} \text{ arc sinh } \left(\frac{N}{2Zn}\right), \quad (1)$$

where k is the Boltzmann constant, T the absolute temperature, and e the elementary electric charge. Conventionally, the Donnan potential is considered to be accompanied by a discontinuous potential gap across the membrane surface (Fig. 1(a)). This is due to the assumption of local electroneutrality, which is used in the classical derivation of Eq. (1). Replacing this assumption by the Poisson-Boltzmann equation, Mauro (1962) has shown that the Donnan potential diffuses over the distances of order $1/\kappa$ (κ is the Debye-Hückel parameter) on both sides of the membrane surface as shown in Fig. 1(b), and has a continuous nature across the membrane surface.

If, on the other hand, it is assumed that all the fixed membrane charges are located only at the membrane surface and the electrolyte ions can not penetrate into the membrane, then the electrical double layer is considered to be formed around the surface (Fig. 1(c)) and the surface potential relative to the bulk solution takes the following form (Verwey and Overbeek, 1948; Davies and Rideal, 1961), which is quite different from Eq. (1):

$$\psi_s = \frac{2kT}{Ze} \text{ arc sinh } \left[\frac{\sigma}{(8n\epsilon_r \epsilon_o kT)^{1/2}}\right], \quad (2)$$

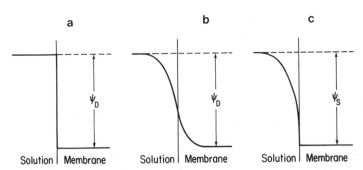

Fig. 1. Schematic representation of the potential distribution near the membrane surface by the hitherto proposed models (a) Classical (discontinuous) Donnan potential. (b) Continuous Donnan potential (Mauro, 1962). (c) Surface potential.

where σ is the surface charge density of the membrane, ε_r the relative permittivity of the solution, and ε_o the permittivity of a vacuum. Several modifications on the surface potential approach have been attempted by allowing the surface region to be permeable to electrolyte ions in relation to cell-cell interactions (Parsegian and Gingell, 1973; Parsegian, 1974) and to cell electrophoresis (Haydon, 1961; Donath and Pastushenko, 1979; Wunderlich, 1982, Levine, et al., 1983).

In spite of the above-mentioned improvements or modifications on each of the two different approaches, there remained ambiguity as to the inter-relation or transition between the Donnan potential and the surface potential.

Ohshima and Ohki (1985) have recently shown that these Donnan and surface potential concepts do smoothly make transition from each other in their potential distributions across a model charge membrane, in which the membrane fixed charges are uniformly distributed through a layer of finite thickness at the membrane surface and the charge layer is permeable to ions. Here, in the first section, we shall review this for the case of uni-univalent electrolyte and non-ion binding. In the following section, we extend the theory to the case where electrolyte ions in the solution can bind to the fixed charge sites of the charge layer and examine how the degree of the ion binding affects the membrane potential profile across such an interfacial region of the charged layer. Since most biological membranes are surrounded by electrolytes containing various monovalent and divalent ions and these ions may bind to the membrane fixed charge sites, the potential distribution across a membrane having a fixed charged layer was studied in terms of ionic concentrations (2,1-1 electrolytes) and ion bindings to the fixed charge sites. A relevant implication of such a theory to biological membrane studies is discussed in regards to the surface charge and the surface potential of cell surfaces obtained from cell electrophoresis.

THEORY

a) 1-1 electrolyte and non-ion binding case

We suppose a planar charge membrane which is in equilibrium with a large volume of a uni-univalent electrolyte solution of concentration n. We choose the x-axis in the direction normal to the membrane surface so that the plane at x = 0 coincides with the left boundary between the membrane and the solution (Fig. 2). We consider the membrane to be composed of three layers: two identical surface layers of thickness d which contain negatively charged groups at a uniform density -eN and are permeable to electrolyte ions ($0 \le x \le d$ and $h + d \le x \le h + 2d$), and one core layer of thickness h which has no fixed charges and is impermeable to ions ($d \le x \le h + d$). Since the membrane considered here is symmetrical with respect to the plane x = h/2 + d, we need to consider only the region $-\infty < x \le h/2 + d$.

We assume that the electric potential $\psi(x)$ at position x in the regions x < 0 and 0 < x < d (relative to the bulk solution (x = $-\infty$)) satisfies the Poisson-Boltzmann equation (in SI units),

$$\frac{d^2\psi}{dx^2} = -\frac{\rho(x)}{\varepsilon_r\varepsilon_o} = -\frac{en}{\varepsilon_r\varepsilon_o}(e^{-\frac{e\psi}{kT}} - e^{\frac{e\psi}{kT}}) = \frac{2en}{\varepsilon_r\varepsilon_o}\sinh\frac{e\psi}{kT}, \quad x < 0, \qquad (3)$$

3

$$\frac{d^2\psi}{dx^2} = -\frac{\rho(x)}{\varepsilon_r'\varepsilon_o} = -\frac{en}{\varepsilon_r'\varepsilon_o}\left(e^{-\frac{e\psi}{kT}} - e^{\frac{e\psi}{kT}}\right) + \frac{eN}{\varepsilon_r\varepsilon_o}, \quad 0 < x < d, \tag{4}$$

where ε_r and ε_r' are the relative permittivities of the solution and of the surface charge layer, respectively. For the region $d < x < h/2 + d$, in which there are no true charges, we have

$$\frac{d^2\psi}{dx^2} = 0 \quad , \quad d < x < h/2 + d. \tag{5}$$

The boundary conditions are:

$\psi(x)$ is continuous at $x = 0$ and $x = d$, \hfill (6)

$$\varepsilon_r \frac{d\psi}{dx}\bigg|_{-0} = \varepsilon_r' \frac{d\psi}{dx}\bigg|_{+0} \quad , \tag{7}$$

$$\varepsilon_r' \frac{d\psi}{dx}\bigg|_{d-0} = \varepsilon_r'' \frac{d\psi}{dx}\bigg|_{d+0} \quad , \quad \text{and} \tag{8}$$

$$\psi(x) \xrightarrow[x \to -\infty]{} 0 \tag{9}$$

where ε_r'' is the relative permittivity of the core layer; and because of symmetry of the membrane we have

$$\frac{d\psi}{dx}\bigg|_{h/2 + d} = 0 \quad . \tag{10}$$

Integrating Eq. (5) and using Eq. (10), we have

$$\frac{d\psi}{dx} = 0 \quad , \quad d < x \le h/2 + d, \tag{11}$$

which can also be obtained directly from the fact that due to symmetry there is no electric field within the core layer. It follows from Eq. (11) that the right-hand side of Eq. (8) becomes zero, and thus the boundary condition (8) may be replaced by

$$\frac{d\psi}{dx}\bigg|_{d-0} = 0. \tag{12}$$

The solution of Eqs. (3) and (4) subject to the boundary conditions (6), (7), (9), and (12) completely determines the potential function $\psi(x)$ in our system.

We now introduce the following dimensionless potential:

$$y = \frac{e\psi}{kT} \quad . \tag{13}$$

Then, Eq. (3) subject to the boundary condition (9) can be easily integrated to give

$$y = 4 \text{ arc tanh } [\tanh \frac{y_0}{4} e^{-\kappa|x|}], \quad x < 0, \tag{14}$$

4

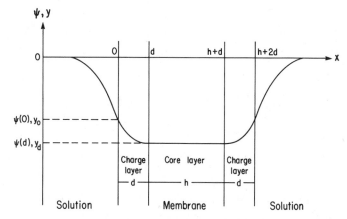

Fig. 2. Schematic representation of the potential distribution
$\psi(x)$ across a membrane with two surface charge layers
(of thickness d) and one core layer (of thickness h).

where κ is the Debye-Hückel parameter of the solution:

$$\kappa = (\frac{2ne^2}{\epsilon_r \epsilon_o kT})^{1/2} \tag{15}$$

and y_0 is defined as

$$y_0 = y(0) = \frac{e}{kT} \psi(0) \tag{16}$$

From eq. (14), we obtain:

$$\frac{dy}{dx}\bigg|_{-0} = 2 \kappa \sinh \frac{y_0}{2} , \tag{17}$$

Integration of Eq. (4) subject to the boundary condition (12) yields

$$\frac{dy}{dx} = \mp\kappa' \ [2(\cosh \frac{y}{y} - \cosh \frac{y_d}{y_d}) + \frac{N}{n} (y - y_d)]^{1/2}, \ 0 < x < d \tag{18}$$

where κ' is the Debye-Hückel parameter of the surface charge layer:

$$\kappa' = (\frac{2ne^2}{\epsilon_r' \epsilon_o kT})^{1/2} \tag{19}$$

and y_d is defined by

$$y_d = y(d) = \frac{e}{kT} \psi(d). \tag{20}$$

Using Eqs. (17) and (18), we obtain

$$2 \sinh \frac{y_0}{2} = - (\frac{\epsilon_r'}{\epsilon_r})^{1/2}[2(\cosh y_0 - \cosh y_d) + \frac{N}{n}(y_0 - y_d)]^{1/2}. \tag{21}$$

Eq. (18) can further be integrated to give

$$\kappa'd = \int_{y_d}^{y_0} \frac{dy}{[2(\cosh y - \cosh y_d) + \frac{N}{n}(y - y_d)]^{1/2}} \quad . \tag{22}$$

LIMITING CASES OF $d \to \infty$ AND $d \to 0$

Case of $d \to \infty$.

This is the case where the membrane itself is a semi-infinite charge layer.

When $d \to \infty$, the region $0 < x < d$ where Eq. (4) is applicable is extended to the region $0 < x < + \infty$, so that

$$\frac{d^2 y}{dx^2} = \kappa'^2 (\sinh y + \frac{N}{2n}), \qquad 0 < x < + \infty \tag{23}$$

and the total potential difference across the membrane $\psi(d)$ (or y_d) becomes equal to $\psi(+ \infty)$ (or $y(\infty)$). Noting that $d^2 y/dx^2 = 0$ at $x = + \infty$, from Eq. (23) we obtain

$$y_\infty = y(\infty) = - \text{ arc sinh } \frac{N}{2n} , \tag{24}$$

which agrees with the Donnan potential (Eq. (1)). We note that Eq. (24) does not depend on ε_r or ε_r'.

Case of $d \to 0$

This is the case where all of the charged sites are located on the membrane surface. Noting that $y - y_d \ll 1$ $(0 < x < d)$ when $\kappa'd \ll 1$, we put $y = y_d + \Delta y$ $(\Delta y \ll 1)$ in Eq. (18) and its integrated form Eq. (22), and linearize them with respect to Δy and solve the differential equation with respect to Δy under the boundary conditions (6), (7), (8). Then, we find

$$\kappa'd = 2 \left[\frac{y_0 - y_d}{2 \sinh y_d + N/(n)} \right]^{1/2}, \qquad \kappa'd \ll 1, \text{ and} \tag{25}$$

$$2 \sinh \frac{y_0}{2} = - \left[(\frac{\varepsilon_r'}{\varepsilon_r})(2 \sinh y_d + \frac{N}{n})(y_0 - y_d) \right]^{1/2}, \text{ respectively.} \tag{26}$$

From Eqs. (25) and (26), we find

$$2 \sinh \frac{y_0}{2} = - \frac{eNd}{(2n \varepsilon_r \varepsilon_0 kT)^{1/2}} - \kappa d \sinh y_d \tag{27}$$

If we take the limit $d \to 0$ in Eq. (27), keeping the product Nd constant, i.e. keeping the total amount of membrane fixed charges - eNd contained in a unit area of the surface charge layer $(0 < x < d)$ constant, then the second term on the right hand side of Eq. (27) becomes negligible and we obtain the following limiting result, which is identical to Eq. (2) (the surface potential):

6

$$y_o = y_d = 2 \ arc \ sinh \ [\frac{\sigma}{(8n \ \epsilon_r \ \epsilon_o kT)^{1/2}}], \qquad (28)$$

where we have defined σ as

$$\sigma = - e \lim_{\substack{d \ \to \ 0 \\ Nd-const.}} (Nd), \qquad (29)$$

which can be interpreted as the surface charge density of the membrane.

TRANSITION BETWEEN THE DONNAN AND SURFACE POTENTIALS

In order to see the transition between these two limiting cases, we consider here the potential distribution for $\kappa'd \leqslant 1$. In order to perform numerical calculations of $y(x)$ ($0 < x < d$) in the range of $\kappa'd \leqslant 1$, we expand $y(x)$ ($0 < x < d$) around $x = d$ in powers of $\kappa'(x-d)$ (The Taylor expansion series).

Since all terms with the odd power of $\kappa'(x-d)$ vanish due to the boundary condition of Eq. (12), $y(x)$ can be written as

$$y(x) = y_d + \sum_{n=1}^{\infty} A_n [\kappa'(x-d)]^{2n}, \qquad 0 < x < d. \qquad (30)$$

where the coefficients A_n can be obtained easily from Eq. (23) and is expressed in terms of N, n and y_d.

Some such expressions for A_n ($n = 1,2,3$) are given below

$$A_1 = \frac{1}{2} B_1$$

$$A_2 = \frac{1}{24} B_1 B_2$$

$$A_3 = \frac{1}{720} B_1 (3B_1 B_3 + B_2^2)$$

where

$$B_1 = sinh \ y_d + \frac{N}{2n}$$

$$B_2 = cosh \ y_d$$

$$B_3 = sinh \ y_d .$$

As $x \to 0$, Eq. (30) becomes

$$y_0 = y_d + \sum_{n=1}^{\infty} A_n (\kappa'd)^{2n} . \qquad (31)$$

By combining Eq. (31) with Eq. (21), it is possible to obtain numerical values of y_0 and y_d, which in turn determine $y(x)$ (or $\psi(x)$) in the regions $x < 0$ and $0 < x < d$, respectively.

Fig. 3 shows numerical results for $\psi(x)$ when d = 0, 5, and 10 $\overset{\circ}{A}$. Here we have varied d, keeping the product Nd constant, i.e., keeping the total amount of membrane fixed charges $-eNd$ per unit area of the surface charge layer (0 < x < d) constant, and using the value $-eNd$ = -0.190 $C \cdot m^{-2}$ so as to give $\psi(d)$ = -120 mV, as d → 0. The values of the other parameters were n = 0.1 M, T = 298 K and $\varepsilon_r'/\varepsilon_r$ = 1 and 0.5, respectively. The value of $1/\kappa'$ then becomes 9.6 $\overset{\circ}{A}$ for $\varepsilon_r'/\varepsilon_r$ = 1 and 6.8 $\overset{\circ}{A}$ for $\varepsilon_r'/\varepsilon_r$ = 0.5.

Fig. 3 shows a strong dependence of $\psi(x)$ on the thickness d of the surface charge layer. We also see from Fig. 3 that the potential drop within the membrane is sharper for smaller ε_r', which gives rise to less negative values of $\psi(0)$ and more negative values of $\psi(d)$.

The calculated potentials at two particular planes x = 0 and x = d, that is, $\psi(0)$ and $\psi(d)$ are plotted in Fig. 4 as a function of d when $\varepsilon_r'/\varepsilon_r$ = 1 for $-eNd$ = -0.190, -0.0842, -0.0421 and -0.0229 $C \cdot m^{-2}$ (which, at d = 0, give $\psi(0)$ = $\psi(d)$ = -120, -80, -50 and -30 mV, respectively), where the values of the other parameters were chosen to be the same as those used in Fig. 3. Fig. 4 shows that the Donnan potential does make a continuous transition to the surface potential as d decreases to zero, and $\psi(0)$ and $\psi(d)$ at d ≳ 5-10 $\overset{\circ}{A}$ take values considerably lower than those which would be predicted if all the surface charges were assumed to be located at x = 0 (i.e., values of $\psi(0)$ and $\psi(d)$ obtained when d = 0).

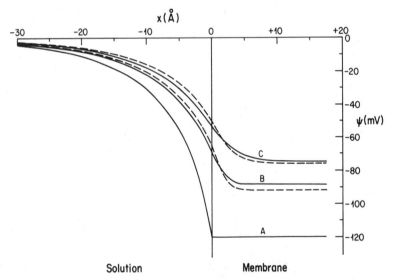

Fig. 3. Potential distribution $\psi(x)$ across the charged membrane with d = 0 , 5 , and 10 $\overset{\circ}{A}$ (curves A, B, and C, respectively). Calculated with T = 298K, n = 0.1 M, $-eNd$ = 0.190 $C \cdot m^{-2}$, $\varepsilon_r'/\varepsilon_r$ = 1 (solid lines) and $\varepsilon_r'/\varepsilon_r$ = 0.5 (broken lines). ε_r = 78.5.

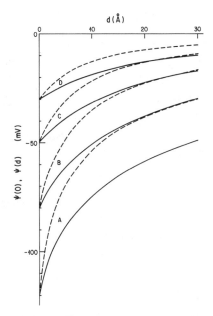

Fig. 4. Potentials $\psi(0)$ (solid curves) and $\psi(d)$ (broken curves) as a function of the surface charge layer thickness d. Calculated with T = 298K, Z = 1, n = 0.1 M, $-e\bar{N}d$ = -0.190, -0.0842, -0.0421, -0.0229 $C \cdot m^{-2}$ (curves A, B, C, and D, respectively) and $\varepsilon_r'/\varepsilon_r$ = 1.

b) 2,1-1 electrolytes and ion binding case

In the previous section, we treated the theory with 1-1 electrolyte and no ion binding to the fixed charge layer. Here, we extend the theory to the case where the solution is composed of a mixture of 1-1 electrolyte of concentration n_1, and 2-1 electrolyte of concentration n_2, and these ions can bind to the fixed charge sites of the surface charge layer. The membrane model to be used is the same as that in the previous section (section a). We assume that each fixed charge group in the membrane surface layer has one negative charge and serves as a site capable of binding either one monovalent cation or one divalent cation.

For a symmetrical membrane, the electric potential $\psi(x)$ at position x relative to the bulk solution phase x = $-\infty$ is related to the space charge density $\rho(x)$ by the Poisson equation,

$$\frac{d^2\psi}{dx^2} = \begin{cases} -\dfrac{\rho(x)}{\varepsilon_r \varepsilon_o}, & -\infty < x < 0, \\[3mm] -\dfrac{\rho(x)}{\varepsilon_r' \varepsilon_o}, & 0 < x < d, \end{cases} \qquad (32)$$

where the space charge density $\rho(x)$ for the region $-\infty < x < 0$ is given by

$$\rho(x) = e[n_1 e^{-y} + 2 n_2 e^{-2y} - (n_1 + 2n_2) e^{y}], \qquad -\infty < x < 0, \qquad (33)$$

9

where $y \equiv e\psi/kT$.

For $0 < x < d$, on the other hand, $\rho(x)$ arises from both the electrolyte ions $(\rho_{el}(x))$ and the membrane fixed charge groups $(\rho_{fix}(x))$;

$$\rho(x) = \rho_{el}(x) + \rho_{fix}(x), \quad 0 < x < d, \tag{34}$$

where $\rho_{el}(x)$ is given by Eq. (33) and $\rho_{fix}(x)$ is written as

$$\rho_{fix}(x) = -e[N - N_1(x) - 2N_2(x)]. \tag{35}$$

Here, $N_1(x)$ and $N_2(x)$ are, respectively, the numbers of monovalent cations and divalent cations bound to the membrane charged groups per unit volume at position x. At chemical equilibrium, we have

$$\frac{N_1(x)}{[N - N_1(x) - N_2(x)]n_1 e^{-y}} = K_1, \quad \text{and} \tag{36}$$

$$\frac{N_2(x)}{[N - N_1(x) - N_2(x)]n_2 e^{-2y}} = K_2, \tag{37}$$

where K_1 and K_2 are, respectively, the equilibrium constants for monovalent cation binding and divalent cation binding. By combining Eqs. (34) – (37), we find that $\rho(x)$ for $0 < x < d$ is given by

$$\rho(x) = e[n_1 e^{-y} + 2n_2 e^{-2y} - (n_1 + 2n_2)e^{y} - \frac{N(1 - K_2 n_2 e^{-2y})}{1 + K_1 n_1 e^{-y} + K_2 n_2 e^{-2y}}], \quad 0 < x < d. \tag{38}$$

The boundary conditions for y are:

y is continuous at $x = 0$, \hfill (39)

$$y \xrightarrow[x \to -\infty]{} 0, \tag{40}$$

$$\varepsilon_r \frac{dy}{dx}\bigg|_{-0} = \varepsilon_r' \frac{dy}{dx}\bigg|_{+0}, \tag{41}$$

$$\frac{dy}{dx}\bigg|_{d-0} = 0. \tag{42}$$

Condition (42) states that there is no electric field within the core layer. Eqs. (32), (33), and (38) subject to boundary conditions (39) – (42) completely determine the potential distribution y(x). For the region $-\infty < x < 0$, Eq. (32) using Eq. (33) can easily be integrated to give explicitly

$$y(x) = \ln\left[\frac{(\frac{Y(x) - 1}{Y(x) + 1})^2 - \frac{n}{3}}{1 - \frac{n}{3}}\right], \quad -\infty < x < 0 \tag{43}$$

where

$$Y(x) = [\frac{1 - ((1 - \frac{\eta}{3})e^{y_o} + \frac{\eta}{3})^{1/2}}{1 + ((1 - \frac{\eta}{3})e^{y_o} + \frac{\eta}{3})^{1/2}}]e^{-\kappa|x|} \quad , \tag{44}$$

$$\eta = \frac{3n_2}{n_1 + 3n_2} \quad , \tag{45}$$

$$y_o = y(0) = \frac{e\psi(0)}{kT} \quad , \tag{46}$$

$$\kappa = [\frac{2(n_1 + 3n_2)e^2}{\epsilon'_r \epsilon_o kT}] \quad . \tag{47}$$

Coupled equations (32), (33) and (38) – (42) for $y(x)$ can be solved numerically by using the above expression for $y(x) (-\infty < x < 0)$ and by expanding $y(x)$ $(0 < x < d)$ around $x = d - 0$ in the Taylor expansion series (Eq. (30) and Ohshima and Ohki, 1985b), viz.

$$y(x) = y_d + \sum_{n=1}^{\infty} A_n[\kappa'(x-d)]^{2n} \quad , \qquad 0 < x < d \tag{48}$$

where

$$y_d = \frac{e}{kT} \psi(d) \qquad \text{and} \tag{49}$$

$$A_n = \frac{1}{(2n)!} \frac{1}{\kappa'^{2n}} \frac{d^{2n}y}{dx^{2n}}|_{d-0} \quad . \tag{50}$$

Here, all the terms with the odd power of $\kappa'(x-d)$ vanish as mentioned before. The coefficients A_n can be obtained from Eq. (38) (n = 1) and by successive differentiation of it (for $n \geq 2$ in terms of y_d, N and η).

NUMERICAL RESULTS

Fig. 5 shows the calculated values of the surface potential for the cases of various ion bindings which occur on the surface charge sites having a constant charge density per unit area of the surface, $-eNd$ (-0.190 c·m^{-2}) which gives $\psi(d) = -120$ mV as $d \rightarrow 0$. Fig. 5-A corresponds to the case of $n_1 = 0.1$ M, $K_1 = 0$ and $n_2 = 0$, $K_2 = 0$; Fig. 5-B the case of $n_1 = 0.1$ M, $K_1 = 0.8$ M^{-1} and $n_2 = 0$; Fig. 5-C the case of $n_1 = 0.1$M, $K_1 = 0.8$ M^{-1}, $n_2 = 10$ mM, $K_2 = 0$, and Fig. 5-D the case of $n_1 = 0.1$ M, $K_1 = 0.8$ M^{-1}, $n_2 = 10$ mM and $K_2 = 10$ M^{-1}. These binding constant values correspond respectively to those for Na$^+$ and Ca^{2+} with a phosphatidylserine bilayer membrane obtained when each ion assumes to bind to one lipid molecule (Nir, et al., 1978; Ohki and Kurland, 1981; Eisenberg, et al., 1979; Ohshima and Ohki, 1985a). From the figure it is evident that the valency and the binding constant of ions greatly affect the magnitude of surface potential.

Some numerical results of potential distribution for the cases where there are ion bindings and various thicknesses, d, of the surface charge layer; 0, 5, or 10 Å, having the same surface charge density of $-eNd =$

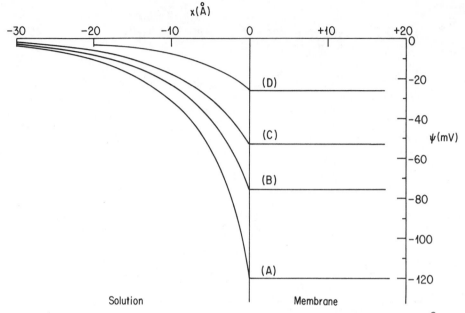

Fig. 5. Potential profiles $\psi(x)$ of a charged membrane with $d = 0$ Å at different divalent cation concentrations. The calculation was done using the following parameters: $T = 298K$, $\varepsilon_r = 78.5$, $-eNd = -0.190$ C·m^{-2}, and the electrolyte solutions (curve A: $n_1 = 0.1$ M, $K_1 = 0.0$, $n_2 = 0$; curve B: $n_1 = 0.1$ M, $K_1 = 0.8$M^{-1}, and $n_2 = 0$; curve C: $n_1 = 0.1$ M, $K_1 = 0.8$ M^{-1}, $n_2 = 10$ mM, $K_2 = 0$; curve D: $n_1 = 0.1$ M, $K_1 = 0.8$ M^{-1}, $n_2 = 10$ mM, $K_2 = 10$ M^{-1}).

-0.190 C·m^{-2}. The other parameters used in the calculation were $\varepsilon_r = \varepsilon_r' = 78.5$, $n_1 = 0.1$ M, $n_2 = 10$ mM, $K_1 = 0.8$ M^{-1} and $K_2 = 10$ M^{-1}. Curves A, B, and C correspond to the cases of $d = 0$, 5, and 10 Å, respectively. While the potential values near the interface are appreciably affected by the presence of divalent cations and the thickness of the charge layer d, the potentials at a distance (greater than $1/\kappa'$) away from the membrane interface are not affected so much by these factors.

The effect of the divalent cation concentration on potential distribution across the membrane interface is shown in Fig. 7 in the cases of 0.1 M monovalent salt having the binding constant of 0.8 M^{-1} and various divalent salt concentrations ($n_2 = 0$ mM, -curve A; 1 mM - curve B; 2 mM - curve C; 5 mM - curve D; and 10 mM - curve E) having the binding constant of 10 M^{-1}. Other parameters needed for calculation were the same as those in Fig. 6. In this case, both the divalent cation concentration and its binding to the fixed charge surface layer affect greatly the potential and magnitude of the potential distribution in the interfacial region.

Fig. 8 shows the calculated surface potential $\psi(0)$ as a function of divalent cation concentration in 0.1 M monovalent salt solution in the case where the thickness of the surface charge layer is 0, 5, or 10 Å. Other parameters are the same as those in Fig. 7 (e.g. $T = 298K$, $\varepsilon_r = \varepsilon_r' = 78.5$, $n_1 = 0.1$ M, $K_1 = 0.8$ M^{-1}, $K_2 = 10$ M^{-1}, $-eNd = -0.190$ C·m^{-2}).

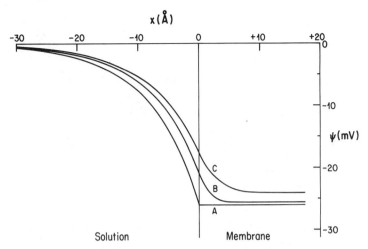

Fig. 6. Potential distribution $\psi(x)$ across a charged membrane with
d = 0, 5 and 10 Å (curves A, B, and C, respectively), cal-
culated with T = 298 K, ε_r = ε_r' = 78.5, $-eNd$ = -0.190 C·m^{-2}
(amount of charges contained in the surface charge layer
per unit area), n_1 = 0.1 M, n_2 = 10 mM, K_1 = 0.8 M^{-1} and
K_2 = 10 M^{-1}.

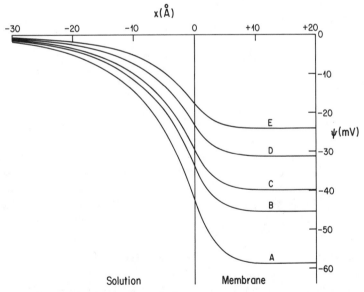

Fig. 7. Potential distribution $\psi(x)$ across a charged membrane with
d = 10 Å for n_2 = 0, 1, 2, 5, and 10 mM (curves A, B, C,
D, and E, respectively). Calculated with T = 298K, ε_r = ε_r'
= 78.5, $-eNd$ = -0.190 C·m^{-2}, n_1 = 0.1 M, K_1 = 0.8M^{-1}, and
K_2 = 10 M^{-1}.

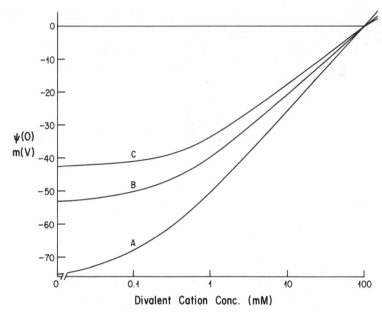

Fig. 8. Surface potential $\psi(0)$ as a function of divalent cation concentration n_2 for d = 0, 5, and 10 Å (curves A, B, and C, respectively). Calculated with T = 298K, $\varepsilon_r = \varepsilon_r' = 78.5$, $-eNd = -0.190$ C·m^{-2}, $n_1 = 0.1$ M, $K_1 = 0.8$M^{-1}, and $K_2 = 10$ M^{-1}.

As expected, the thicker the surface charge layer is, the larger (the magnitude is smaller) is the surface potential, $\psi(0)$ and as the concentration of divalent salt increases the magnitude of the surface potential is monotonically reduced in all cases. The surface potential, $\psi(0)$, becomes zero at a certain divalent cation concentration, and the sign of the surface potential is altered when the divalent salt concentration is further increased.

DISCUSSION

In the above study, we have shown that the potential distribution obtained undergoes a smooth transition from the Donnan potential (d → ∞) to the surface potential (d → 0), and that the apparent difference between these two concepts is not of a fundamental nature (Fig. 4).

The potential profile near the solution/membrane interface is continuous and changes in a diffused manner to both sides of the membrane interface (Figs. 3, 5, 6, 7). The potential distribution near the interface depends strongly on the thickness, d, of the surface charge layer when the layer thickness d is smaller than the Debye-Hückel length (1/κ') in the charged layer region. It is interesting to note that the different dielectric constants in the solution and the surface charge layer do not appreciably effect the potential distribution across the membrane interface (Fig. 3).

The magnitude of the surface potential $\psi(0)$ is reduced as the thickness of the surface charge layer (Figs. 3,4) increases. For example, the

magnitude of surface potential $\psi(0)$ for the thickness $d = 5$ Å is reduced
to about 40% of that for $d = 0$; the magnitude of the surface potential $\psi(0)$
for $d = 10$ Å is reduced to 60% of that for $d = 0$. Such a large reduction
in the surface potential strongly suggests the necessity of correction for
the estimation of the surface potential and surface charge density obtained
from the zero thickness assumption of the surface charge layer. As mentioned in the Introduction section, several authors (Haydon, 1961; Donath
and Pastushenko, 1979; Wunderlich, 1982; Levine, et al., 1983; and McLaughlin,
et al., 1985) have already pointed this possibility out. Since most biological membranes may possess fixed charges which may be distributed within
a certain depth (e.g., surface layer) of about the same order of magnitude
as the Debye-Hückel length ($1/\kappa' \sim 5 \sim 10$ Å in usual physiological conditions) from the membrane surfaces, the conventional method to determine
the surface charge density and surface potential of biological membranes
from cell electrophoresis measurements (Seaman and Cook, 1965; Sherbet,
1978) should give under-estimated values of those obtained from the membrane model taking into account the thickness for the surface charge layer.
This effect is especially enhanced when electrolyte ions are able to bind
to the membrane fixed sites as shown in the above (Figs. 5,6,7).

Although the introduction of such surface charge layers should be an
improvement over the conventional surface potential treatment for biological membrane surface studies, there are still several difficulties
that remain to be solved in order to have correct information about the
potential profile at the membrane interface as well as the amounts of
ion adsorption onto the membrane surface regions. For instance, the
estimation of the correct thickness of the surface charge layer, the amount
of the fixed charges in the charge layer, the distribution of such fixed
charges in the charge layer, etc.

In the present paper, we have treated a symmetrical membrane with
respect to the surface charge layers and the electrolyte solutions surrounding the membrane. When the membrane is asymmetrical (it is usually
the case for most biological membranes) with respect to the above two
aspects, in general, there exists electrostatic fields inside the core
layer ($d < x < h + d$) and thus, the boundary condition Eq. (12) no longer
holds. If, however, the right-hand side of Eq. (8) is negligibly small,
then the boundary condition of Eq. (12) holds approximately well and the
results obtained here may be still applied to such a membrane system.

In spite of these complications associated with the interfacial
electrical potential distribution, the above theoretical treatment should
be valuable in the future for the analysis and determination of the
electrokinetic potentials derived from cell electrophoresis studies,
the studies of ion adsorption and binding onto the charged membranes
influenced by electrical potential profile in the membrane surface
regions. The above treatment would also relate to the study of the
transmembrane potential profile across cell membranes.

ACKNOWLEDGEMENTS

This work was partially supported by a grant from the U.S. National
Institutes of Health (GM24840).

REFERENCES

Davies, J.T., and E.K. Rideal. (1961) Interfacial Phenomena. Academic
    Press, New York and London. 75.
Donath, E., and V. Pastushenko. (1979) Bioelectrochem. Bioenerg. 6:543.
Eisenberg, M., Gresalfi, T., Riccio, T. and McLaughlin, S. (1979)
    Biochemistry 18:5213.
Haydon, D.A. (1961) Biochim. Biophys. Acta 50:450.
Levine, S., M. Levine, K.A. Sharp, and D.E. Brooks. (1983) Biophys. J.
    42:127.
Mauro, A. (1962) Biophys. J. 2:179.
McLaughlin, S., D. Brooks, M. Eisenberg, R. McDaniel, A. McLaughlin,
    L. Pasquale, K. Sharp, and A. Winiski (1985) personal communication.
Nir, S., C. Newton and D. Papahadjopoulos (1978) Bioelectrochem. Bioenerg.
    5:116.
Ohki, S. (1965) J. Phys. Soc. Japan 20:1674.
Ohki, S., and R. Kurland (1981) Biochim. Biophys. Acta 695:170.
Ohshima, H. and S. Ohki (1985a) J. Colloid Interface Sci. 103:85.
Ohshima, H. and S. Ohki (1985b) Biophysical J. 47:673.
Parsegian, V.A. and D. Gingell (1973) in "Recent Advances in Adhesion",
    L. H. Lee, Ed., Gordon & Breach, New York, p. 153.
Parsegian, V.A. (1974) Ann. N.Y. Acad. Sci. 238:362.
Seaman, G.V.F. and V.F. Cook (1965) in "Cell Electrophoresis", E.J. Ambrose,
    Ed., Little Brown, Boston, Mass., p. 78.
Sherbet, G.V. (1978) in "The Biophysical Characterization of the Cell
    Surface". Academic Press, New York.
Teorell, T. (1953) Progr. Biophysics and Biophysic. Chem. 3:305.
Verwey, E.J.A., and J. Th. G. Overbeek (1948) in "Theory of the Stability
    of Lyophobic Colloids". Elsevier, Amsterdam, p. 25.
Wunderlich, R.W. (1982) J. Colloid Interface Sci. 88:385.

AN EMPIRICAL RELATION FOR THE SURFACE POTENTIAL OF PHOSPHATIDIC ACID

MONOLAYERS: ITS DEPENDENCE ON CALCIUM AND THE ROLE OF DOUBLE LAYER

THEORY

John A. DeSimone, Gerard L. Heck, and Shirley K. DeSimone

Virginia Commonwealth University
Medical College of Virginia
Department of Physiology and Biophysics
Richmond, VA 23298

INTRODUCTION

The physical chemistry of the phospholipid-aqueous interface has been a subject of interest for many years. The comparative ease with which phospholipid monolayers, bilayers, and vesicles can be prepared, along with their unique properties, has made them attractive as models for biomembranes. The actual "work" that biomembranes do depends invariably on the specific nature of their integral proteins. However, these are thought to be modulated in function, to some extent, by the phospholipid backbone comprising the basic bilayer structure. The relative fluidity of the phospholipid matrix is regarded as important in this respect. Accordingly the crystal-liquid crystal phase transition of phospholipids has been much studied (e.g. Trauble and Eibl, 1974; Jacobson and Papahadjopoulos, 1975; Galla and Sackmann, 1975; MacDonald et al., 1976; Trauble et al, 1976; Disalvo, 1983). In the case of acidic phospholipids, an isothermal phase transition from the crystal to the liquid-crystal state can be induced by increasing the monovalent cation concentration, or by increasing the pH. An increase in the concentration of divalent ions has the opposite effect. Gouy-Chapman double layer theory has often been invoked to advance an explanation of these phenomena (e.g. the fluidizing effect of monovalent cations on the acidic phospholipids in the protonated state, c.f. Trauble and Eibl, 1974).

There is by no means unanimity in the application of the Gouy-Chapman theory to phospholipid assemblies. Indeed the theory is often justifiably criticized as simplistic, and a variety of useful corrections have been offered (e.g. Haydon and Taylor, 1960). One of its attributes is that it provides a simple framework upon which to sort out stoichiometric from nonstoichiometric interactions in the micro-enviroment. This distinction is often overlooked and has led to overestimates of the intrinsic stability constants between acidic phospholipids and metal ions (e.g. Abramson et al., 1966). More recently attempts have been made to extract the intrinsic, or stoichio-metric ion binding constant by first correcting the binding data for electrostatic screening (e.g. McLaughlin et al., 1971; Nir et al., 1978). However, any of these approaches must begin by making a few basic assumptions about the nature of the surface stoichiometric re-

17

actions. For monovalent cations including protons, there is no reason to assume other than simple mass action relations between the phospholipid head groups and the cations in solution. These invariably lead to simple Langmuir adsorption relations (Payens, 1955; Ninham and Parsegian, 1971; DeSimone, et al., 1980). The Langmuir equation assumes that binding is noncooperative. In most instances this is born out. DeSimone et al. (1980) have shown that the dependence of the surface pressure of phosphatidic acid monolayers on the NaCl concentration in the subphase is consistent with the surface charge density of the monolayer being determined through a Langmuir relation involving the surface hydrogen ion concentration. There was no indication that Na ion binding was a significant factor. These results are similar to those of Sacre and Tocanne (1977) on the monolayer properties of phosphatidylglycerols. They also observed film expansion as the subphase monovalent cation concentration was raised. In several cases they also observed the condensing effect of high salt concentrations. Some discrimination among alkali metal ions could be seen although these effects were slight and only apparent at concentrations above 100 mM. The same effects are observed with phosphatidic acid at both the air/aqueous and the oil/aqueous interface (DeSimone et al. 1980).

The problem of the surface stoichiometry of reactions involving divalent cations with phospholipid anionic sites is considerably more formidable. The very fact that the binding ligand is divalent suggests that site-site interactions will be the rule rather than the exception. In treating divalent cation binding to phosphatidylserine, McLaughlin et al. (1971) and Nir et al. (1978) assumed that Ca ions bound two neighboring phosphate groups. No 1:1 Ca-phospholipid complexes were assumed to occur. On the other hand Toko and Yamafugi (1980) treated Ca binding to phosphatidylglycerol monolayers as 1:1 complexes, but allowed for nearest neighbor interactions on a one-dimensional Ising lattice. Interactions were assumed to be attractive if the neighboring site was vacant, and repulsive if the neighboring site was occupied by Ca. More recently Cohen and Cohen (1984) have considered reaction schemes which permit complexes of divalent cations with both 1:1 and 1:2 stoichiometric ratios, the two forms being interconvertible. Given that divalent cations may bind to more than one site, the probability that a site will be occupied will increase if a nearest neighbor site contains a 1:1 complex, i.e. some form of positive cooperativity is likely to exist.

In an earlier study we used a relaxation method for continuously changing the subphase salt concentration beneath a phospholipid monolayer (DeSimone et al., 1980). In that study we demonstrated that a monolayer of dipalmitoylphosphatidic acid (DPPA) expands as the NaCl concentration is raised from 0 to approximately 200 mM. Above that concentration it condenses. In the work reported herein we have used the same method to record simultaneously changes in both the surface potential and the surface tension of DPPA monolayers at the air/aqueous interface as a function of a changing NaCl or $CaCl_2$ concentration. In this case the surface potential increases as the NaCl concentration increases in the subphase, and the magnitude of the changes are in accord with the predictions of the Gouy-Chapman theory assuming that only H^+ ions bind specifically to the phospholipid. Under the conditions of the experiments herein the surface tension change always indicated film expansion.

In the case of $CaCl_2$ we show that, at a fixed concentration of 0.1 mM NaCl, a change of just a few micromolar Ca causes an increase in surface potential equivalent to that produced by NaCl 1000 to 10,000 times higher in concentration.. We establish that the relation between the surface potential and the calcium concentration satisfies Hill's equation up to about 4 μM Ca. Between 4 μM and 20 μM the potential increases logarithmically with a slope of kT/e. These empirical relations derived from the data allow for the calculation of the surface charge density if one assumes the validity of the Gouy-Chapman theory. This is equivalent to the adsorption isotherm for DPPA monolayers with calcium ions for the special conditions of the experiments. Accordingly any assumed mass action relations must satisfy the experimentally derived adsorption isotherm. Our results therefore circumscribe the set of allowed stoichiometric surface reaction schemes. Simultaneous measurements of the surface tension show that, in the case of calcium infiltration of the double layer, only film condensation is detected although in theory there should also be a slight expansion during the earliest phase of infiltration.

EXPERIMENTAL

L-α-phosphatidic acid, dipalmitoyl (DPPA) was obtained from Sigma Chemical Company; it was routinely acid washed according to the method of Rathbone (1962) to remove latent calcium. Washes were monitored for calcium with an IL-353 atomic absorption mass spectrometer. The purity of a DPPA sample was ascertained by two-dimensional thin layer chromatography. All chemicals were reagent grade. Sodium chloride was recrystallized according to the method of Betts and Pethica (1956). DPPA was spread from a solution of 9:1 chloroform-methanol.

Monolayers of DPPA were formed on subphases of 0.1 mM NaCl and subsequently transferred to a new subphase containing 0.1 mM NaCl plus a certain concentration of $CaCl_2$. The new subphase equilibrated with the film by diffusion and the consequent changes in the surface potential and surface tension recorded simultaneously. In experiments where increasing subphase sodium concentration was under investigation, the monolayer was formed on a subphase of 0.01 mM and subsequently transferred to a subphase containing: 0.01 M, 0.1 M, 0.5 M 1 M, or 2 M NaCl. In all cases the transfer of the phospholipid film was carried out by first forming a monolayer on one section of a teflon trough divided into two adjacent compartments. Each compartment was 1 cm deep and had an area of 97.1 cm^2. The compartment upon which the film was formed was fitted with a teflon collar which conformed to the dimensions of the compartment. Typically the film rested about 1 mm above the surface of the compartment level where it was maintained, because of the negative meniscus, within the perimeter of the collar. With the collar held in position, the trough could be moved beneath it so as to expose the film and about 1 mm of the original subphase to the new subphase in the adjacent compartment. The new subphase equilibrated with the film by diffusion and changes in the surface potential and tension were measured. The surface potential was measured with an Am^{241} electrode positioned 1 mm above the film. Potentials were referenced to a saturated calomel electrode in contact with the interior of the subphase and measured with a high input impedence electrometer. The electrometer out put was then fed to one channel of a dual channel chart recorder. The surface tension was measured with the Wilhelmy plate method. The plate was thin platinum 1 cm wide and was suspended from a Cahn Electrobalance (model RG). The balance output was amplified and fed to the second channel of the recorder. All experiments were done at room temperature (22-23°C). All salt solutions were unbuffered. Measured pH values varied between 5.5 and 6.0.

The thickness of the diffusion zone was measured as described previously (DeSimone et al., 1980). Briefly, a lamina of water was transferred to a compartment containing concentrated NaCl (1 or 2 M).

The thickness of the lamina was estimated by the extent the solution was diluted. Calibrations of this sort gave a thickness of about 1 mm. Generally the increase in surface tension was recorded as the NaCl diffused. The trace was an accurate representation of the NaCl concentration at the surface because the surface tension was a linear function of the NaCl concentration (cf. DeSimone et al., 1980, fig. 6). Assuming the same diffusional relaxation occurred in the presence of a film, we could infer a certain concentration of diffusing cation at the surface just outside the double layer by solving the diffusion equation for the salt in question given the appropriate diffusion coefficient and diffusion distance. Thus time could be eliminated as a parameter in favor of concentration.

THEORY

We begin by examining the Gouy-Chapman theory. At high surface density of monolayer it seems reasonable to consider the monolayer to be of infinite extent, i.e. we shall consider the monolayer to be free from inhomogenieties in the plane. Accordingly, we regard the electric field established by the distribution of surface bound anions (ionized DPPA) as directed normal to the plane of the monolayer. We further regard the surface to be in equilibrium with the solution several Debye lengths away. This will be true even when the subphase is changing composition by diffusion because the relaxation within the double layer (over a few Debye lengths) is so much faster than the process of diffusion in the 1 mm zone. We can consequently ignore perturbations of the structure of the electrical double layer and proceed as if the double layer were always in equilibrium with its current subphase. As such the potential gradient can be obtained as the first integral of the Poisson-Boltzmann equation, viz.:

$$\left(\frac{d\phi}{dx}\right)^2 = \frac{8\pi Ne^2}{1000\,\epsilon\,kT} \sum_i c_{i\infty} (e^{-z_i\phi} - 1) \tag{1}$$

where we have used the condition of vanishing field and potential at infinity. Here ϕ is the normalized potential $e\psi/T$, where ψ is the potential in the x-dimension normal to the surface, e the protonic charge, k Boltzmann's constant and T the temperature. N is Avogadro's number, ϵ the dielectric constant of the aqueous medium, and $c_{i\infty}$ is the molar concentration of the ith species in the electroneutral phase whose valence is z. This concentration governs the structure of the double layer at the surface and is regarded as the concentration just outside the double layer at a plane where deviations from electoneutrality are negligible. The surface charge density σ, is given by the boundary condition at the surface (x=0), viz.

$$\sigma = -\frac{kT}{4\pi e}\left.\frac{d\phi}{dx}\right]_0 \tag{2}$$

Specializing to the case of a mixture of a di-univalent and a uni-univalent salt of the type BX_2 and AX, we obtain

$$\sigma = \sqrt{\frac{\epsilon NkTa}{500\pi}}\ \sinh\frac{\phi_o}{2}\left[1 + \frac{b}{a}(2 + e^{-\phi_o})\right]^{1/2} \tag{3}$$

where a is the bulk concentration of monovalent salt and b is the concentration of divalent salt. This relation can be put in the more familiar form:

$$\sigma = \sqrt{\frac{\varepsilon\, NkTc}{500\,\pi}} \quad \sinh \frac{\phi_o}{2} \tag{4}$$

if the effective concentration is defined as:

$$c = a + b(2 + e^{-\phi_o}) \tag{5}$$

Thus the effect of adding divalent cations is substantially different depending on the sign of the potential. If the surface is positive, the effect of adding the salt BX_2 is roughly additive with that of AX, viz.

$$c \cong a + 2b \tag{6a}$$

But if the surface is highly negative, the concentration of BX_2 is amplified by the exponential factor, and

$$c \cong a + be^{|\phi_o|} \tag{6b}$$

Thus even low divalent ion concentration may have major effects on the surface charge density.

Equation (3) cannot be used directly to find the surface charge density when specific ion binding occurs. This is because neither σ nor ϕ_o is constant for arbitrary changes in a or b. An additional relation is required to specify the surfaces potential uniquely.

If the mechanism of ion binding is known a priori, an additional relation can be deduced from:

$$\sigma = -e\,\Gamma_1 \; -2e\,\Gamma_2 \tag{7}$$

where Γ_1 and Γ_2 are the surface concentrations of monovalent and divalent phosphate anions respectively. In the absense of divalent cations, specific adsorption to DPPA seems to be limited mainly to protons, and the adsorption isotherm is simply:

$$\sigma = -e\,\Gamma \; \frac{K_1/He^{-\phi_o} + 2K_1 \, K_2/H^2 e^{-2\phi_o}}{1 + K_1/He^{-\phi_o} + K_1 \, K_2/H^2 e^{-2\phi_o}} \tag{8}$$

where Γ is the surface concentration of DPPA in all its forms, H is the bulk hydrogen ion concentration and K_1 and K_2 are the two proton dissociation constants. Putting equation (8) equal to equation (3) with b = 0, determines ϕ_o uniquely.

The surface pressure has an electrostatic component and this is given by the well-known Davies equation (Davies, 1951)

$$\Delta\Pi = \sqrt{\frac{\varepsilon\, Nk^3T^3 a}{125\,\pi e^2}} \left[\cosh \frac{\phi_o}{2} - 1 \right] \tag{9}$$

provided b = 0. These equations have been tested for the case of DPPA in the absence of divalent cations. They correctly predict the nonmonotonic character of $\Delta\Pi$ as a function of the NaCl concentration and the dependence of $\Delta\Pi$ on pH (DeSimone et al., 1980).

In the presence of divalent cations equation (9) takes the form (DeSimone, 1978).

$$\Delta \Pi = \sqrt{\frac{\varepsilon Nk^3 T^3 b}{2000 \pi e^2}} \left[\eta (2 + e^{-\phi_0}) + r \; ctnh^{-1} \eta \; - 3\sqrt{3 + r} \right.$$

$$\left. - r \; ctnh^{-1} \sqrt{3 + r} \right] (10)$$

where

$$\eta = \sqrt{(2 + r)e^{\phi_0} + 1}$$

and $r = a/b$.

The number of possible surface reactions and the high probability of cooperative interactions makes guessing the form of the surface stoichiometry speculative at best. The alternative is to determine the surface charge density from an experimentally derived relation between the surface potential and the divalent ion concentration. If this is done by means of the diffusional relaxation method outlined above, we require a single additional relation, the solution of the diffusion equation describing the surface concentration of divalent ions control-ling the structure of the double layer. This allows us to assign a value for the divalent ion concentration to each surface potential point as it evolves in time. Thus the change in ϕ_0 as a function of ion concentration is established over a range of concentrations in any given experiment. Assuming reflecting boundary conditions we can write:

$$c(t) = c_\infty \left[1 - \frac{4}{\pi} \sum_{n=0}^{\infty} \frac{(-1)^n}{2n + 1} e^{-(2n + 1)^2 \pi^2 \xi/4} \right] \quad (11)$$

where $c(t)$ is the concentration of the diffusing species at the surface, but beyond the double layer, c_∞ is the bulk phase concentration, and $\xi = Dt/1^2$, where D is the diffusion coefficient, t is time and 1 is the diffusion path length.

RESULTS

In all cases, DPPA films were formed to an initial starting surface pressure of 8 dynes/cm. Since the DPPA surface pressure versus area curve at the air/aqueous interface is condensed, this starting surface pressure roughly corresponds to a specific area of 40-50 A^2/molecule. Figure 1 shows the typical behavior of a DPPA monolayer when there is an increase in NaCl concentration in the absence of divalent cation. In this particular instance the film was transferred from a subphase of 10^{-5} M NaCl to 0.1 M NaCl. We observed an immediate rise in the surface potential followed by a slower rise toward an asymptotic value. Note that the surface tension decreased (which is of course equivalent to an increase in the surface pressure). The film clearly becomes more expanded in keeping with the model outlined above and the results of DeSimone et al. (1980). Although not done so here, it is possible to account for the increase in surface potential accurately by computing the concentration of NaCl in equilibrium with the film at any time using equation (11) and by computing the surface potential from equations (3) and (8) with b = 0. Note that it requires a 10,000 fold increase in the NaCl concentration to produce these changes in potential and surface tension.

The situation with CaCl$_2$ is quite different. Figure 2 shows the

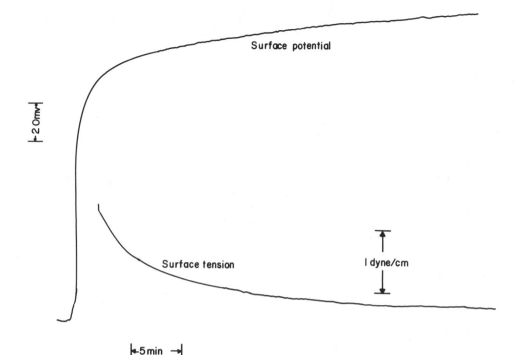

Figure 1.   Time course of the surface potential and surface tension
after transfer of a DPPA monolayer from 0.01 mM NaCl to
0.1 M NaCl. The starting surface pressure was 8 dynes/cm.
Note that while the surface potential depolarizes, the
monolayer expands.

result of a film transfer from 0.1 mM NaCl to a new subphase containing
0.1 mM NaCl plus 4 μM $CaCl_2$. There was a sustantial increase in surface
potential, and rather then expand, the film condensed. As might be
expected an increase in the concentration of $CaCl_2$ to 20 μM resulted in
both a faster and a larger response. This is shown in figure 3. It is
clear that 20 μM $CaCl_2$ has about the same effect on the potential as 0.1
M NaCl. The lower trace in figure 3 shows the concentration profile of
$CaCl_2$ computed from equation (11) with $D = 1.16 \times 10^{-5}$ cm^2/sec and 1 =
mm. Together the measured potential and the computed concentration
profile determine the concentration dependence of the surface potential.
This data reduction is displayed in figure 4. There are several notable
features in this dependence. First the initial increase in potential is
nonlinear. This is followed by a steep rise which is nearly linear out
to about 4 μM. Following this the potential rises more slowly.

The form of the early nonlinear part of the response can be seen
more convincingly by examining the dependence of the potential on the
concentration for the case of 4 μM $CaCl_2$. As seen in figure 5, the
potential is initially an accelerating function of the calcium concentration. A log-log plot of the data in figure 4 shows that the poten-

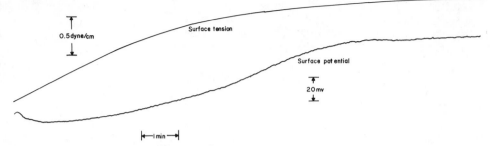

Figure 2.   Time course of the surface potential and surface tension
             after transfer of a DPPA monolayer from a 0.1 mM NaCl
             subphase to 0.1 mM NaCl plus 4 μM CaCl$_2$. In this case
             depolarization of the surface potential is accompanied by
             film condensation. Starting surface pressure was 8
             dynes/cm.

tial, ψ (in mV) satisfied the following empirical relation with [Ca]
expressed in micromolar units:

$$\psi = 13.15\ [Ca]^{1.32} \qquad\qquad 0 < [Ca] < 3\ \mu M \qquad\qquad (12)$$

This can be extended out to 4 μM by modifying it slightly in the form of
Hill's equation, viz:

$$\psi = \frac{13.15\ [Ca]^{1.32}}{1 + 0.01[Ca]^{1.32}} \qquad\qquad (13)$$

From 4 μM out to 20 μM, the relation which best describes the data is:

$$\psi = 23.88\ \ln\ [Ca] + 52 \qquad\qquad (14)$$

It should be noted the empirical constant in equation (14) is
approximately equal to kT/e, and that the empirical constant in equation
(12) is roughly kT/2e.

DISCUSSION

The initial dependence of the surface potential on the calcium ion
concentration suggests the appearance of the first quantities of calcium
in the microenvironment of the film set up conditions which promote
further calcium uptake, i.e. a cooperative interaction. The electrostatic forces within the double layer are central in establishing the
appropriate conditions for cooperativity. A calculation of the surface
charge density as calcium approaches the interface shows that there is
an initial increase in negative surface charge before appreciable
calcium binding occurs. The mechanism behind this probably depends on
proton release which may occur as follows: In the absence of calcium

24

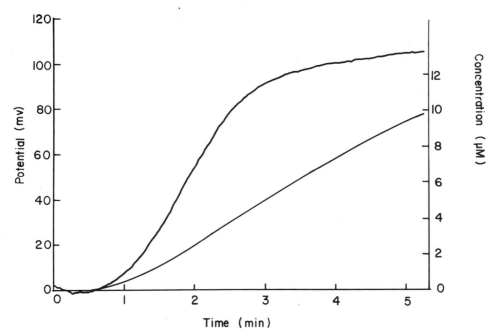

Figure 3.   The top curve is the measured time course of the change
              in surface potential after transfer of a DPPA film from
              0.1 mM NaCl to 0.1 NaCl plus 20 μM $CaCl_2$. Starting film
              surface pressure was 8 dynes/cm. Lower curve is the
              computed time course of the surface concentration of
              $CaCl_2$ corresponding to the potential change.

ion the principally bound ion is the hydrogen ion. At 0.1 mM, there is
no evidence suggesting that sodium ions have more than a negligible
affinity for the phospholipid. As calcium enters the double layer a
more effective screening occurs and the surface potential becomes more
positive. This results in a decreased surface hydrogen concentration
and by mass action the monolayer begins to give up hydrogen ions. The
net effect is an increased surface charge density, and at least
initially the more calcium ions entering the double layer, the more
calcium binding sites become available. At the initially low calcium
ion concentration more sites are created than are bound. This formation
of additional negative charge retards the decrease in potential until
significant ion binding can occur; thus the sigmoidal change in the
surface potential with calcium concentration. Also contributing to the
cooperative nature of the binding is the possibility of intermolecular
links between calcium and the phospholipid at the surface. Before
saturation occurs, a bound calcium ion is in a favorable position to
bind also to vacant neighboring sites. Thus the binding of one site
increases the probability of binding to nearby sites. This early phase
is halted when proton dissociation is complete.   In these experiments

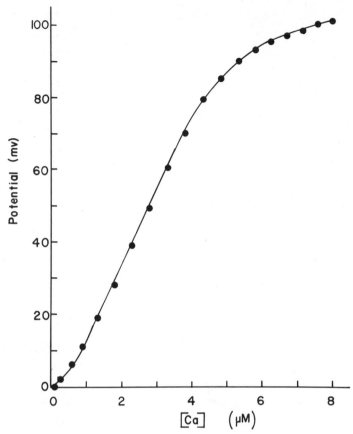

Figure 4.   The empirical relation between the change in surface
             potential and the subphase calcium ion concentration
             derived from the data of figure 3 by eliminating time
             betwen the two curves. Note the early nonlinear
             dependence of the potential on concentration followed by
             a quasi-linear phase. The logarithmic dependence begins
             at about 4 μM and continues out to 20 μM.

this seems to happen when the bulk concentration of calcium is about
3-4 μM. Above 4 μM the potential becomes a logarithmic function of the
calcium concentration.

    It is noteworthy that the slope of the potential for a unit log
increase in the concentration of calcium is kT/e rather than kT/2e.
This suggests that the calcium ions in the microenvironment of the film
can, under some conditions, behave as though they were singly charged
entities. This possibility has been previously discussed by Hauser et
al. (1976) and Nir et al. (1978).

Unlike the case of sodium ions entering the double layer treated in

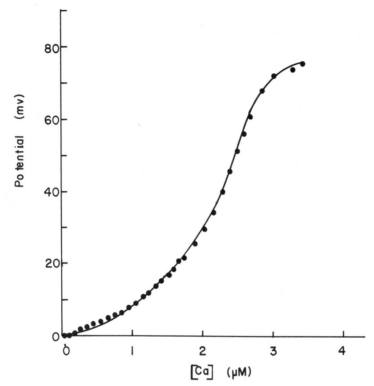

Figure 5.   An amplification of the early nonlinear phase of the
              dependence of the surface potential on subphase calcium
              ion concentration. The curve was obtained by eliminating
              time between the surface potential and concentration time
              courses for the case of a 4 μM $CaCl_2$ subphase (such as
              that in figure 2).

figure 1, calcium ions always result in film condensation. This is also
the major predicted effect according to theory. For example, if we
assume an initial surface potential of about -165 mV on the subphase of
0.1 mM NaCl, calculations using the equations above indicated an
increase in surface tension of about 0.8 dyne/cm when the film is
transferred to the 4 μM $CaCl_2$ subphase. Film transfer to a 20 μM $CaCl_2$
subphase results in a predicted surface tension change of about 0.9
dyne/cm. This is in good agreement with observation. In theory there
should be a slight film expansion during the earliest phases of the
diffusional relaxation process coinciding with the rapid increase in
surface charge density. This is never observed. This may, however, be
for technical reasons because mechanical artifacts, inherent in the
transfer of the film between subphases, tend to obscure this early
period in the relaxation process. Even if it could be observed
calculations show that the expansion would be slight. Again assuming a
starting potential of -165 mV, a concentration of 0.5 μM calcium ion
outside the double layer would result in doubling the surface charge

density, but would only increase the surface pressure by about 0.3 dyne/cm. However, the ensuing phase of condensation is easily measured and most of the increase in surface tension coincides with the cooperative phase of calcium binding. This too is in agreement with calculations based on double layer theory as outlined herein.

In summary, an experimental determination of the change in surface potential and surface tension of DPPA monolayers as a function of the subphase calcium concentration indicates that calcium binding is a cooperative process. The surface potential satisfies Hill's equation at low calcium concentrations. The origin of the cooperativity is most likely electrostatically promoted dissociation of protons from the DPPA thereby exposing new negatively charged binding sites. Any mechanism of ion binding based on an application of the law of mass action in the microenvironment of the monolayer must accordingly include both calcium and proton binding reactions. The empirical relations developed herein provide a criterion which any theoretical model of calcium binding must meet. Any theoretically derived relation between the surface potential and the calcium ion concentration must ultimately also satisfy the empirical relations developed herein. These relations should not only be useful in judging the validity of various mass action formulations but should also aid in the scaling of key parameters such as disso-ciation constants between ions and the monolayer. These results also show that the theoretical relation for the effect of divalent ions on the surface pressure (DeSimone, 1978) accurately describes the depen-dence of the surface pressure of DPPA monolayers on the calcium ion concentration.

ACKNOWLEDGMENT

This work was supported by NIH Grant NS 13767.

REFERENCES

Abramson, M.B., Katzman, R., Gregor, H., and Curci, R., 1966, Biochemistry, 5:2207

Betts, J. J., and Pethica, B.A., 1956, Trans. Faraday Soc. 52:1581

Cohen, J.A., and Cohen, M., 1984, Biophys. J. 46:487

Davies, J.T., 1951, Proc. Roy. Soc. Lond., 208A:224

DeSimone, J.A., 1978, J. Coll. Interface Sci., 67:381

DeSimone, J.A., Heck, G.L., and DeSimone, S.K., 1980, in: "Bioelectrochemistry: Ions, Surfaces, Membranes", M. Blank, ed., Advances in Chemistry Series, No. 188, Washington, D.C.

Disalvo, E.A., 1983, Bioelectrochem. Bioenerg. 11:145

Galla, H.-J., and Sackmann, E., 1975, Biochim. Biophys. Acta 401:509

Hauser, H., Darke, A., and Phillips, M.C., 1976, Eur. J. Biochem. 62:335

Haydon, D.A., and Taylor, F.H., 1960, Phil. Trans. A, 253:31

Jacobson, K., and Papahadjopoulos, D., 1975, Biochemistry 14:152

MacDonald, R.C., Simon, S.A., and Baer, E., 1976, Biochemistry, 15:885

McLaughlin, S.G.A., Szabo, G., and Eisenman, G., 1971, J. Gen. Physiol.,
58:667

Ninham, B.W., and Parsegian, V.A., 1971, J. Theoret. Biol., 31:405.

Nir, S., Newton, C., and Papahadjopoulos, D., 1978, Bioelectrochem.
Bioenerg., 5:116

Payens, Th. A. J., 1955, Philips Research Report 10:425

Rathbone, L., 1962, Biochem. J., 85:461

Sacré, M. M., and Tocanne, J. F., 1977, Chem. Phys. Lipids, 18:334

Toko, K., and Yamafugi, K., 1980, Chem. Phys. Lipids, 26:79

Träuble, H. and Eibl, H., 1974, Proc. Nat. Acad. Sci. 71:214

Träuble, H., Teuber, M., Woolley, P., and Eibl, H., 1976, Biophysical
Chem. 4:319

A FUNDAMENTAL QUESTION ABOUT ELECTRICAL POTENTIAL PROFILE IN

INTERFACIAL REGION OF BIOLOGICAL MEMBRANE SYSTEMS

V.S. Vaidhyanathan
Department of Biophysical Sciences
School of Medicine, State University of
New York at Buffalo, Buffalo, New York, 14221

INTRODUCTION

It is generally presumed, that if a plane plate is positively charged, its potential is positive and that if a plate contains an excess of negative charges, its potential is negative. One may assume that the distribution of positive and negative ions, near a surface with an excess of fixed negative charges, is similar to the schematic representation presented in Figure 1 a. In Figure 1 b, are presented the corresponding plots of charge density, and a function $Y(x) = -4\pi e \sum_{\sigma} Z_\sigma C_\sigma(x)$, where Z_σ is the signed valence charge number of ions of kind σ, and $C_\sigma(x)$ is its number density (concentration in a small volume element) at location x. e is the protonic charge and $Y(x) = \epsilon(x)\phi''(x) + \epsilon'(x)\phi'(x)$, where $\epsilon(x)$ is the value of the dielectric coefficient at x, and $\epsilon'(x) = [d\epsilon/dx]$ is its first differential with respect to position variable x. $\phi(x)$ is the value of electric potential felt by a unit charge placed at x. $\phi'(x)$ and $\phi''(x)$ are respectively the first and second differentials of ϕ, with respect to x. If the above stated statements are valid, then one may presume that the electric potential profile in interfacial regions, will be similar to the schematic plot $\phi(x)$, presented in Figure 1 c. This schematic profile is drawn under the assumption, that if a thin parallelepiped layer of aqueous electrolyte contains an excess of positive ions than negative ions, then the potential of this thin layer should be positive and not negative.

This conclusion is in disagreement with the results suggested by the classical limiting expression[1,2] for electrochemical potential, μ_σ, of ions of kind σ,

$$\mu_\sigma(x) = \mu_\sigma^*(T,P) + kT \ln C_\sigma(x) + Z_\sigma e \, \phi(x) \qquad (1)$$

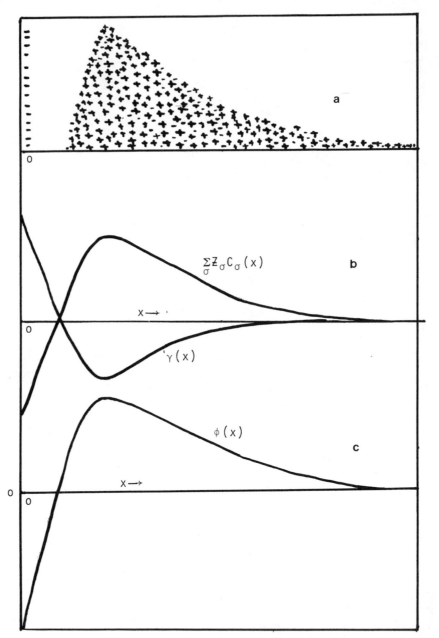

FIGURE 1. Schematic representation of distribution of charges, charge density and electric potential profiles, near a surface with fixed negative charges.

In equation (1), T is temperature in Kelvin scale, P is pressure and k is the Boltzmann constant. $\mu_\sigma^*(T,P)$ is the composition and hence position independent part of the partial molar free energy of ions of kind σ in the system. x is the position variable, defined normal to the yz plane of the interface. $\emptyset(x)$ is the electric potential. Equation (1) is valid for extremely dilute solutions, and is obtained from computation of the ideal entropy of mixing equation of statistical mechanics[3]. Equation (1) requires that if there is an excess of positive ions near the interface, $C_+(x)$, decreasing monotonically to its bulk average concentration, $C_+(d)$, at distances far from the interface, then the electric potential $\emptyset(x)$, in this region must increase monotonically from some negative value $\emptyset(o)$ at x = 0, to the value zero, at large distances away from the interface. The conclusion that when $C_+(x)$ decreases, $\emptyset(x)$ must increase, follows from equation (1), since at equilibrium the chemical potential of a species should be the same everywhere[4]. One should recall that in obtaining equation (1) from the ideal entropy of mixing expression, the mole fractions were replaced by the concentrations of solutes in dilute solutions and that the contribution of electric potential to chemical potential was phenomenologically added. The contributions to chemical potential arising from ion-ion and ion-dipole interaction energy terms were neglected. The schematic forms of concentrations of ions, the negative of local charge density $Y(x)$ and electric potential profile, as dictated by equation (1) are presented in Figure 2.

It is evident that the surface must contain negative charges in order that the concentrations of mobile positive ions exceed the concentrations of mobile negative ions, in regions near the interface. Thus, the electric potential profile depicted in Figure 2, describes essentially, the potential arising from the negatively charged surface. However, this potential profile ignores the contribution to potential at x, $\emptyset(x)$, from other ions present in solution around the region x. The sign and magnitude of the total potential felt by a unit charge at x is open to question.

Equation (1) leads to the equilibrium Nernst expression, which relates the concentrations of specified ion at two different locations, with the difference in electric potential at these two locations.

$$Z_\sigma \beta_{Nernst} = \ln \left\{ [C_\sigma(x_1)/ C_\sigma(x_2)] \right\} = (Z_\sigma e/kT) [\emptyset(x_2) - \emptyset(x_1)]$$

(2)

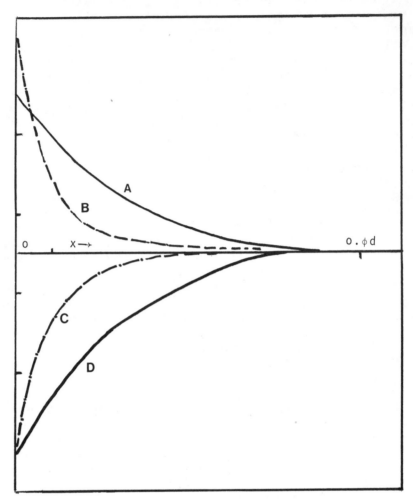

FIGURE 2. Concentration profiles, $C_+(x)$ [Plot B], $Y(x)$ [Plot C] and electric potential profile [Plot D] as given by equations (1) and (2) are presented. Plot A, representing $\beta(x)$ was assumed and other plots were computed. Y-axis scales vary for different plots.

The Nernst-Planck equation and the Poisson-Boltzmann equation[5,6], which form the basis for most quantitative approaches to electrical phenomena in biology, are based on the validity of equation (1). It is well known that these equations are nonlinear and several reviews concerning the solutions of these equations[7,8] are available. Most of these attempts invoke the convenient additional assumption that the positional dependence of the dielectric coefficient can be ignored.

The structure of the double layer, resulting from equation (1), is presented in the treatise of Kruyt[9] and the review of Grahame[10]. Grahame has included several criticisms and limitations of the classical theory of the electrical double layer. The nonvanishing nature of the electric potential and its gradient for finite distances away from the interface, imply that the inhomogeneous interfacial region is almost infinite in extent. On the other hand, the interfacial inhomogeneous region near a biological interface is finite, whose magnitude d, is determined by the system in a self-consistant manner[11]. In the inhomogeneous part of interfacial region, the concentrations of the ionic species, electric potential and dielectric coefficient are functions of the position variable. At locations for values of x greater than d, these become independent of x. The dielectric coefficient of aqueous electrolyte solution is about 80.36 at 293.15° K. The value of the dielectric coefficient near the interface is not precisely known. The usual speculation is that the value is about 2 to 6. If indeed, due to finite size of ions and water molecules, positive ions are precluded from being in close proximity with fixed negative charges of the surface, one may even speculate the value of dielectric coefficient, very near the surface will be unity.

The value of the dielectric coefficients of hydrocarbons is of the order 2. It is known that the value of the dielectric coefficient is a function of the local electric field[12]. Since the electric field near the interface is not a constant, even as solution of the Poisson-Boltzmann equation, the study of inhomogeneous region must include the positional dependency of the dielectric coefficient. If one deems that the dielectric coefficient is essentially a fudge factor, in reduction of the local intercharge interactions, then the local value of the dielectric coefficient is a measure of the influence of the uncharged species present in effecting this reduction.

THE RIDDLE

There arises a number of questions, when one examines critically this classical concept of the electrical double layer, as applied to biological membrane systems. The concentrations of ions in aqueous solutions of biological systems are of the order, uninormal. One is therefore inclined to study the role played by higher concentrations of ions, on distributions of ions and electric potential profile, in interfacial region, than the concentrations at which equation (1) is valid. For values of electrical potential difference of about 100 millivolts, (a value that one normally encounters in biological systems), one obtains from equation (2), that $C+(o) = 54.6 \, C_{+}(d)$ for univalent ions and that $C_{\delta}(o) = 2981 \, C_{\delta}(d)$, if δ is a divalent ion. Evidently these values are too high, for ions to be physically accommodated at or near the surface. The second question that arises is to determine the factors which stipu-

lates the magnitude of the extent of the interfacial inhomogeneous region, d. The third question that one should investigate is the approximations involved in the neglect of the existance of a dielectric profile.

Assuming either the concentration profile of ions, or the
electric potential profile in interfacial region, one can com-
pute the values of $\beta(x)$, $C_\sigma(x)$, $\emptyset(x)$ and $\emptyset''(x)$ for various
values of x, in interfacial region, using equations (1) and
(2). The knowledge of composition of electrolyte will enable
one to compute $Y(x) = -(4\pi e) \sum_\sigma C_\sigma(x) Z_\sigma$, for various values
of x. If indeed the classical equations are correct, and the
assumption of constant dielectric coefficient is valid, then
it is imperative that the values of $Y(x)$ are proportional to
the corresponding values of $\emptyset''(x)$. One may verify by simple
calculations that this conclusion is not borne out by equations
(1) and (2).

One of the unsolved problem of the interfacial region
system, consisting of a surface with charges, and an electrolyte of known composition, is the lack of a reliable expression
existing between the electric potential difference and the
surface charge density. An expression for this relation, known
as Gouy equation is presented in the review of Grahame[10].
Gouy equation is obtained from Gauss theorem, and its validity
is restricted to a symmetrical 1-1 electrolyte system, provided
that the dielectric coefficient can be regarded as a constant.
A general equation valid for unsymmetrical multi-ion system
as one encounters in biology, is not available. Since the
magnitude of the electric field near a biological surface is of
the order of 10^5 volts/cm, or larger, the effect of the electric field on dielectric coefficient cannot be ignored[13].

In an attempt to answer some of these questions, at least[14]
partially, the contributions from ion-ion interaction energy
terms, which become significant at higher concentrations, were
included by me, as a correction to the classic limiting expression for chemical potential of ions in concentrated solutions.
The main elements of this approach are briefly presented in the
appendix. Our analysis has led to certain surprising conclusions and have raised a fundamental question, which forms the
main subject of this paper. The fundamental question is that
whether the electric potential profile and concentration profiles of positive ions in the interfacial region should be
schematically similar or dissimilar. Our analysis, based on
the assumed validity of equations of appendix, leads to the
conclusion that a dielectric profile must exist in the interfacial region and that the ratio of $[\emptyset'(x)/Y'(x)]$, where
$Y'(x) = [dY/dx]$, in the interfacial region must be negative
for all values of x. This conclusion implies that the electric potential profile and concentration profile of positive
ions, must be schematically similar as presented in Figure 3.
Evidently this contradicts the conclusions that one obtains
from equations (1) and (2).

36

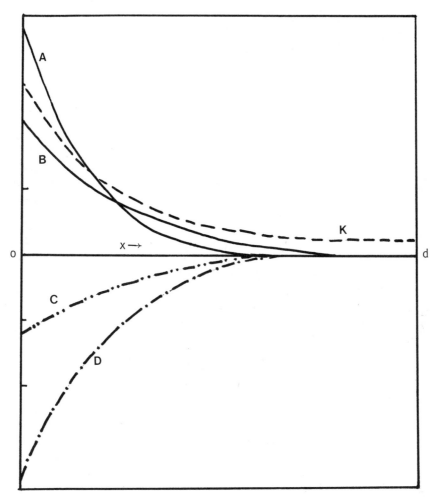

FIGURE 3. The various profiles in interfacial regions
as expressed by equations (3) and (4). Plot A denotes
the electric potential profile $\emptyset(x)$, while plot D denotes its derivative, $\emptyset'(x)$. Plot C denotes computed
values of $Y(x)$ while plot B represents its derivative
$Y'(x)$. $K(x) = \varkappa^2(x) \in (x)$ computed for the three ion
system, presented in the text.

These conclusions are based on the results, which are
exact, dependent on the validity of equations of appendix, viz.,

$$\beta(x) = (e/kT)\Delta\emptyset(x) + [He/4\pi kT] \, Y(x) \tag{3a}$$

$$= (e/kT)\Delta\emptyset(x) + \left\{ [\sum_{\sigma} z_{\sigma}c_{\sigma}(x)]/[\sum_{\sigma} z_{\sigma}^2 c_{\sigma}(d)] \right\} \tag{3b}$$

$$\beta(x) = (e/kT)\Delta\emptyset(x) + \sinh \beta(x), \text{ for } 1-1 \text{ ions.} \tag{3c}$$

$$\varkappa^2(x) \in (x)\emptyset'(x) = Y'(x) \left\{ 1 - [K(x)/K(d)] \right\} \tag{4a}$$

$$[4\pi/H] = -K(d) \qquad (4b)$$

$$K(x) = \varkappa^2(x)\,\epsilon(x) = (4\pi e^2/kT)\sum_\sigma z_\sigma^2 C_\sigma(x) \qquad (4c)$$

$K(x)$ is the product of local values of dielectric coefficient and Debye-Huckel parameter of strong electrolyte theory.
One may verify that for an electrolyte of any composition, the
value of $K(x)$ is always greater than the value of $K(d)$. [This
statement is exact for symmetrical electrolyte system. An exception to this statement occurs for 2-1 electrolyte system,
for values of β lying in the range, $-1 < \beta < 0$, and for
unsymmetrical ion system such an exception occurs for smaller
range of negative values for β.] Therefore, when equation
(4a) is valid, it follows that $[\emptyset'(x)/Y'(x)]$ in interfacial
regions must be negative definite. Equation (1) on the other
hand, suggests that $[\emptyset'(x)/Y'(x)]$ should be positive definite
for all values of x. When dielectric coefficient is not a
function of position variable, $Y'(x) = \emptyset'''(x)$, where
$\emptyset''' = [d^3\emptyset/dx^3]$. Debye-Huckel linearized Poisson-Boltzmann[15]
equation, requires that $\emptyset'$ and $\emptyset'''$ must have similar signs.
Equation (3a) essentially states that the concentration distributions of ions in interfacial region, is determined by the
sign and magnitude of local charge density $Y(x)$, in addition
to the difference in electric potential $\emptyset(x)$. This perturbation due to ion-ion interaction energy terms, to Nernst expression, results in equation (3). Equation (3a) is valid in general, and equation (3b) is valid, when equation (4b) is valid,
for many ion unsymmetrical electrolyte system. Equation (4b)
can be shown to be exact for a 1-1 electrolyte system, provided
that a dielectric profile exists, and is valid in general, if
an additional boundary condition that $\emptyset'''(d) = 0$. Equation
(3c) is derived from (3b), when electrolyte consists of univalent ions only. Since the magnitude of β is always less
than the magnitude of $\sinh\beta$, when β is nonzero, equation
(3) implies that the electric potential must have the same
sign as the sign of local charge density. This riddle, leading
to profiles of Figure 3, needs to be solved.

Since most applications of electrochemistry to biological
phenomena are based on the assumed validity of equation (1),
which is evidently approximate and inadequate, it is important
that one obtains an answer to the fundamental question raised.
The correction to equation (2) presented in equation (3), involves the deviation from microscopic electroneutrality, that exist
locally. Since most experiments are carried out under conditions
where electroneutrality is valid, it is difficult, if not impossible, to devise a suitable experiment to verify the validity
of one and nonvalidity of another of the two diverse conclusions.

The problem of variation of chemical potential of a species, in solution, with composition is an yet unsolved fundamental problem in physical chemistry. [In literature, two theories of solutions, one attributed to Mayer and McMillan[16] and another to Kirkwood and Buff[17], which are stated to be exact, address these problems. These cannot as yet explain the dependence of excess free energy of mixing on composition in an unsymmetrical manner. Additional remarks about these are extraneous to our current discussion.] The familiar statement that equation (1) is valid for concentrated solutions, by replacement of concentrations by activities, does not enlighten the solution to the problem. It is true that one can measure the activity coefficient experimentally from one colligative property and can utilize this value to compute other colligative property. However, it should be remarked that the activity coefficient is just a measure of our ignorance factor, and one does not as yet have a chemical potential meter. In order to answer the fundamental question raised, an analysis independent of the validity of equations (3) and (4) is presented in the following section.

A SIMPLE ANALYSIS

With the objective of finding an answer, assume that in the interfacial region, of extent d, the electrical potential decays from a value $\phi*$ at $x = 0$, to a value zero at $x = d$, in a monotonic manner. When the electric field is constant, the electric potential profile will be linear. The presence of ions in interfacial region, distorts this linear profile and one may express this perturbation as

$$\phi(x) = \phi_o + \phi_1 x + \phi_2 x^2 \; ; \quad \phi_o = \phi* \qquad (5)$$

Equation (5) may be regarded as a Taylor series, where only the leading three terms have been retained. If $\phi(d) = 0$, $\phi'(d) = 0$, one obtains $\phi_1 = - (2\phi*/d)$ and $\phi_2 = (\phi*/d^2)$. If the magnitude of $\phi* = 100$ mV $= (3.3137333/10000)$ esu/cm, and $d = 2 \times 10^{-7}$ cm, ϕ_1 will equal $- 3313.733$ esu/cm^2, when equation (5) is valid, and ϕ_1 will equal $- 1658.7$ esu/cm^2, when field is constant. Thus, the value of electric field very near the interface has doubled in value, when ϕ_2 is nonzero. When the electric potential is linear, $Y(x) = 0$, and $\epsilon'(x)$, the gradient of dielectric coefficient is zero. This is the situation of constant dielectric coefficient. Since $Y(d) = 0$, $\phi''(d)$ should equal zero, when ϕ_2 does not necessarily equal to zero. Thus, one should retain at least the leading four terms in the Taylor expansion of the electric potential profile, to satisfy the condition that $Y(d) = 0$, and $Y(x) \neq 0$.

One may express the electric potential profile in interfacial region in a Taylor series.

$$\phi(x) = \sum_{i=0} \phi_i x^i \; ; \quad (i!) \; \phi_i = [d^i\phi/dx^i]_{x=0} \qquad (6)$$

It is shown elsewhere[18], that retention of only the leading six terms of equation (6) leads to the result that $d = 19.38 \times 10^{-8}$ cm, for a three ion system, with concentrations, $C_-(d) = 5 \times 10^{-5}$, $C_+(d) = 3 \times 10^{-5}$ and $C_\xi(d) = 1 \times 10^{-5}$ (moles/cm^3) $[Z_+ = - Z_- = 1; \; Z_\xi = 2]$. When concentrations of ions in electrolyte are of this order, one may assume that $d = 2 \times 10^{-7}$cm, a value we shall employ. For every tenfold decrease in ion concentrations, the value of d increases by a multiplicative factor of $(10)^{1/2}$.

If one assumes that a dielectric profile exists and that it can be expressed in a Taylor series, with truncation of terms of order x^4 and higher order, one has

$$\epsilon(x) = \epsilon_0 + \epsilon_1 x + \epsilon_2 x^2 + \epsilon_3 x^3$$

$$\epsilon'(d) = 0 = \epsilon''(d) \; ; \quad \epsilon_3 d^3 = \Delta\epsilon = \epsilon(d) - \epsilon(0)$$

$$\epsilon_1 d = 3 \Delta\epsilon = -\epsilon_2 d^2 \qquad (7)$$

In this manner one may specify the electric potential and the dielectric profiles. Specification of the values of $\epsilon(0)$ and $\epsilon(d)$, yields the Taylor coefficients and one can compute $\epsilon'(x)$ and $\epsilon(x)$ for any value of x. $[0 < x < d]$. When $\phi* = 100$ mV, and $\phi_5 d = - 0.2 \phi_4$ say, the values of Taylor coefficients are known and one can compute $\phi'(x)$ and $\phi''(x)$ for any value of x, in interfacial region.

One may now compute, the Taylor coefficients of the function $Y(x)$, using the relations,

$$Y(x) = \epsilon(x)\phi''(x) + \epsilon'(x)\phi'(x) = \sum_{i=0} Y_i x^i$$

$$Y_0 = 2 \epsilon_0 \phi_2 + \epsilon_1 \phi_1$$

$$Y_1 = 6 \epsilon_0 \phi_3 + 4\epsilon_1 \phi_2 + 2 \epsilon_2 \phi_1$$

. . . .

. . . . $\qquad (8)$

Since, one has in addition

$$Y(x) = - (4\pi e) \sum_\sigma Z_\sigma C_\sigma(d) \exp [Z_\sigma \beta(x)] \qquad (9)$$

the values of $C_\sigma(0)$ and $\beta(0)$ corresponding to the value of Y_0 can be computed. These calculations does not involve the validity of equations (1), (3) and (4). They involve only

the assumptions that an electric potential profile and a dielectric profile exist in the interfacial region and that
these are continuous analytic functions of position variable.
The validity of equations (8) and (9) are assumed. Assuming
that the value of ϵ_o equals 10, and that $\epsilon(d) = 80.36$,
$d = 2 \times 10^{-7}$ cm, the computed values of various quantities,
obtained by assuming that the Taylor expansion of functions
may be truncated, retaining only the leading $(n + 1)$ terms,
are presented in Table 1. In these calculations it is assumed
that $\emptyset* = 3.3137333 \times 10^{-4}$ esu/cm and that the electrolyte
contains 3 ions of concentrations specified. Utilization of
the relation,

$$(kT/e)\ \beta(o)\ =\ (H/4\pi)\ Y_o\ -\ \emptyset* \tag{10}$$

one may compute the value of ion-ion interaction energy contribution term, $(H/4\pi)$. The values of $(H/4\pi)$, obtained in this
manner are also presented in Table 1. These values are negative, for all values of n, and have magnitudes in the order
of 10^{-16} cm^2. The value of $[1/K(d)]$ equals 1.9308×10^{-16}
cm^2, for the electrolyte with three ions at $20^{\circ}C$. The agreement between the orders of computed values of $(H/4\pi)$ and $[K(d)]^{-1}$
indicate the plausible validity of equation (4b).

TABLE 1. Computed values of various quantities, for various
orders of truncation of Taylor series of equation (6) and assuming the validity of equation (7). $\epsilon(o) = 10$; $\epsilon(d) = 80.36$.

Term	n = 1	n = 2	n = 3	n = 4	n = 5
Y_o	− 174.87	− 332.17	− 474.9	− 516.6	− 706.6
$\beta(o)$	1.4514	1.8050	1.9996	2.0452	2.2152
Y_1	174.86	720.49	1548.9	1932.1	4123.1
$\emptyset_1$	− 1658.7	− 3313.7	− 4970.6	− 5522.9	− 8284.3
$K(o)$	374.96	720.41	1040.4	1134.6	1570.5
$[H/4\pi]$	− 1.2142	− 0.5987	− 0.4170	− 0.3740	− 0.2646

$\emptyset* = 100$ mV, $d = 2 \times 10^{-7}$cm and $K(d) = 51.79 \times 10^{-14}$ cm^{-2}.
The values of Y_o are in 10^{10} esu/cm^3 and the values of $\emptyset_1$
are in 10^3 esu/cm^2. The values of Y_1 are in 10^{17} esu/cm^4
and the values of $K(o)$ are in 10^{14} cm^{-2}. The values of $[H/4\pi]$
are in 10^{-16} cm^2.

The computed values presented in Table 1, demonstrate that $[\emptyset^*/Y_o]$ and $[\emptyset_1/Y_1]$ are negative for all values of n. These negative values support the conclusions obtained on the basis of equations (3) and (4). The magnitudes of various quantities computed depend on the assumed value of $\varepsilon(o)$ and the validity of equation (7a) for the dielectric profile. If one now assumes different values of $\varepsilon(o)$, assuming the validity of equation (7) and computes the value of $(H/4\pi)$ in each case, retaining the leading seven terms of equation (6), one obtains the result that $\varepsilon(o)$ should equal 29.82, in order that the computed value of $[4\pi/H]$ equals $- K(d)$.

SUMMING UP

Our analysis of distribution of ions in interfacial inhomogeneous region, is based on the assumed validity of three postulates. These are: 1. The chemical potential of an ionic species in the interfacial region is adequately described by equation (A.2) of the appendix, where certain interaction contributions arising from the presence of nonelectrolyte molecules, (H* terms) have been neglected. 2. The finite extent, d, of the inhomogeneous region is determined by the system. [The value of d, determines the ratio of $\Delta\emptyset$ and Y_o, as well as the ratio $[\emptyset_1/Y_1] = [K(d) - K(o)]/[K(d)K(o)]]$. 3. In the interfacial region, concentrations, electric potential and dielectric coefficient are functions of position variable.

Thus, our basic postulates, which include ion-ion energy interaction contribution to chemical potential, require that the system parameters determine the electric potential profile and dielectric profile. If one knows or specifies the electric potential profile together with the concentrations of ions in aqueous solution, one should be able to specify the Taylor coefficients of dielectric profile. The results of such computation, assuming that the electric potential decreases monotonically from the value of 100 mV at $x = 0$, is presented in Figure 3. In these calculations it is assumed that $d = 2 \times 10^{-7}$ cm and that the electrolyte consists of three ions with concentrations $C_- = 5$, $C_+ = 3$ and $C_\delta = 1$ $(\times 10^{-5}$ moles/cm$^3)$. For this electrolyte system, the values of $\beta(x)$ are given by the solution of the equation,

$$12\,\beta(x) - 3\,e^\beta - 2\,e^{2\beta} + 5\,e^{-\beta} = -12(e/kT)\emptyset(x) \qquad (11)$$

where $\emptyset(d) = 0$. Equation (11) is obtainable from (3b). The values of $Y(x)$ are known from the relation,

$$Y(x) = -(4\pi eN)\left\{3\,e^\beta + e^{2\beta} - 5\,e^{-\beta}\right\} \times 10^{-5} \text{ (esu/cm}^3)$$

$$N = 6.02486 \times 10^{23}/\text{mole;} \quad e = 4.80298 \times 10^{-10} \text{ esu} \qquad (12)$$

Retention of the leading six terms of equation (6), with $\emptyset(d) = 0$, yields,

$$\emptyset_o = \emptyset_4 d^4 (1 + 4R) \quad ; \quad \emptyset_1 = - \emptyset_4 d^3 (4 + 15R)$$

$$\emptyset_2 = \emptyset_4 d^2 (3 + 10R) \quad ; \quad \emptyset_3 = - \emptyset_4 d (4 + 10R)$$

$$R = [\emptyset_5 d / \emptyset_4] = -0.2, \text{ when } [d^4 \emptyset / dx^4]_{x=o} = 0. \tag{13}$$

The plots presented in Figure 3, were obtained using equations (10), (11) and (13), with the assumption that $\emptyset_o = 100$ mV. One may verify that for any multi-ion system, β has the same sign as $\emptyset_o$ and an opposite sign of Y_o.

However, one is now confronted with two new problems. One may notice that the profiles of $\emptyset(x)$ and $Y(x)$ are not quite akin to the schematic plots presented in Figure 1. The dielectric profile specified by the system, should in principle include specification of the value of $\varepsilon(o)$. Assuming the value of $\varepsilon(o)$ and computation of Taylor coefficients of dielectric profile, when $\emptyset(x)$ is monotonic always exhibits an extremum value. The presence of such an extremum in dielectric profile is not physically realistic. The dielectric profile, when it exists in the interfacial region should be monotonic.

The value of dielectric coefficient $\varepsilon(o)$, is given by the Gauss integral,

$$\int_o^d Y(x) \, dx = - \varepsilon(o)\emptyset'(o)$$

$$Y(x) = (d/dx) [\varepsilon(x)\emptyset'(x)] \tag{14}$$

The value of the integral of $Y(x)$, is negative as shown in Figure 3. Since $\emptyset'(o)$ is negative, the validity of equation (14) requires that the value of $\varepsilon(o)$ is negative, if the profiles of $\emptyset(x)$ and $Y(x)$ are similar to those presented in Figure 3.

Since the value of the integral presented in equation (14) is negative definite and the value of dielectric coefficient at $x = 0$ is positive definite, the only manner in which one can overcome this dilemma, is to insist that $\emptyset'(o)$ is positive definite. The electric potential profile satisfying all these conditions is presented in Figure 4. The Taylor expansion coefficients utilized to compute the profile of Figure 4, are listed in Table 2. The computed value of the profile of $Y(x)$ using the profile of Figure 4 and equations (11) and (12) are presented in Figure 5.

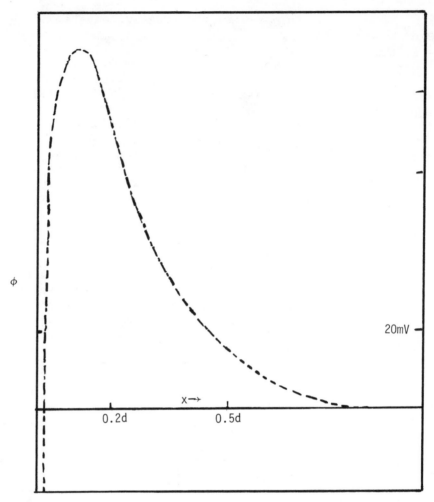

FIGURE 4. Assumed electric potential profile, computed with the values of Table 2. $\emptyset(o) = -100$ mV.

TABLE 2.
Computed values of Taylor expansion coefficients of electric potential profile depicted in Figure 4.

$\emptyset_o = -3.3137333 \times 10^{-4}$ esu/cm. $\quad d = 2 \times 10^{-7}$ cm.

$\emptyset_1 d = -48.59\ \emptyset_o$ $\quad ;\quad \emptyset_1 = +80.5071 \times 10^3$ esu cm^{-2}

$\emptyset_2 d^2 = 467.21\ \emptyset_o$ $\quad ;\quad \emptyset_2 = -387.052 \times 10^{10}$

$\emptyset_3 d^3 = -2285.51\ \emptyset_o$ $\quad ;\quad \emptyset_4 d^4 = 6748.234\ \emptyset_o$

$\emptyset_5 d^5 = -12650.39\ \emptyset_o$ $\quad ;\quad \emptyset_6 d^6 = 15118.13\ \emptyset_o$

$\emptyset_7 d^7 = -11136.97\ \emptyset_o$ $\quad ;\quad \emptyset_8 d^8 = 4602.65\ \emptyset_o$

In addition to the values listed in Table 2, one has $\phi_9 d^9 = -815.7192\phi_0$. Equation (3) yields that $\beta(o) = -2.78217$ and that $Y_O = 293.02445 \times 10^{10}$ esu/cm^3.

One obtains the value of the integral, $\int_0^d Y(x)\,dx$, presented in Figure 6, as equal to -81.825×10^3 esu/cm^2. The value of $\phi'(o)$ equals $+80.507 \times 10^3$. Therefore, when equations (3), (4), (6), (8), (9) and (14) of this paper are valid, the value of the dielectric coefficient at interface containing fixed charges, equals UNITY.

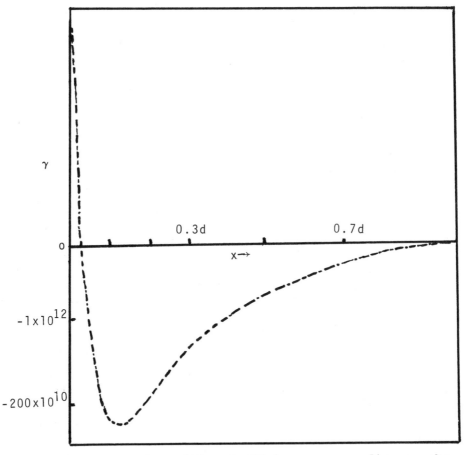

FIGURE 5. Computed profile of Y(x), corresponding to the electric potential profile of figure 4, using equations (3) (9) and (11).

In Figure 6, two possible dielectric profiles, assuming that $\epsilon(d) = 80.36$ and $\epsilon(o) = 4$, using equation (7) and equation of the type,

$$\epsilon(x) = \sum_{i=0}^{m} \epsilon_i x^i \qquad (15)$$

with $m = 7$, are exhibited. When one insists that the dielectric profile is monotonic and approaches $\epsilon(d)$ in an asymptotic manner, one may impose the conditions that $\epsilon'(d) = \epsilon''(d) = 0$, $\epsilon'''(d) = 0$, etc. These conditions lead to the result,

$$\epsilon_1 d = [m \Delta \epsilon] ; \qquad \Delta \epsilon = \epsilon(d) - \epsilon(o) \qquad (16)$$

where m is the integer where the Taylor expansion of dielectric profile is truncated. Since $\emptyset_1$, $\emptyset_2$, $\epsilon(d)$, Y_o and ϵ_1 are known, one may solve for the value of dielectric coefficient, $\epsilon(o)$, using the relation,

$$Y_o - \emptyset_1 \epsilon(d) = \epsilon(o) [2 \emptyset_2 + (m \emptyset_1/d)] \qquad (17)$$

One obtains utilizing the values presented, that $\epsilon(o)$ equals respectively, 7.23 ($m = 2$), 10.517 ($m = 3$) and 21.168 ($m = 7$). One may observe from figure 6, that for the distance of separation of 2×10^{-8} cm, the dielectric coefficient has attained the value of 22.9, when $m = 3$. The basic reason for the disagreement in the computed values of $\epsilon(o)$ by this method and unity, is due to the assumption of arbitrary electric potential profile, with an arbitrary location at which the extremum occurs for the system, with specified concentrations. The system in principle decides all these factors.

If $\epsilon(o)$ equals unity, and if one assumes that equation (13) is valid, one has

$$[\emptyset_o/Y_o] = \left\{ [4(3+10R) - 79.36 \; m \; (4+15R)]/[d^2(1+4R)] \right\}$$

$$[\emptyset_1/Y_1] = \left\{ [K(d) - K(o)]/[K(o)K(d)] \right\} \qquad (18)$$

For the electrolyte under consideration, $K(d) = 51.79 \times 10^{14}$. If $R = -0.2$, one obtains the results that $d = 36.38 \times 10^{-8}$cm from equation (18a) and that $d = 49.64 \times 10^{-8}$ using equation (18b), when $\emptyset_o = -3.313733 \times 10^{-4}$, and equation (7) adequately describes the dielectric profile.

If instead of stipulating the value of R, one computes R using equation (18b) and (7) for the system, one obtains, $R = -0.277$, while equation (18a) yields, $R = -0.299$. The extremum value in electric potential profile is evidently determined by the ratio $[\emptyset_1/\emptyset_2]$, which needs to be computed more accurately.

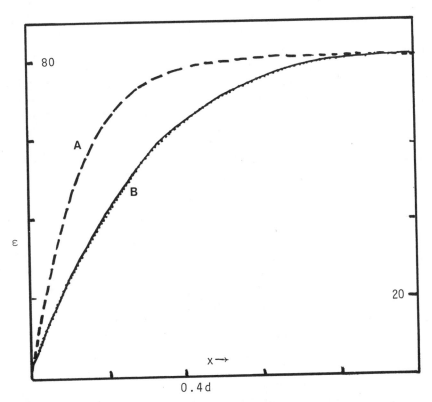

FIGURE 6. Dielectric profiles in inhomogeneous region computed with the use of equations (15) and (16). Plot A denotes profile obtained when $m = 7$ ($\epsilon_1 = 267.26 \times 10^7$) and plot B denotes the profile obtained with $m = 3$. ($\epsilon_1 = 114.54 \times 10^7$, $\epsilon_2 = -57.27 \times 10^{14}$ and $\epsilon_3 = 9.545 \times 10^{21}$).

CONCLUSION

It is shown that inclusion of the contribution to chemical potential of an ion from ion-ion interaction energy terms leads to three main conclusions. 1. In interfacial region, ion distributions, $\beta(x)$, are determined by magnitude and sign of local charge density (deviation from electroneutrality), in addition to electrical potential difference. 2. The electric potential profile and concentration profile of positive ions are schematically similar, exhibiting an extremum value, as shown in Figure 4. (The magnitude of local charge density dominates over the contribution to local potential from surface of the interface). 3. A dielectric profile exists. Two other plausible conclusions are that the value of dielectric coefficient at the interface is almost unity and that the magnitude of region d, is determined by electrolyte composition and property of surface.

The results presented in the section entitled 'A Simple Analysis' requiring only that a dielectric profile and an electric potential profile exist in the interfacial region, lend support to the conclusions obtained when equations (3) and (4) are valid. One must therefore reexamine the applicability and validity of equations (1) and (2) for the study of biological interfacial systems. The use of Maxwell's osmotic balance equation (A.5) and equations (3) and (4) have been shown to yield more reasonable values of resting potentials of various axon systems,[19,20] than classical Nernst expression using potassium concentrations.

APPENDIX: BASIC EQUATIONS

When the system contains aqueous solution with electrolytes in significant concentrations, it is assumed that the expressions for the chemical potentials of ionic species, σ, and nonionic species j, are

$$\mu_\sigma(x) = \mu_\sigma^*(T,P) + Z_\sigma e \, \emptyset(x) + kT \ln C_\sigma(x)$$
$$+ \sum_\eta C_\eta(x) H_{\sigma\eta} + \sum_j C_j(x) H_{\sigma j} \qquad (A.1\ a)$$
$$\mu_j(x) = \mu_j^*(T,P) + kT \ln C_j(x) + \sum_k C_k(x) H_{jk}$$
$$+ \sum_\sigma C_\sigma(x) H_{j\sigma} \qquad (A.1\ b)$$

$H_{\sigma\eta}$ are molecular integrals, of potential energy of interactions weighted by probability functions, of σ and η -th kind of ionic species. $H_{\sigma j}$ and H_{kj} are similar molecular integrals of ion-uncharged molecule interactions and dispersion force interactions contributions to chemical potentials. The greek subscripts denote charged species and the roman subscripts indicate uncharged species present in the system. $H_{\sigma\eta}$, denote the contribution of chemical potential of σ -th kind of ion, by the presence of η -th kind of ions. Since $H_{\sigma\eta}$, $H_{\sigma j}$ and H_{jk} are molecular integrals over the whole volume of space, one may assume that these are position independent, and that one can represent, $H_{\sigma\eta} = Z_\sigma Z_\eta e^2 H$ and $H_{\sigma j} = Z_\sigma e H^*$, where H and H* are charge independent portions. With the use of Poisson equation, one may now write the gradients of chemical potentials as

$$\mu_\sigma'(x) = kT [d \ln C_\sigma /dx] + Z_\sigma e \left\{ F'(x) + H^* \sum_j C_j' \right\}$$
$$\mu_j'(x) = kT [d \ln C_j/dx] + \sum_k H_{jk} C_k' + H^* e \sum_\sigma Z_\sigma C_\sigma'$$
$$F(x) = \emptyset(x) - (H/4\pi) Y(x) \qquad (A.2)$$

Under equilibrium conditions, the gradients of chemical potentials, μ_σ' and μ_j' are zero. Under stationary state conditions, the gradients of chemical potentials are the mean force acting on species, and these can be related to partial frictional coefficients. With H* term neglected, one obtains the Nernst-Planck analog equation, and further neglect of H term and positional dependence of frictional coefficient, one obtains the Nernst-Planck equation of electrodiffusion.[21]

Provided that ion-neutral molecule term, H*, is nonvanishing, when concentration gradient and flux of any permeant uncharged molecule vanishes, it can be shown that the third derivative of electric potential, under stationary state conditions should be a constant, independent of position variable x.

$$\emptyset'''(x) \;=\; - \,(4\pi/\text{H}^*\epsilon)\; \sum_\sigma J_\sigma \zeta_{\sigma j} \qquad (A.3)$$

where J_σ is the flux of ionic species σ, and $\zeta_{\sigma j}$ is the partial frictional coefficient representing the interaction between j and σ kind of molecules. Equation (A.3) implies that under these conditions, the electric potential will be a third order polynomial in x, in the inhomogeneous regions. Recognizing the constant field approximation, equation (A.3) suggests that the electric potential profile in general case may be expressed in various order polynomials in x. This is the basis for the adoption of Taylor series expansion of the electric potential profile.

Equations (3) and (4) of the main text follows from equation (A.2), when one neglects H* term, and confines attention to equilibrium situations. The resultant differential equations, relating gradients of concentrations of ionic species, with gradients of electric potentials and the gradients of dielectric profile, can be written as leading to equation (4a) and to equation (A.4) which is a generalized form of Maxwell's Osmotic balance equation.

$$4\pi kT \quad K(x)A'(x) \;=\; Y'(x)Y(x)$$

$$A(x) \;=\; \sum_\sigma C_\sigma(x) \;;\; A'(x) = [dA/dx] \qquad (A.4)$$

One may verify that for an electrolyte of any composition, $A(x)$ is always greater than $A(d)$ in interfacial regions. Equation (A.4) is obtained by summing the differential equation for concentration gradient of each ionic species, over all ionic species. Equation (4a) is obtained by multiplying each equation for species σ, by $Z_\sigma e$, and summing over all species. Thus, equations (3), (4) and (A.4) are exact, subject only to the validity of equations (A.2) and neglect of H* terms. Since $F(x)$ replaces $\emptyset(x)$ of equation (1), by the inclusion of energetic interactions, the condition of constant chemical potential, requires that $F(x)$ and $C_+(x)$ be schematically dissimilar in interfacial region.

REFERENCES

1. J.G.Kirkwood and I. Oppenheim, 'Chemical Thermodynamics'
    McGraw Hill Book Co. New York, (1961).

2. T.L.Hill, in 'Membrane Phenomena' Disc. Faraday Society,
    (1956) p.31.

3. R.H.Fowler and E.A.Guggenheim, 'Statistical Thermodynamics'
    Cambridge University Press, New York, (1939).

4. E.J.W.Verwey and Th.G.Overbeek, 'Theory of Stability of
    Lyophobic Colloids' Elsevier Publishing Co. New York (1945)
    p. 31.

5. N.Lakshminarayaniah, 'Transport Phenomena in Membranes'
    Academic Press, New York, (1969).

6. S.Rice and M.Nagasawa, 'Polyelectrolyte Solutions'
    Academic Press, New York, (1961).

7. D.Agin, in 'Foundations of Mathematical Biology' R.Rosen
    editor, Academic Press, New York, (1971) vol. 1.

8. H.R.Leuchtag and H.M.Fishman, in 'Structure and Function
    in Excitable Cells' edited by Chang, Tasaki, Adelman and
    Leuchtag, Plenum Publishing Copn. New York,(1983).

9. H.R.Kruyt, 'Colloid Science' Elsevier Publ.Corpn. Amster-
    dam, (1952).

10. D.C.Grahame, Chem. Reviews, $\underline{41}$ 441 (1947).

11. V.S.Vaidhyanathan, Colloids and Surfaces, $\underline{6}$ 291-306 (1983).

12. F.J.Booth, J.Chem. Phys., $\underline{19}$, 39, 327 (1951).

13. V.S.Vaidhyanathan, J. Biol. Phys., $\underline{10}$, 153,167 (1982).

14. V.S.Vaidhyanathan, in 'Topics in Bioelectrochemistry and
    Bioenergetics' G.Milazzo, editor, John Wiley & Sons,
    London, (1976) vol.1,p.287-378. also, Bioelectrochemistry
    and Bioenergetics, $\underline{5}$ 754-775 (1978).

15. T.L.Hill, 'Introduction to Statistical Thermodynamics'
    Addison Wesley Publishing Co., Reading, Mass. (1960).

16. W.G.McMillan and J.E.Mayer, J. Chem. Phys., $\underline{13}$, 276 (1945).

17.   J.G.Kirkwood and F.P.Buff, J. Chem. Phys.,19 774 (1951).

18.   V.S.Vaidhyanathan, Bioelectrochemistry and Bioenergetics, 12, 105-118 (1984).

19.   V.S.Vaidhyanathan, Bioelectrochemistry and Bioenergetics, 7, 25-30 (1980).

20.   V.S.Vaidhyanathan, in 'Bioelectrochemistry:Ions, Surfaces and Membranes' M.Blank, editor, Advances in Chemistry Series, 188, 313-336 (1980).

21.   V.S.Vaidhyanathan, Bull. Math. Biology, 41, 365-385 (1979).

(The references on the main subject of this paper is vast. Only the references, which the author feels adequate are listed.)

THE EFFECT OF WATER POLARIZATION AND HYDRATION ON THE PROPERTIES OF

CHARGED LIPID MEMBRANES

I.S. Graham[1], A. Georgallas[2] and M.J. Zuckermann[3]

[1]Department of Physics, McGill University, Montreal
Quebec, Canada H3A 2T8
[2]Department of Physics, Queen's University, Kingston
Ontario, Canada K7L 3N6
[3]Department of Physics and Center for Polymer Studies
Boston University, Boston, MA 20215 (Permanent address:
Department of Physics, McGill University, Montreal
Quebec, Canada H3A 2T8)

Research supported in part by the NSERC of Canada and Les Fonds Formation
de Chercheurs et Action a la Recherche du Quebec and the Queen's University
Advisory Research Council.

ABSTRACT

We describe the physical properties of a planar interface between an
electrolyte and a charged lipid membrane. The model differs from the
usual Gouy-Chapman-Debye-Huckel theory for electrolytes in that the dipolar nature of the water molecules and the hydration force at the
interface are explicitly included. The dielectric constant and
potential of the electrolyte are calculated as functions of distance
from the interface for surface charge densities corresponding to anionic
lipid bilayers. Finally we look at the electrostatic properties of two
bilayers separated by a distance d.

1. INTRODUCTION

The object of this paper is to examine the electrostatic properties of
a set of ionic charges in a polar liquid near a charged surface. An
example of current experimental interest is that of charged lipid bilayer
membranes in an electrolytic solution. Clearly a theoretical understanding of the electrostatics is essential when interpreting data for
these systems.

The electric double layer which forms at the interface of charged
surfaces and electrolytic solutions has been the subject of much research
since the appearance of the Gouy-Chapman-Debye-Huckel (GCDH) theory
(Gouy (1910); Chapman (1913)) which in its turn has been modified in a
number of ways. For example, Stern (1924) introduced the concept of an
adsorption layer of counterions on the charged surface and pointed out
that the dielectric constant of the electrolyte above the charged
surface must be a function of the local electric field.

The GCDH theory has been used by several authors to describe the properties of charged lipid bilayers in contact with an electrolytic solution. This includes a calculation of the repulsive potential between charged bilayers (Lis et al (1982)) examination of the adsorption isotherms in the liquid crystalline state (Cohen and Cohen (1981)), an analysis of the experimentally determined zeta potential (McLaughlin (1979)) and an investigation of the behaviour of the temperature, T_f, of the main gel to liquid crystalline phase transition as a function of ionic strength and pH (Traüble et al (1976)). In these investigations the water is treated as a continuum with a dielectric constant of $\simeq 80$.

The question then arises: how are the properties of charged bilayers affected by a spatially varying water polarization and hydration forces? This question was addressed by Zuckermann et al (1985), who showed that, for a 1-1 electrolyte the inclusion of water polarization in the absence of a hydration force leads to a decrease in the dielectric constant and an increase in the predicted value of the electrostatic potential at the surface of the liquid bilayer (this work will be referenced as ZGP in the present paper). ZGP treated the water solvent as a system of permanent dipoles. The resulting dipole moment was described by a Langevin function and the hydrogen bonding matrix included by a method due to Booth (1951), in which the free dipole moment, p_o, inside the Langevin function was replaced by a moment $p \sim 4\,p_o$, giving a very approximate description of hydrogen bonding in water. However, the dipole-dipole interaction between the solvent molecules was not directly included in the formalism.

In the present paper we extend the treatment of ZGP to include both dipole interactions of the water molecules and the hydration force which decays exponentially away from the charged surface with a decay length ξ. (For a review of the hydration force see Rand and Parsegian (1984)). In section 2.1 we include these terms in a free energy and use a variational approach to derive two coupled differential equations for the electric potential, V, and the electric polarization, P, at any point in the 1-1 electrolytic solution.

A formalism involking ice-like Bjerrum defects to explain the hydration force has been investigated by Gruen and Marcelja (1983). We show in section 2.1 that our differential equations are identical to theirs in the linear limit of low potential and electric polarization.

We proceed to apply this formalism to the case of a single bilayer in section 2.2, where we calculate the values of the dielectric constant and potential near the surface of the bilayer and compare to the values obtained by GCDH and ZGP.

In section 2.3 we look at the case of two bilayers at a separation d and calculate effective electrostatic pressure and the ionic concentration at the midpoint. Our results are compared to the predictions of GCDH and to a recent Monte Carlo study by Jönsson et al (1980).

A discussion of the results follows in section 3.

2. THEORY

2.1 General Formalism

We have developed a statistical theory of a polar liquid (water) containing monovalent ionic charges which is bounded by a negatively charged planar surface. Because of the planar symmetry it is most convenient to work with the quantity $f(x)$ which is the free energy density at a distance x from the surface. In general

$$f(x) = f_c(x) + f_E(x) \tag{2.1}$$

where we have decomposed the free energy into a configurational and electrostatic contribution. The total free energy F is then

$$F = \int_0^\infty f(x)\,dx \tag{2.2}$$

Since we are concerned with the limit of low salt concentrations, the configurational free energy density is given by (Gruen and Marcelja (1983), ZGP):

$$f_c(x) = kT \{c^+(x)\ln c^+(x) - c^-(x)\ln c^-(x)$$
$$-(c^+(x) + c^-(x) - 2c_o)\} \tag{2.3}$$

where $c^\pm(x)$ are the concentration of positive and negative monovalent ions at x and c_o is the bulk salt concentration.

We use an expression for $f_E(x)$ which has been derived using the mean spherical approximation in the limit of interacting point charges and dipoles with zero ionic radius. The full derivation will be reported elsewhere by the authors. The expression is generally:

$$f_E(x) = \frac{\varepsilon_\infty E^2(x)}{8\pi} + [E(x) + \frac{\gamma}{2} P(x)]P(x)$$

$$- \frac{kT}{V_o} \ln Z(E,P) - \frac{\varepsilon_o}{\chi} \frac{\xi^2}{\varepsilon_\infty} \left(\frac{dP}{dx} \right)^2 \tag{2.4}$$

where Z(E,P) is the dipolar partition function

$$Z(E,P) = \frac{\sinh Y}{Y} \tag{2.5}$$

with

$$Y = \left\{ E + \gamma P + \frac{\varepsilon_o}{\chi} \frac{\xi^2}{\varepsilon_\infty} \frac{d^2P}{dx^2} \right\} \frac{p_o}{kT} \tag{2.6}$$

The symbols have the following meanings: P(x) is the water polarization, E(x) the electric field and T the absolute temperature. p_o and V_o are the dipole moment and specific volume of a water molecular and χ is the bulk water polarizability. ε_o is the bulk static dielectric constant of water and ε_∞ the dielectric constant above the first dispersion (Fröhlick (1958)). The decay length of the hydration force ξ has been estimated at $\sim 3\mathring{A}$ (Rand and Parsegian (1984)). The constant γ which arises through the mean spherical approximation (see Augousti and Rickayzen (1984)) can be related to ε_o through

$$\varepsilon_o = [\varepsilon_\infty + \frac{4\pi p_o^2}{3kTV_o} (1 - \frac{p_o^2}{3kTV_o} \gamma)^{-1}] \tag{2.7}$$

which is obtained by the requirement that in the absence of hydration forces and in the linear limit (2.4) must reduce to the GCDH expression.

We now have from (2.1) (2.3) and (2.4) an expression for F of (2.2) which must be minimized with respect to c^+ and c^- subject to the following electrostatic boundary conditions:

$$E(x) = -\frac{dV(x)}{dx} \tag{2.8}$$

$$\frac{dD(x)}{dx} = 4\pi q(c^+(x) - c^-(x))$$

$$D(x) = \varepsilon_\infty E(x) + 4\pi P(E)$$

$$D(0) \equiv \varepsilon_\infty E(0) + 4\pi P(0) = 4\pi\sigma_0$$

where D is the displacement vector, $V(x)$ is the potential, q is the ionic charge and the surface charge density $\sigma_0 = -q/A$ where A is the area associated with a single charge at the surface. A straightforward though tedious minimization gives

$$c^+(x) = c_0 \exp \{-qV(x)/kT\}$$

$$c^-(x) = c_0 \exp \{qV(x)/kT\} \tag{2.9}$$

and

$$\varepsilon_\infty \frac{d^2V}{dx^2} - 4\pi \frac{dP}{dx} = 8\pi q c_0 \sinh(qV/kT) \tag{2.10}$$

$$P = \frac{P_0}{V_0} \mathcal{L}(Y) \tag{2.11}$$

where Y is given in (2.6) and $\mathcal{L}(Y)$ is the Langevin function i.e.

$$\mathcal{L}(Y) = \coth Y - 1/Y \tag{2.12}$$

Equations (2.10) and (2.11) are the main equations used in this work and reduce to GCDH theory in the limit of small polarization and the absence of a hydration force. More generally we define the dielectric constant, $\varepsilon(x)$ for all values of the polarization to be:

$$\varepsilon(x) = D(x)/E(x) \tag{2.13}$$

For small values of P and V (i.e. in the limit $| qV(x)/kT | \ll 1$, $|p_0 P(x)/kT| \ll 1$) equations (2.10) and (2.11) can be linearized to give:

$$\varepsilon_\infty \frac{d^2V}{dx^2} - 4\pi\frac{dP}{dx} = \frac{8\pi q^2 c_0}{kT} V \tag{2.14}$$

$$P = \frac{p_0^2}{3kTV_0} \left[-\frac{dV}{dx} + \gamma P + \frac{\varepsilon}{\chi}\frac{\xi^2}{\varepsilon_\infty}\frac{d^2P}{dx^2} \right] \tag{2.15}$$

which can be rewritten as

$$\varepsilon_\infty \frac{dE}{dx} + 4\pi \frac{d^2P}{dx^2} = \frac{\varepsilon_0 E}{\lambda^2} \tag{2.16}$$

$$-\frac{\varepsilon_0 \xi^2}{\varepsilon_\infty} \frac{d^2P}{dx^2} + P = \chi E \tag{2.17}$$

where λ is the Debye-Hückel screening length. These are identical to

56

the equation of Gruen and Marcelja (1983) for inhomogeneous water
polarization in the linear limit. This implies that our linearized
theory is analogous to that of Gruen and Marcelja and applies to the
case where the solvent is described by a system of permanent interacting electric dipoles. Returning to the full equations (2.10) and
(2.11) we may consider the absence of hydration forces by taking the
decay length ξ to be zero. In this case we obtain

$$\epsilon_\infty \frac{d^2 V}{dx^2} - 4\pi \frac{dP}{dx} = \frac{\epsilon_o}{\lambda^2} \; \sinh(qV/kT) \qquad (2.18)$$

$$P = \frac{p_o}{V_o} \mathscr{L} \left[\frac{p_o}{kT} \left[- \frac{dV}{dx} + \gamma P \right] \right] \qquad (2.19)$$

Equation (2.18) was obtained previously by ZGP for the absence of both
hydration forces and dipolar interactions. However the equation corresponding to (2.19) was derived as

$$P = \frac{p_o}{V_o} \mathscr{L} \left[- \frac{\bar{p}}{kT} \frac{dV}{dx} \right] \qquad (2.20)$$

where $\bar{p} \simeq 4p_o$ and describes the constraint imposed on a given water
molecule by the presence of hydrogen bonds to its neighbours (Booth
(1951)). The solution arising from these equations will be compared
in the next section.

2.2 Single Bilayer

We consider in this section the properties of a single bilayer
in the absence of the hydration force ($\xi = 0$). We wish to evaluate the
dielectric constant $\epsilon(x)$ at the surface (as a function of area per unit
charge) and as a function of distance x away from the surface (for a
given value of surface charge). A second quantity that is of interest
is the potential $V(x)$ which we wish to evaluate under the same conditions.

Evaluation of these quantities involves the numerical solution of
(2.8), (2.13) and (2.19) to give $\epsilon(o)$ and a numerical integration of
(2.18) and (2.19) for $V(o)$, $V(x)$ and $\epsilon(x)$.

In all cases the following values of the physical parameters were
assigned: The absolute temperature was taken as 300K and the bulk
dielectric constant of water ϵ_o = 76.82. The dipole moment of a single
water molecule p_o = 2.35 x 10^{-18} esu and its specific volume V_o = 29.8$\overset{o}{A}^3$.
The bulk ionic concentration c_o was chosen as 1.2 x 10^{19} ions cm^{-3} (which
corresponds to 0.02M). Finally for any given value of ϵ_∞ (which is an
adjustable parameter of our model) that of γ may be obtained using (2.7).

Firstly in fig. 1 we present the results for the dielectric constant
at the surface $\epsilon(o)$ as a function of area per unit charge ($1/\sigma(o)$).
These results are presented for ϵ_∞ = 1 and 4, together with the corresponding GCDH and ZGP results for comparison. Both the ZGP and present
work show a significant reduction of $\epsilon(o)$ from the bulk value, but the
dipole coupling in the present work extends the range of this reduction
to higher values of area/surface charge.

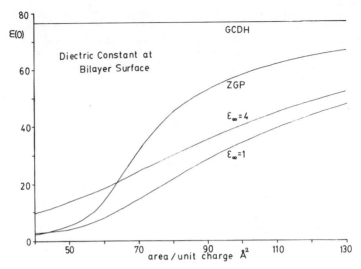

Figure 1: Dielectric constant at the bilayer surface as a function of area/unit charge.

Figure 2 shows how $\varepsilon(x)$ varies with distance x away from the surface. The area/surface charge was chosen 40 $Å^2$/unit charge, which corresponds to a gel state bilayer. Again a similar reduction is seen, this reduction being almost independent of the choice of ε_∞. We find that $\varepsilon(x)$ reaches 90% of ε_o at 10 Å whereas in the ZGP treatment it reaches this value at only 3 Å. This long range variation of $\varepsilon(x)$ will play a significant role when two bilayers are brought close together.

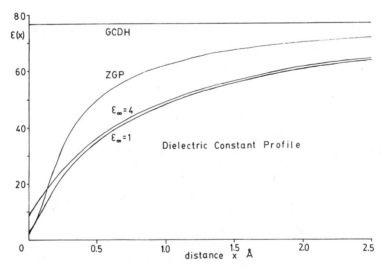

Figure 2: Dielectric constant as a function of distance from the bilayer surface.

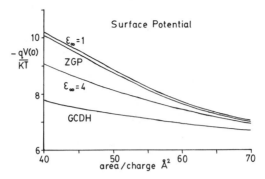

Figure 3: Surface potentional as a function of area/unit charge.

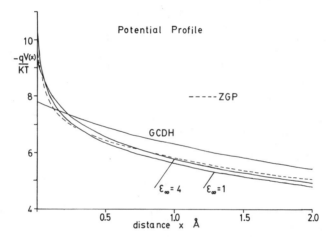

Figure 4: Potential as a function of distance from the bilayer surface.

The surface potential is shown in fig. 3. We see that at low values of area/unit charge (high values of surface charge density) both the ZGP and present work predict significantly higher values of V(o). The implications of this to experimental studies of surface binding are discussed in section 3.

Figure 4 shows V(x) as a function of x. It is seen that as we move away from the surface this enhancement is reversed whilst at large distance (off) scale) all theories give the same behaviour.

2.3 Two Bilayers

In section 2.2 we considered the case of a single bilayer by solving the differential equation of section 2.1 subject to certain boundary conditions given in equations (2.8). These same differential equations can be used to model the electrostatic properties of two bilayers separated by a distance d. The difference in the two physical systems is reflected in the boundary conditions. For the case of two bilayers symmetry about the mid-point leads to the requirement that

$$-\frac{dV}{dx} = E(x) = 0$$
$$\text{at } x = d/2 \qquad\qquad (2.21)$$
$$P(x) = 0$$

The quantity of interest is the electrostatic pressure, calculated as the osmotic pressure at the mid-point. Following Verwey and Overbeek (1948) we write

$$P = 2\, c_o kT \left\{ \cosh\left(\frac{qV(d/2)}{kT}\right) - 1 \right\} \qquad\qquad (2.22)$$

Clearly by calculating $V(d/2)$ for the GCDH theory we obtain a pressure P_{GC}. In our present theory we calculate $V(d/2)$ by numerical integration of (2.18), (2.19) subject to the boundary conditions (2.21). This has been done in two ways:

(i)   The surface charge density $\sigma(o)$ is maintained constant, the value chosen being 70 Å^2/unit charge corresponding to a fluid bilayer.

(ii) The surface potential $V(o)$ is kept constant at $qV/kT = -7.597$, which corresponds to $V(o)$ as calculate for a single bilayer with $1/\sigma(o) = 70$ Å^2/unit charge (see fig. 3).

It is useful to compare the value of the pressure P obtained using the present formulas with P_{GC} obtained from the GCDH theory. Defining

$$\Delta P = P_{GC} - P \qquad\qquad (2.23)$$

we show in fig. 5 both $\ln\Delta P$ and the fractional difference $\Delta P/P_{GC}$ as functions of the separation d for both the cases (i) and (ii) outlined above.

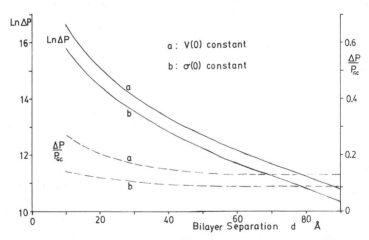

Figure 5:   The electrostatic pressure as compared to the GCDH theory as a function of bilayer separation d.

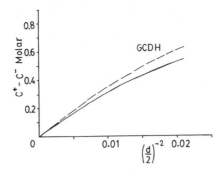

Figure 6: The charge concentration at the mid-point as a function of separation.

We can see that for all bilayer separations and under both surface constraints the present model predicts a significantly weaker electrostatic pressure than GCDH (approx. 10 - 12%). At large separations this pressure difference appears to vary exponentially with separation, following the equation.

$$\Delta P = P_o e^{-d/\lambda*} \qquad (2.24)$$

where from the graph we estimate $\lambda* = (21.5 \pm .2)\overset{o}{A}$. This is very close to the value of the Debye-Hückel screening length for a 0.02M monovalent ionic situation $\lambda = 21.3 \overset{o}{A}$. This is significant as at large separations it has been shown (Verwey and Overbeek, (1948)) that

$$P_{GC} = P_o e^{-d/\lambda} \qquad (2.25)$$

where λ is the Debye-Hückel screening length. It may also be shown that values of ε_∞ other than $\varepsilon_\infty = 1$ do not change the decay length $\lambda*$.

The value of $V(d/2)$ also determines the ionic concentrations at the mid-point $c^\pm(d/2)$, a quantity which has been studied using Monte Carlo computer simulations by Jönsson et al (1980). Figure 6 shows the values of $c^+(d/2) - c^-(d/2)$ as a function of the corresponding GCDH value. It is seen that the present theory shows a marked decrease from that predicted by GCDH, as a result also found in the Monte Carlo study.

3. DISCUSSION

In most studies of multilamellar lipid systems the electrostatic contribution is derived from standard GCDH theory. This can be used to calculate an effective (repulsive) electrostatic pressure, P_{GC}, between opposing bilayer as a function of separation d. A Van der Waals interaction free energy can also be calculated leading to an effective (attractive) pressure, P_{vw}, also a function of d. When the bilayer system is put under an external pressure, P_E, the system comes to equilibrium with a bilayer separation d_w. For equilibrium we must have

$$P_{GC}(d) + P_{vw}(d) + P_E = 0 \quad \text{for } d = d_w \qquad (2.26)$$

In experimental studies this equality is usually satisfied for low surface charge densities and large values of d_w. However it is found that for small values of d_w or large surface charge density (2.26) cannot be balanced. This necessitates the introduction of a phenomenological term $P_x(d)$ such that

$$P_{GC}(d) + P_{vw}(d) + P_E + P_x(d) = 0 \quad \text{for } d = d_w \qquad (2.27)$$

Such an analysis has been used to demonstrate strong short range repulsive forces, the so-called hydration force (Rand and Parsegian (1984), Marra and Israelachvili (1985)). Similarly, forces due to mechanical fluctuations of the bilayers have also been proposed (Ostrowsky and Sornette (1984)).

The validity of such an analysis depends on the applicability of P_{GC} calculated from GCDH theory. In the limit of low surface charge densities (such as those found in neutral lipid bilayers) the GCDH and present theories are equivalent. At high surface charge densities (such as charged anionic lipid bilayers) our model predicts a repulsive force smaller than predicted by GCDH. This may be of considerable relevance to the type of pressure studies of interacting charged bilayer systems outlined above.

Finally we shall address the question of the Stern layer. As mentioned in the introduction Stern (1924) introduced the concept of an adsorption layer of counterions of the charged surface and pointed out that the dielectric constant of the electrolyte above the charge surface must be a function of the local electric field. It is clear from figures 2 and 4 that our model predicts just such a layer. However we stress that our model is a continuum model and that the question of ion binding is more appropriately tackled using discrete mechanics.

REFERENCES

Augousti, A.T. and Rickayzen, G. 1984, J. Chem. Soc., Faraday Trans. 2 80, 141.
Booth, F. 1981, J. Chem. Phys. 19, 391.
Chapman, D.L. 1913, Phil. Mag. 25, 475.
Cohen, J.A. and Cohen, M. 1981, Biophys. J. 36, 623.
Fröhlich, H. 1958, 'Theory of Dielectrics: dielectric constant and dielectric loss', 2nd ed. (Clarendon Press: Oxford).
Gouy, G. 1910, J. Phys. 9, 457.
Gruen, D.W.R. and Marcelja, S. 1983, J. Chem. Soc., Faraday Trans. 2, 79, 211.
Jönsson, B., Wennerström, H. and Halle B. 1980, J. Phys. Chem. 84, 2179.
Lis, L.J., McAlister, M., Fuller, N., Rand, R.P. and Parsegian, V.A. 1982, Biophys. J. 37, 667, and references therein.
Marra, J. and Israelachvili, J. 1985, Biochemistry (in press).
McLaughlin, S. 1979, Biochemistry 18, 5213.
Ostrowsky, N. and Sornette, D. 1985, Chemica Scripta (in press).
Rand, R.P. and Parsegian, V.A. 1984, Can. J. Biochem. and Cell Biol. 62, 752.
Stern, O. 1924 Electrochem. 30, 508.
Träuble, H., Teubner, M., Wooley, P. and Eible, H. 1976, Biophys. Chem. 4, 319.
Verwey, E.J.W. and Overbeek, J. Th. G. 1948, 'Theory of the Stability of Lyophobic Colloids', (Elsevier Publishing Co. Inc.: New York).
Zuckermann, M.J., Georgallas, A. and Pink, D.A. 1985, Can. J. Phys. (in press).

INFLUENCE OF THE SURFACE CHARGE DISTRIBUTION AND WATER LAYERS ON THE

PERMEABILITY PROPERTIES OF LIPID BILAYERS

E. Anibal Disalvo and Laura S. Bakás

Instituto De Investigaciones Fisicoquimicas Teóricas y
Aplicadas (INIFTA) C.C. 16-Suc 4-1900 La Plata
Argentina

ABSTRACT

The barrier properties of a biological membrane are mainly due to the stabilization of the lipid components in a bilayer conformation. The overall permeation phenomena for polar molecules involves the interaction of the permeant at the membrane/solution interface and its penetration into the hydrocarbon core. The hydration shell of the phospholipid molecules contributes to the permeability barrier as determined by equilibrium dialysis experiments. In this paper, it is shown that the adsorption of Ca^{+2} on the inner and the outer interfaces of a sonicated vesicle bilayer of egg phosphatidylcholine can affect the barrier properties to non-electrolytes. In addition, ^{31}P nuclear magnetic resonance shows that the adsorption of Ca^{+2} on the inner interface of the vesicle promotes a conformational change of the bilayer affecting the adsorptive properties of the outer interface.

This transbilayer effect is related to the changes in the order parameter of the hydrocarbon chains of the bilayer as demonstrated by fluorometric experiments. A semi-quantitative analysis of the conformational change and the stoichiometry of the Ca^{+2} adsorption is discussed.

INTRODUCTION

In the context of an electrochemical approach the properties of the aqueous membrane interfaces and those of the electrical double layer in particular are relevant for the permeation process in lipid bilayers.

It has been very useful for the studies of the permeability, structural and stability properties the utilization of experimental model systems of lipid bilayers (Bangham et al., 1974). These systems can be obtained in the laboratory from pure phospholipids isolated from natural membranes or prepared by chemical synthesis. Phosphotidylcholines dispersed in aqueous solutions above their transition temperatures stabilize in a bilayer in which the polar head groups are oriented towards the aqueous phase and the hydrocarbon residues are segregated from the water. Such a conformation of the lipids is one of the main features of biological membranes. It constitutes the principal barrier of permeability for different biologically relevant substances such as ions and non-electrolytes (Bangham et al., 1967; Blok et al., 1976; Disalvo and de Gier, 1983). The bilayer determines three regions of different physicochemical properties: two membrane-solution interfaces where polar groups, water and ions are located and the non-polar region of the fatty acid residues (Figure 1). The magnitude of the interfaces amount to nearly 40% of the total thickness of the bilayer and therefore their influence on the permeability barrier properties of the membrane cannot be ignored. Several formalisms for interpreting the permeation of water and non-electrolytes describe the phenomena as an activated process in which the barrier of energy is located at the membrane interfaces. The overall permeation phenomena for polar molecules has been described as a two step process: dehydration of the molecule at the water-membrane interface and penetration into the hydrocarbon phase (de Gier, et al, 1971). The phenomena occurring at the polar head group regions will be related to the surface properties in general, and the electrical double layer in particular of the membrane. However, the processes occurring at the membrane interface (i.e. adsorption) are not independent from those determining the phase penetration. It is a general criteria to describe the membrane as a phase

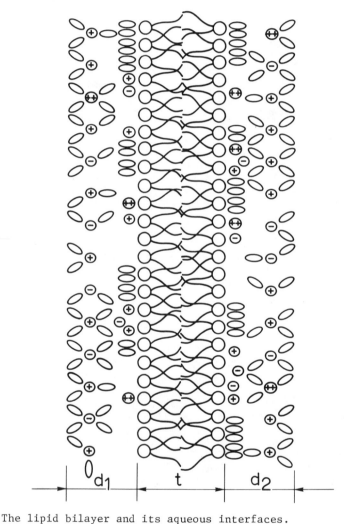

Fig. 1. The lipid bilayer and its aqueous interfaces.
t = bilayer thickness
d_1 and d_2 = aqueous layer interfaces where ions and water molecules are located.

interposed between two aqueous compartments. Hence, at a given temperature and pressure the composition of the membrane should be accurately defined. However, as lipid membranes are only stable in aqueous solutions, the composition of the membrane depends on that of the aqueous phases via the changes in the interfaces. As a consequence changes in the aqueous bulk phases may affect the properties of the interfaces (water and ions distribution) and of the lipid one (liquid crystalline-gel state). The purpose here is to analyze the interrelations of the permeability of a lipid bilayer with those properties. One of the most commonly used experimental model systems for structural and permeability studies are the sonicated vesicles. Upon sonication of a phospholipid dispersion, high curvature bilayered particles (diameter c.a. 300 Å) are formed (Huang, 1969). Considering the high surface/volume ratio, it is possible to ask which are the thermodynamic properties of those particles in relation to the adsorption and the phase penetration of water, ions and non-electrolytes. The sonicated vesicles are stable structures, above the transition temperature of the phospholipid, as a consequence of the intermembrane lipid-lipid interaction (mainly hydrophobic and Van der Waals forces) and the interaction of the polar head groups among them and the aqueous phase (Israelachvili et al., 1976). When a polar permeant approaches a lipid/water interface, several phenomena may occur: adsorption, redistribution of the water molecules and ions, disruption of the electrical double layer, phase transition etc. In addition, non-permeant ions can penetrate the outer and the inner Helmholtz plane modifying one or several of the above named processes. In connection with these possibilities it is of importance to know the thickness of the water layers adjacent to the lipid bilayer contributing to its barrier properties, the effects of the adsorption of non-permeant ions on the inner and the outer monolayer of the vesicle and if the hydrocarbon core properties are changed by the presence of ions at the membrane interface. It would be also important to know if the interfacial properties of one side of the membrane depends on the properties of the opposite interface. For a complete analysis of the permeation phenomena the interaction of the permeant molecules, such as non-electrolytes, with the interfacial and the bulk region of the lipid membrane and its interrelation must be taken into account. The composition at the interfaces can be, in a simple way, splitted in two contributions: the amount of water hydrating the polar head groups and the distribution of charges. The changes in some of these contributions can affect not only the interfacial properties but also the membrane behavior at the hydrocarbon phase and at the opposite interface.

It is the purpose here to analyze the influence of the distribution of water and Ca^{+2} ions on both sides of a vesicle bilayer of egg phosphatidylcholine in relation to the permeability of non-electrolytes. Several methodologies can be used for those purposes. In this paper, data obtained by means of excluded volume determination, ^{31}P nuclear magnetic resonance (^{31}P NMR) and fluorometric measurements are analyzed.

As a consequence, it can be concluded that the properties of the regions determining the barrier properties of a bilayer are not independent from each other and one of the principal parameters affecting their properties is the distribution of Ca^{+2} at the interfaces.

MATERIALS AND METHODS

Phosphatidylcholine (PC) was extracted from egg yolks using standard techniques and further purified by column chromatography of silica gel (van Kessel et al., 1981). The purity was checked by thin layer chromatography and ultraviolet measurements.

Sonicated vesicles were obtained by sonication of a lipid dispersion at 90-100 watts at 0°C and under N_2, during 20 min at intervals of 20s.

After the sonication the solution was centrifuged at 17500 rev/min during 20 min. To build up the asymmetric atmosphere at each side of the bilayers, vesicles prepared in a solution of a given ion were dialyzed against a solution of the same isotonicity of other ions. The equilibrium dialysis experiments were carried out in a Dianorm equilibrium dialyzer, which consists of a cellulose acetate membrane (molecular weight cut off ∿5000) separating a Teflon dialysis cell into two halves, each one of approx. 200 μl capacity. The distribution of ^{14}C-erythritol was measured at pH:7 (buffer TrisHCl/10mM) and at 25°C in all cases.

The ^{31}P nuclear magnetic resonance measurements were done in a Bruker WH-90 spectrometer operating in the Fourier transform mode at 36.4 MHz. The spectrometer was interfaced with a Bruker DS computer and equipped with temperature control, broad band proton decoupling (3W input power) and field stabilization via a deuterium lock.

Accumulated free induction decay were obtained using a 1200 Hz sweep width and 4K data point from 500 transients with 9 μs 90° radio frequency pulses with an interpulse time of 1.7 s. The free induction decay was exponentially filtered resulting in a 2 Hz line broadening. Measurements of line width have been corrected for this line broadening. Triphenylphosphine in chloroform was used as an external standard.

The fluorometric measurements were done in a Aminco-Bowman spectrofluorimeter at an excitation wave length at 355 nm, using 1,6,diphenil-1,3,5,hexatriene (DPH) as a probe. The anisotropy fluorescence parameter was calculated by

$$r = \frac{I_{||} - I_{\perp}}{I_{||} + 2I_{\perp}}$$

Where $I_{||}$ and $I_{\perp}$ are the intensities obtained when the analyzer is parallel ($I_{||}$) or normal ($I_{\perp}$) to the direction of the polarization of the excitation beam.

The ratio PC/DPH was 10^5 μg/μmol. In all experiments, corrections for the light scattering were done.

RESULTS

The thickness of the permeability barrier of a lipid membrane for a non-electrolyte

It has been discussed elsewhere that by means of equilibrium dialysis experiments it is possible to obtain the excluded volume of a lipid bilayer (Disalvo and de Gier, 1983). The distribution of a labeled permeant molecule, such as, ^{14}C erythritol, between two aqueous compartments separated by a dialysis membrane is measured at the equilibrium. Hence, the excluded volume can be obtained from the plot of the percentage of permeant between both compartments as a function of the lipid concentration in one of them. The experimental values found depend on the type of membrane particles (i.e. vesicles or planar bilayers). The experimental excluded volume can be defined in a naive way as the regions of the bilayer which is not accessible to permeant molecules. It can be described as a contribution of two terms. One of them corresponds to the intrinsic volume (v_L) of the phosphatidyl-choline molecules. It can be calculated from geometrical considerations, monolayer experiments and density determination of vesicles that it amounts to 0.84-0.77 l/mol of phospholipid (Hauser, 1975; Huang, 1969; Phillips, 1972). The other contribution corresponds to the nonsolvent water

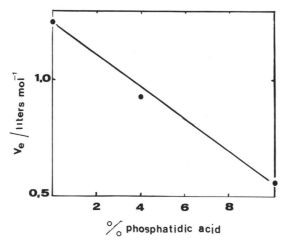

Fig. 2.   Effect on the excluded volume of lipid bilayers of sonicated
vesicles of the percentage of phosphatidic acid incorporated to the
membrane.

associated with the polar head groups of the phospholipid. It is about
0.41-0.48 1/mol and it fits well with differential scanning calorimetry and
hydrodynamic methods (Chapman et al., 1967; Hauser, 1975). As a consequence,
the amount of non-solvent water contributing to the barrier membrane properties can be distributed in two or three water layers (average thickness 9-
10Å) at the water/membrane interface. In addition, it must be pointed out
that similar experiments showed that the amount of nonsolvent water in
planar bilayer is around 11-15 moles H_2O/mole of lipid as it has been also
confirmed by partition techniques and DSC determinations (Chapman et al.,
1967; Katz and Diamond, 1974). It can be concluded therefore that the
barrier properties of a lipid bilayer for a non-electrolyte such as [14]C
erythritol is determined both by the lipid molecules and the water sequestered by the polar head groups.

In addition to water, membrane interfaces can also be composed of
ions and counterions. Therefore it was of interest to determine the effect
of the presence of net charges at the membrane interphase on the values of
excluded volume.

Figure 2 shows that there is a decrease in the excluded volume of a
lipid bilayer of sonicated vesicles as a function of the percentage of
charged lipids.

However, in this type of experiment it is difficult to conclude if
such variations are due to the charged head groups at the interfaces or to
the hydrocarbon chains of the phosphatidic acid molecules inserted in the
bilayer. Furthermore, the distribution of the phosphatidic acids between
the inner and the outer monolayer of the vesicle bilayer is not well known.

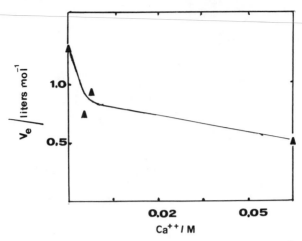

Fig. 3. Effect on the excluded volume of lipid bilayers of sonicated
vesicles of the Ca^{+2} concentration on the outer solution.
Vesicles were prepared as in Fig. 2.

Table I

Effect of Ca^{+2} on the Excluded Volume of Planar and Curved Lipid Bilayers

	Ca^{+2} Concentration (M)	Interface	Lipid in the Monolayer (%)	Decrease of Excluded Volume (%)
curved bilayers		outer	67	60
(sonicated vesicles)	0.05			
		inner	35	33
planar bilayers		outer	50	0
(liposomes)	0.05			
		inner	50	0

With the purpose to differentiate the effects of the charges at the
interfaces from those promoted by the insertion of lipid molecules in the
membrane phase, experiments with Ca^{+2} were performed.

Figure 3 shows that the excluded volume for ^{14}C erythritol decreases
with the addition of Ca^{+2}. Furthermore, as shown in Table I, the values for

68

the excluded volume in the presence of Ca^{+2} depend on if the Ca^{+2} is present on the inner or the outer membrane/solution interface. In all cases, except that for planar bilayers, Ca^{+2} decreases the excluded volume. Interestingly, the percentage of the decrease is directly related to the percentage of phospholipids on each monolayer of the bilayer. The inspection of Table I and its comparison with the values obtained for the excluded volume in the absence of Ca^{+2} (see above), shows that this decrease cannot be attributed only to a change in the amount of non-solvent water at the interfaces.

In other words, Ca^{+2} is affecting the barrier properties for the non-electrolyte by changing other properties of the interfacial region and/or the hydrocarbon core. These effects are investigated below by means of ^{31}P nuclear magnetic resonance and fluorometric determination.

The effect of the adsorption of Ca^{+2} on the interfacial and bulk properties of a lipid membrane

^{31}P nuclear magnetic resonance (^{31}P NMR) provides a suitable means to measure the effect of ions on the interfacial groups in lipid bilayers. By this technique the chemical shift of the phosphate groups located at the inner or at the outer monolayer of the vesicle bilayer can be determined as a function of the composition of the aqueous solution. Several works have been done in order to determine the size of the vesicle, the distribution of the lipids between the inner and the outer monolayer, the adsorption of ions and the amount of associated water (Phillips, 1972; de Kruijff et al., 1976; Hauser and Phillips, 1979; Disalvo, 1983). The chemical shift of the phosphate groups located at the inner monolayer of a vesicle bilayer are displaced to higher fields in the presence of Ca^{+2} in the inner aqueous solution (Figure 4). On the contrary, when Na^+ is inside the vesicle the addition of Ca^{+2} to the external medium has no effect on the chemical shift of the outer phosphates. From these experiments it can be concluded that Ca^{+2} adsorption is more probable on the inner than on the outer monolayer of a vesicle bilayer.

In addition to these results, when an egg phosphatidylcholine vesicle contains Ca^{+2}, the addition of Ca^{+2} to the outer solution displaces the chemical shift signal of the outer phosphates to higher fields (Figure 5).

It is concluded, therefore, that the presence of Ca^{+2} on the inner monolayer affects the properties of adsorption of Ca^{+2} on the outer monolayer of the vesicle.

In this sense. a transversal perturbation of Ca^{+2} from the inner to the outer monolayer can be inferred. With the purpose to corroborate such effect with the changes occcuring in the hydrocarbon core of the lipid membrane fluorometric determinations were carried out.

The effect of Ca^{+2} on the order parameter of the lipid bilayer

The introduction of a lipid soluble fluorometric probe into the lipid bilayer such as 1-2 di phenil hexatriene (DPH) allows to obtain information on the physical state of the hydrocarbon core (Shinitzky and Barenholz, 1978). The results of Figure 6 show that the anisotropy of fluorescence, which is a measure of the viscosity of the environment where the probe is, increases with the Ca^{+2} concentration. The increase of the anisotropy by Ca^{+2} can be interpreted as a promotion of an ordered state in the bilayer. Table II shows the magnitude of anisotropy or order parameter with Ca^{+2} at one or both interfaces of the vesicle bilayer. Ca^{+2} inside promotes an increase in the order parameter in comparison to the Na^+-containing vesicles. In addition, external Ca^{+2} promotes an additional increase on Ca^{+2}-containing vesicles. The findings of fluorometry parallels those obtained with

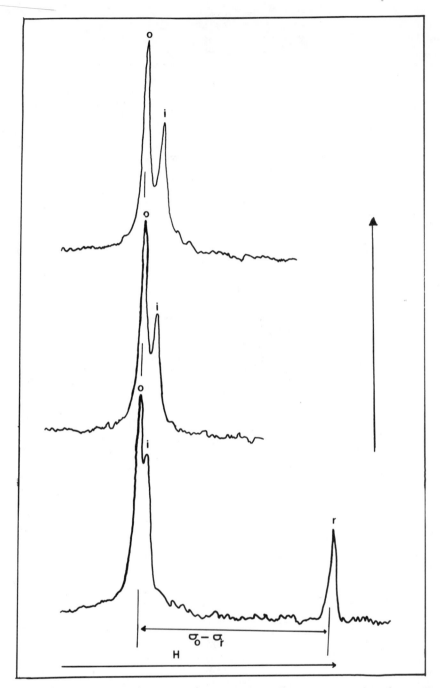

Fig. 4. Displacement of the chemical shift signal of the phosphate groups on the inner monolayer of a vesicle bilayer for different concentrations of Ca^{+2} inside the vesicle, σ_i and σ_o represent the signal position corresponding to the inner and the outer phosphates respectively with respect to an external reference (σ_r) (triphenylphosphine in chloroform). The arrow shows the increment of the internal Ca^{+2} concentration. In all cases, the spectra were obtained in the absence of external Ca^{+2}.

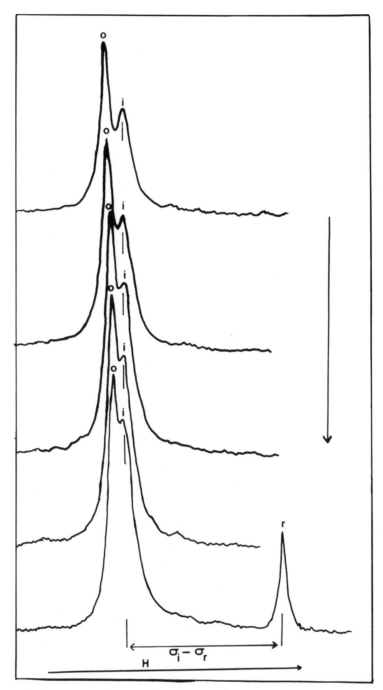

Fig. 5. Displacement of the chemical shift signal corresponding to the phosphate groups laying on the outer monolayer of a vesicle bilayer for different external Ca^{+2} concentrations, σ_o, σ_i, and σ_r have the same meaning as in Figure 4. The inner Ca^{+2} concentration was fixed in 0.5 N in all cases. The increments of external Ca^{+2} is indicated by the arrow (see experimental details in Materials and Methods).

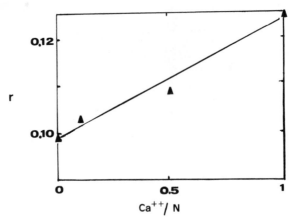

Fig. 6. Anisotropy fluorescence parameter $r = (I_{||} - I_{\perp})/(I_{||} + 2I_{\perp})$ as a
function of the Ca^{+2} concentration of the inner and the outer
solution of sonicated vesicles of phosphatidylcholine.

Table II

Effect of the Symmetric and Asymmetric Na^+ and Ca^{+2} Distributions on the
Fluorescence Anisotropy (r) of Vesicle Bilayers

Inner Solution	Outer Solution	r
Na^+ · 1N	Na^+ 1N	.099
Na^+ 1N	Ca^{+2} 1N	.107
Ca^{+2} 1N	Ca^{+2} 1N	.119
Ca^{+2} 1N	Na^+ 1N	.110

^{31}P NMR. However, the external Ca^{+2} also affects the anisotropy of the
hydrocarbone core of a vesicle containing Na^+ which is not in correspondence
with the NMR experiments.

The results of fluorometry show that vesicles containing Na^+, adsorbs
Ca^{+2} without perturbing the interfacial phosphate groups. However, the
presence of Ca^{+2} on the inner interface would promote a change on the external interface, via the ordering of the hydrocarbon chains, as a result of
which the phosphate groups become affected by the Ca^{+2} action. This effect
can be interpreted as a conformational change of the lipid bilayer and it is
discussed in the next section.

DISCUSSION AND CONCLUSIONS

The binding of Ca^{+2} to phosphatidylcholine monolayers and bilayers has been studied by different methodologies. Although there is a general acceptance that the adsorption in planar bilayers is very low, enhancement of the interaction has been found below the phase transition temperature (Lau et al., 1980). However, it is still polemic if Ca^{+2} is able to compete with water for the same binding sites (Cerbon, 1967; Simon et al., 1975).

The interaction of Ca^{+2} with the PC molecules of the lipid interface can be described stoichiometrically by

$$PC + nCa^{+2} \rightleftharpoons PC(Ca^{+2})_n \qquad (1)$$

The equilibrium constant (K) for the Ca^{+2}/phosphatidylcholine interaction can be written as:

$$K = \frac{|PC(Ca)_n|}{(PC) \ |Ca|^n} \qquad (2)$$

When additional changes in free energy ($\Delta G'$) occur as a result of conformational changes during binding, the gradual change in affinity is given by

$$K'_{(\theta)} = K \ exp[\frac{\Delta G'}{RT}]$$

K is the equilibrium constant when $\Delta G' = 0$ (Blank, 1972).

The constants K' and K can be written as a function of the Ca^{+2} concentration in the aqueous solution adjacent to the bilayer as follows.

$$K' = \frac{\theta}{\theta-1} \ \frac{1}{C} \qquad (3)$$

and

$$K = \frac{\theta}{1-\theta} \ \frac{1}{C^n} \qquad (4)$$

where θ is the degree of coverage of the interface. As the equilibrium constant can be written as a function of the standard free energy of the process of adsorption

$$\Delta G^\circ = -RT \ ln \ K$$

the logarithm of equations (3) and (4) can be derived with respect to ln C. The resulting expressions can be substracted one from another yielding:

$$-\frac{1}{RT} \ \frac{\partial \Delta G'}{\partial lnC} = (n-1) \qquad (5)$$

Equation (5) denotes that the exponent n affecting the Ca^{+2} concentration in the stoichiometric relation Ca^{+2}/PC (equation 1) is connected with a surface excess concentration as described by the Gibbs adsorption isotherm.

Expressing now the degree of coverage as the ratio between the chemical shift displacement at a given Ca^{+2} concentration ($\Delta\sigma$) and the maximal chemical shift displacement ($\Delta\sigma_{max}$),

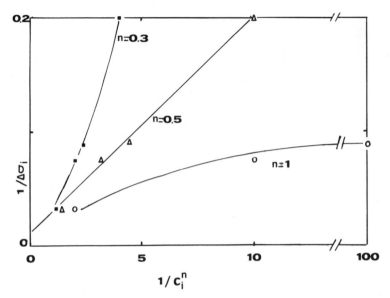

Fig. 7. Double reciprocal plot of the separation of the peaks corresponding to the chemical shift signals of the inner and the outer phosphate groups of a vesicle bilayer as a function of the reciprocal of the inner Ca^{+2} concentration for different values of n.

$$\theta = \Delta\sigma/\Delta\sigma_{max}$$

the isotherm in (4) can be represented in a double reciprocal plot.

As it is shown in Figure 7, the value n = 0.5 linealizes the results of ^{31}P NMR for which the stoichiometry of the interactions of Ca^{+2} with the inner interface is one Ca^{+2} per two phospholipids. Furthermore, the interpretations of a n-coefficient less than 1 corresponds to, according to equation (5), a finite change in the conformational state of the bilayer.

In this way, it can be concluded that Ca^{++} adsorbs on the outer monolayer of a vesicle bilayer by interacting with the external phosphate if there is Ca^{+2} inside the vesicle. The adsorption of Ca^{+2} on the inner interface promotes, then, the adsorption of Ca^{+2} on the outer interface affecting the phosphate groups.

As a consequence of the interaction of Ca^{+2} on one or both interfaces of the lipid bilayer, the barrier properties for non-electrolytes are affected. The changes are found both in the aqueous interfaces and in the hydrocarbon core of the membrane.

ACKNOWLEDGMENTS

The INIFTA is a research institute jointly established by the Universidad Nacional de La Plata, the Consejo Nacional de Investigaciones

Cientificas y Tecnicas and the Comision de Investigaciones Cientificas de la Provincia de Buenos Aires.

E.A.D. is a member of the career of investigator of the Consejo Nacional de Investigaciones Cientificas y Tecnicas of the Argentine Republic.

L.S.B. is recipient of a fellowship of the Comision de Investigaciones Cientificas de la Provincia de Buenos Aires (Argentina).

The authors are grateful to Prof. J. de Gier and Prof. B. de Kruijff of the Biochemistry Laboratory of the State University of Utrecht, The Netherlands, for the discussion and facilities for the equilibrium dialysis and ^{31}P NMR experiments and to Dr. Brenner from the INIBIOLP (La Plata, Argentina) for the facilities to perform the fluorometric measurements.

REFERENCES

Bangham, A. D., Hill, M. W. and Miller, N. G., 1974, Preparation and use of liposomes as models of biological membranes, Methods. Membr. Biol., 1:1.

Bangham, A. D., de Gier, J. and Greville, G. D., 1967, Osmotic properties and water permeability of phospholipid liquid crystals, Chem. Phys. Lipids, 1:225.

Blank, M., 1972, Cooperative effects in membrane reactions, J. Colloid. Int. Sci., 41:97.

Blok, M. C., van Deenen, L. L. M. and de Gier, J., 1976, Effect of the gel to liquid crystalline phase transition on the osmotic behavior of phosphatidylcholine liposomes, Biochim. Biophys. Acta, 433:1.

Cerbon, J., 1967, NMR studies on the water immobilization by lipid systems "in vitro" and "in vivo," Biochim. Biophys. Acta, 144:1.

Chapman, D., Williams, R. M. and Ladbrooke, B. D., 1967, Physical studies of phospholipids VI thermotropic and lyotropic mesomorphism of some 1,2-diacyl-phosphatidylcholines (lecithins), Chem. Phys. Lipids, 1:445.

de Gier, J., Mandersloot, J. G., Hupkes, J. V., McElhaney, R. N. and van Beek, W. P., 1971, On the mechanism of non-electrolyte permeation through lipid bilayers and through biomembranes, Biochim. Biophys. Acta, 233:610.

de Kruijff, B., Cullis, P. R. and Radda, G. K., 1976, Outside-inside distributions and sizes of mixed phosphatidylcholine-cholesterol vesicles, Biochim. Biophys. Acta, 436:729.

Disalvo, E. A. and de Gier, J., 1983, Contribution of aqueous interphases to the permeability barrier of lipid bilayers, Chem. Phys. Lipids, 32:39.

Disalvo, E. A, 1983, Divalent cations promote transversal perturbations in lipid bilayers of sonicated vesicles, Bioelectrochem. Bioenerg., 11:145.

Hauser, H., Lipids, in: "Water: A Comprehensive Treatise," F. Franks, ed., Plenum Press, New York (1975).

Hauser, H. and Phillips, M. C., 1979, Interactions of the polar groups of phospholipid bilayer membranes, Prog. Surf. Membr. Sci., 13:297.

Huang, C., 1969, Studies on phosphatidylcholine vesicles, formation and physical characteristics, Biochemistry, 8:134.

Israelachvili, J. N., Mitchell, D. J. and Nimham, B. N., 1976, Theory of self-assembly of lipid bilayers and vesicles, Biochim. Biophys. Acta, 470:185.

Katz, Y. and Diamond, J. M., 1974, Nonsolvent water in liposomes, J. Membr. Biol., 17:87.

Lau, A. L. Y., McLaughlin, A. C., McDonald, R. C. and McLaughlin, S. G. A., 1980, The adsorption of alkaline earth cations to phosphatidylcholine bilayer membranes: A unique effect of calcium, <u>Adv. Chem. Ser.</u>, 188:49.

Phillips, M. C., 1972, The physical state of phospholipids and cholesterol in monolayers, bilayers, and membranes, <u>Prog. Surf. Membr. Sci</u>, 5:139.

Shinitzky, M. and Barenholz, Y., 1978, Fluidity parameters of lipid regions determined by fluorescence polarization, <u>Biochim. Biophys. Acta</u>, 515:367.

Simon, S. A., Lis, L. J., Kauffman, J. W. and McDonald, R. C., 1975, A calorimetric and monolayer investigation of the influence of ions on the thermodynamic properties of phosphatidylcholine, <u>Biochim. Biophys. Acta</u>, 375:317.

van Kessel, W. S. M. G., Tieman, M. and Demel, R. A., 1981, Purification of phospholipids by preparative high pressure liquid chromatography, <u>Lipids</u>, 16:58.

FACILITATION OF ION PERMEABILITY OF BILAYER MEMBRANES AND THEIR PHASE TRANSITION

Israel R. Miller

Dept. of Membrane Research, The Weizmann Institute of Science
Rehovot, Israel

ABSTRACT

The rate of relase of Tl+ from phospholipid vesicles of different composition was measured by pulse polarography as a function of temperature or in
the presence of valinomycin, tetraphenyl boron (TPB⁻) or dipicylamine (DPA⁻)
as transport facilitators. The release from pure dipalmitoyl phosphatidylcholine (DPPC) vesicles increased abruptly around the pretransition temp-
earture. The steepness of the increase decreased with the width of the
transition peak. Valinomycin TPB⁻ (tetraphenyl boron) and DPA⁻ (dipicryl
amine) facilitate release of Tl⁺ from unilamellar vesicles above their
phase transition temperature with a first order release rate constant.
They do not facilitate release below the phase transition. Bursts of release were observed upon their addition to the vesicles but after annealing, which was completed within less than a minute, the vesicles were re-
sealed. The rate of phase transitions, their reversibility or irreversi-
bility and hysteresis with concomitant existence of metastable states,
will be discussed.

1. INTRODUCTION

Unilamellar lipid bilayer vesicles are a convenient tool for investigation
of lipid membrane permeability under different conditions and under the
influence of different agents penetrating the membrane, modifying its
structure and eventually facilitating the ion transport. The polarographic method or the voltametric methods in general, using different types
of electrodes (e.g. rotating electrodes) besides the dropping mercury
electrode are suitable for measuring the rate of release from vesicles
(Sequaris and Miller, 1984). They all measure very accurately the concentration in the external solution and are not affected by the substance
entrapped within the vesicles. Similarly to some fluorescence methods
(Wilshut and Paphadjopoulos, 1979, Lelkes et al., 1982; Bentz et al.,
(1983) the pulse polarographic method permits measurement of outflow kinetics in real time. Its advantage is the wide choice of ionic and non-
ionic permeants, amenable for investigation. We have chosen for our present
investigation Tl⁺, which shows strong resemblance to K⁺ with respect to
ionic radius, hydration and binding (Eisenman et al., 1975); Urban et al.,
1980) and is reduced around 0.65 V relative to N Calomel electrode. We
measured the release rate as a function of temperature and of surface
change with particular consideration of events occuring during phase
transitions. The effectiveness of transport facilitators like valinomycin

or hydrophobic anions depends very strongly on the state of the lipid in the liquid crystalline or in the rigid crystalline phase and on the rate of its transition between different states. Many sphingolipids show hysteresis phenomena when the temperature of their aqueous dispersions is scanned in the differential scanning calorimeter (Bunov, 1979; Estep et al., 1980; Freire et al., 1980). Acidic phospholipids behave likewise when crosslinked by multivalent cations (Elamrani and Blume, 1984). It is assumed that in all these cases temperature dependent potential barriers exist between the different states. Thus in some temperature regions two or more phases can coexist, when only one of the states is stable and all the rest are metastable. In the discussion part of this paper I shall consider transition rates and hysteresis phenomena occuring during the transition between a liquid crystalline and one or two rigid crystalline states.

2. EXPERIMENTAL

2.1 Materials

The lipids used were dipalmitoylphosphatidylcholine (DPPC) purchased from Dr. Berchtold's CH-3007 Bern, Switzerland, and egg phosphatidylcholine and bovine brain phosphatidylserine purchased form Lipid Products, Nutfield, England. The salts K_2SO_4, Na_2SO_4, Tl_2SO_4 were all pro analysis grade and used without further purification.

Valinomycin, tetraphenyl boron (TPB$^-$) and dipicrylamine (DPA$^-$) were purchased from Sigma.

2.2 Vesicle Formation

The vesicles were formed by the Deamer and Bangham procedure (1976). Briefly, 3 ml of phospholipid solution 1.5-2 mg/ml in diethylether/methanol (4/1, v/v) were injected into 4 ml of an equeous solution of 0.1 M Tl_2SO_4 + 0.1 M K_2SO_4 or Na_2SO_4 at 60°C and at a rate of 0.2 ml/min. A slow stream of N_2 was then bubbled through the solution, at this temperature, for an additional ten minutes to remove most of the organic solvent. The rest of the organic solvent was removed together with the untrapped Tl+ by consecutive dialysis against 0.5 l of 0.1 M K_2SO_4 or Na_2SO_4 respectively, changed every hour for 5 hours or every half an hour for 3 hours at 40°C. The concentration of Tl^+ in the extravesicular solution was reduced to less than $2 \cdot 10^{-15}$ M or $5 \cdot 10^{-5}$ M respectively and after dilution the initial concentration in the polarographic cell was below $4 \cdot 10^{-7}$ M or 10^{-6} M. The exception were egg PC vesicles which had even at 4 ^{o}C an outflow half time of about 50 hours.

The average radius Ri of the vesicles was derived from the entrapped volume V_i(1) by the relation:

$$V_i = v_p\ Ri/3d$$

where v_p is the molar volume of the phospholipid molecule and d is the thickness of the bilayer. The total entrapped volume V, was determined polarographically after release of the vesicular content with 0.1% Triton X100 or by sonication. The entrapped volume of the different preparations varied between 6 and 81 per mole lipid and the radius varied between 1100A and 1500A. This is an underestimate since 10%-20% of the entrapped volume was in multilamellar vesicles. The entrapped volume in the unilamellar vesicles was derived from the maximal facilited release by valinomycin or the organic anions.

2.3 Polarographic Determination

We used pulse polarography in the measurements employing PAR 170 with a
drop timer. It had the advantage that time dependence could be recorded
directly as follows: After completing the polarogram, the recorder was
transferred to the time base and thus the sampled pulse current at the
final set potential was measured as a function of time. Then either the
temperature was scanned at a certain rate or different additives like
valinomycin, orgainc anions etc. were injected and the current was measured
isothermically, as a function of time. The concentration/current relation
was obtained by running the polarograms at different Tl^+ concentrations
and by changing the Tl^+ concentration when the recorder was on time base
at constant potential as demonstrated in Fig. 1. At room temperature,
increments of 10^{-7} M in concentration of Tl^+ produced an increase of
1.8 nA in pulse polarographic current at 2 sec drop time at a potential of
0.9 V relative to N-Ag/Ag Cl electrode.

The rate of release is considered to be a first order reaction and the reverse reaction could be neglected since the outer concentration of the de-
polarizer was always by at least two orders of magnitude lower than its
intravesicular connection.

3. Results

3.1. The temperature dependence of the release rate

The rate of the release at 4 $^{\circ}$C was extemely slow if the lipids were
below the phase transition temperature like in the case of DPPC or
mixtures of DPPC with egg PC at ratios 70/30 and 50/50 or even a mixture
of 50% DPPC, 30% egg PC and 20% PS (all the mixtures had broad phase transition between 10°C-37°C) (Davis et al., 1980). Vesicles formed from DPPC
or from DPPC with 30% PC released less than 2% of their content per day
at 40C. The release from vesicles of the (DPPC, egg PC, PS) mixture was
up to 5% per day, while the release from egg PC vesicles reached 20%-25%
per day at this temperature. The rate of release increases with temperature. In Fig. 2 the time dependence of the pulse polarographic current
is presented. The temperature which is gradually increased is

Fig. 1.
Calibration of pulse polarographic current of Tl^+. After
potential scan in absence of
Tl^+ till 0.9 V, the recorder
is changed to time base. Subsequent amount of Tl^+ added
as indicated by arrows. The
concentration of Tl^+ noted
near each current line.

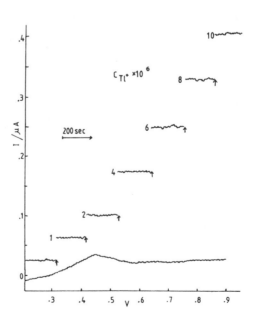

79

marked along the scan. The corrected current for a constant temperature (20°C) which is proportional to the Tl$^+$ concentration in the extra-vesicular solution, can be derived from the presented curves with the aid of the measured variation of the current with temperature for any fixed concentration of Tl$^+$.

It is evident from the curve in Fig 2 that there is an abrupt release from the vesicles of DPPC around 35°C which is the temperature region of the pre-transition. This differs from the reported maximal leakiness at the transition temperature which was obtained after abrupt increase of tempe-rature from below the phase transition to different temperatures around and above the phase transition (Sequaris and Miller, 1985; Papahadjopou-los et al, 1973; Blok et al, 1976). Indeed if the temperature is changed abruptly from 4°C to 37°C only about 15% of the inner content is released in the first instant and the release continues very slowly after annealing of the vesicles at this temperature.

No abrupt increase in release at any particular temperature is observed from vesicles made of mixtures of DPPC and egg PC at ratios 70/30 (and 50/50 and of 20% PS 30% PC and 50% DPPC having broad phase transitions. Curves b and c corresponding to release from negatively charged vesicles show a distinct lowering of the rate of release increase with temperature, if K$_2$SO$_4$ is replaced by Na$_2$SO$_4$ as the supporting electrolyte in the outer solution.

3.2. Facilitated transport of Tl$^+$.

The transport of Tl$^+$ could be facil-itated by valinomycin and by the organic anions tetraphenylboron (TPB$^-$) or dipicryl amine (DPA$^-$). TPB$^-$ and DPA$^-$ are not known as car-riers but the organic anions and their neutral complexes with cations are more permeable than inorganic ions (Lauger et al, 1981). Neither of these agents were active way below the phase transition.

Very reproducible results could be obtained above the phase transition as in the case of egg PC or even within the broad phase transition region as in the case of the lipid mixture (DPPC,PC,PS) at 25°C and of DPPC/PC 70/30 and 50/50 at 30°C. Unfortunately in the case of the PC vesicles only about 50%-60% of the Tl$^+$ expected to be in the vesi-cles remained entrapped. The rest leaked out during the dialysis and the incubation at 4°C. The release of Tl$^+$ was enhanced when the vesi-cles were inserted into the 0.1M K$_2$SO$_4$ at 24°C (Fig 3). The rate of

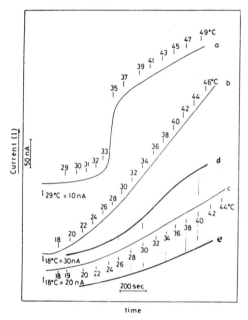

Fig. 2. Temperature scan. After completing the pulse polarogram bet-ween -0.2V and -0.9V relative to N(Ag/AgCl) electrode at 29°C for DPPC vesicles loaded with 0.05M K$_2$SO$_4$ and 0.05M Tl$_2$SO$_4$, and at 18°C for vesi-cles from the lipid mixtures the temperature was increased at nearly constant rate - as indicated. The currents above the background at the initial temperature are presented on each scan. Scans: a-DPPC - support-ing electrode K$_2$SO$_4$; b-(DPPC+PC+PS) - supporting electrolyte K$_2$SO$_4$; c-(DPPC +PC+PS) - supporting electrolyte Na$_2$SO$_4$; d-(70DPPC/30PC); e-(50DPPC/ 50PC).

release is further augmented by addition of valinomycin, dipicrylamine (DPA$^-$) or tetraphenylboron (TPB$^-$) at concentrations of the order of 10^{-7}M. In all the cases the rate of release increases with concentration of the consecutively added facilitators.

Similar results were obtained with the lipid mixture (DPPC,PC,PS) in the presence of K_2SO_4 as supporting electrolyte. When K_2SO_4 was replaced by Na_2SO_4 the behaviour of valinomycin still remained the same but TPB$^-$ and particularly DPA$^-$ behaved differently (Fig 4). Added

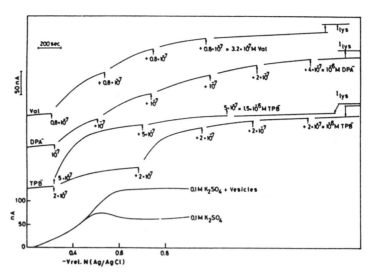

Fig. 3. Kinetics of facilitated release of Tl$^+$ from egg PC vesicles. Supporting electrolyte 0.1M K_2SO_4. The charging current background at the start of the time scan varied between 60 and 70 nA. The name of the facilitator is given at the start and at the end of each run. Their addition times and concentrations are indicated. Current after lysis with Triton X100 - I$_{lys}$ is given for each run. Low inset: Pulse polarograms of supporting electrolyte alone and after addition of Tl$^+$ vesicles loaded with Tl_2SO_4 and K_2SO_4.

DPA$^-$ and also TPB$^-$ at its lowest concentration caused a burst of release during the first few seconds after addition. The membranes seem to become leaky upon the addition of the lipid soluble anions but anneal rapidly and the rate of release decreases to a value slightly above the initial rate. This behaviour is in keeping with the stabilising and possibly rigidifying effect of Na$^+$ on the membrane which upon impact of the interacting additive undergoes a local structural perturbation with transient perforation

Fig. 4. Kinetics of facilitated release of Tl$^+$ from mixed lipid vesicles (50% DPPC, 30% PC, 20% PS) at 24°C. Initial net current was around 70 nA. Supporting electrolyte was 0.1M Na_2SO_4. The facilitators then addition and concentration indicated.

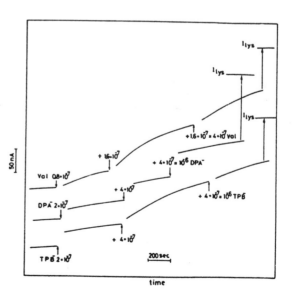

and leakiness. The annealing of the membrane is accomplished by rearrangement of the lipid molecules around the interacting additive resealing the perforation.

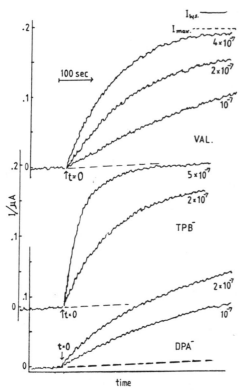

Fig. 5. Facilitated release of Tl^+ from 50/50 DPPC/egg PC. The name of the facilitator noted below the group of current/time curves it generated, its concentration noted on each curve.

To carry out accurate kinetic analysis only one concentration of facilitator was added to each lipid vesicle sample and the release was measured for a longer time. In Fig 5 and 6 the measured current is plotted as a function of time for different concentrations of facilitators (valinomycin DPA⁻ or TPB⁻) added to vesicles produced from the phosphatidylcholine mixtures DPPC/egg PC 50/50 (Fig 5) and 70/30 (Fig 6). Here more diluted solution of facilitators were added under vigorous stirring of the vesicle suspension to assure their even distribution. It is evident from these figures that the rate of increase in polarographic current which represents the rate of release is proportional within experimental accuracy to the concentration of the facilitator added. The difference between the different facilitators was not very large. For 50/50 DPPC/PC their relative efficiencies were as follows: TPB⁻ valinomycin DPA⁻.

4. Discussion

The kinetics of carrier facilitated transport has been discussed in detail and extensively reviewed by many authors (Lauger et al, 1981; Laugher and Stark, 1970; Markin and Liberman, 1973; Ciani et al, 1973; Hladky, 1979). The treatment dwelled mainly on transport through planar lipid bilayers with potential control. The release kinetics from vesicles be it in the form of charged ions or neutral complexes (Sequaris and Miller, 1984; Gutknecht, 1983) can be described as a homogeneous first order reaction:

$$- \frac{dc_i}{dt} = kc_i \qquad (1$$

Where c_i is the intravesicular concentration. The back reaction is neglected since the outer concentration is by orders of magnitude lower. It can also be described as a heterogeneous reaction.

$$\frac{dn}{dt} = \frac{-dc_i}{dt} V_i = Ak_p c_i \qquad (2$$

with a permeability rate constant k_p which is the product of the complexation and of the complex translocation rate constants as well as of the surface concentration of the carrier. V_i being the entrapped volume and A the surface area of the vesicle. Thus:

$$k_p = k \cdot \frac{V_i}{A} = \frac{kR}{3} \qquad (3$$

R is the mean radius of the vesicles. Solution of equation (1 in
the absence of complexation or at
very fast complex dissciation when
the internal concentration of the
active permeant is proportional to
its concentration after its out -
flow, renders:

$$\ln\frac{I_{\infty}-I_o}{I_{\infty}-I_t} = kt \qquad (4$$

where I_{∞} is the current measured
at infinite time after the facilitator has been added. This is usually lower by 10-20 % than the
current measured after lysis of
the vesicles by Triton X 100. I_o
and I_t are the currents at the moment of injection of the carrier
and at the time t.
Since Tl^+ forms an ion pair complex $TlSO_4^-$ with a stability constant $K_s=23$ (Smith& Martell 1976)
the concentration of free Tl^+
which can interact with the carier
or the hydrophobic anion is in the
presence of excess of SO_4 approximately:

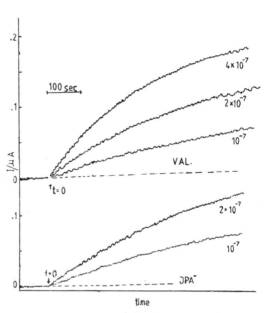

Fig.6 Facilitated release of Tl^+ from
70/30 DPPC/egg PC. Facilitators and
their concentrations indicated.

$$c_{Tl^+}^{free} = c_{Tl^+}^{tot} /K_s(c_{SO_4}+1/K_s-c_{Tl^+}^{tot}) \qquad (5$$

and the solution of equation 2 becomes:

$$kt = \ln\frac{(I_{\infty}-I_0)(B-I_{\infty}+I_t)}{(I_{\infty}-I_t)(B-I_{\infty}+I_0)} \qquad (6$$

$$B= (c_{SO_4}+1/K_s)x \text{ current per } c_{Tl^+}$$

The choice Of equation 6 or 4 for the description of the kinetics of facilitated release of Tl^+ from the vesicles will depend on the rate of dissociation of the ion pair $TlSO_4^-$. If the dissociation is very fast and the
binding of Tl^+ to the fa-ilitators and their translocation across the membrane is rate controlling, then equation 4 is valid. In the oposite case
it will be equation 6. Applying the two equations tothe data presented in
Figs. 3,4,5 and 6 shows that equation 4 represents faithfully the unimole-
cular kinetics, indicating that the dissociation of $TlSO_4$ is, as expected
a much faster process than the Tl^+ translocation through the membrane.
   In Fig. 7 we plotted $\ln(I_{\infty}-I_{\phi})/(i_{\infty}-I_t)$ against time according to
equation 4, for the release of Tl^+ from 50/50 DPPC/egg PC vesicles facili-

tated by valinomycin, DPA⁻, and by TPB⁻. The release rate constant is
nearly proportional to the concentration of the facilitator. At 30^0, it
is about $1.2x10^{-3}$ sec⁻¹ for 10^{-7} M DPA⁻, about $1.6x10^{-3}$ sec⁻¹ for 10^{-7} M Valinomycin and $2.3x10^{-3}$ sec⁻¹ for 10^{-7} M TPB⁻. These values correspond to
the respective heterogeneos rate constants: k_p= $4x10^{-9}$, $5.3x10^{-9}$ and $7.7x$
10^{-9} cm.sec⁻¹ respectively for a vesicle radius of 100 nm. The rate constants in vesicles made of DPPC/eggPC mixtures at aratio 7/3 is about half
of those in 1/1 DPPC/eggPC mixtures. theefficacy of the different facilitators is a function of the distribution coefficient of the facilitator
between theaqueos solution and the membrane and of their binding, translocation and dissociation rates.

The transition between two phases

The translocation rate in the phase
transition region is clearly related
to thelipid transition between the
two states. A tridimensional diagram
depicting the variation of the potential profile with temperature representing a thermotropic behavior with
hysteresis during the transition between the two states is presented in
Fig.8. This diagram is drawn following the potential profile suggested
by Everett(1978) explaining hysteresis.

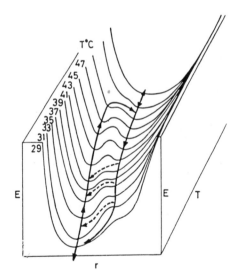

Tridimensional diagram of potential
barrier profiles for a two state
transition, changing with temperature.

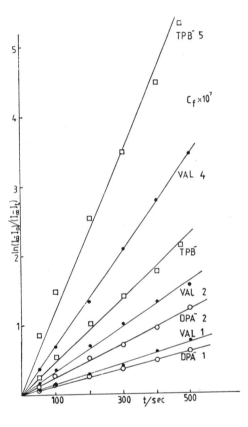

Fig. 7: First order kinetic analysis
ofTl⁺ release from 50/50 DPPC/eggPC
vesicles $\ln(I_\infty - I_0)/(I_\infty - I_t)$ plotted
against time in the presence of:
Valinomycin o-o; DPA⁻ o-o; TPB⁻ - .
The concentration of the facilitators
indicated.

To derive amore quantitative expression for this behavior let us consider
only the lower, approximately parabolic part of the potential profile of
the molecules in each state. In the rigid crystalline state the area per
lipid molecule is about 40 A^2 while in the liquid crystalline state it
is about 64 A^2. The intermolecular distance varies thus during melting from
6.3 to 8 A. In the rigid structure the deviation from equilibrium is more

costly t-an in the liquid crystalline one. The parabolic equation for the
relation between the energy increment A and the deviation from the equilibrium molecular distance r_s in each stateis: $A_r= a r_s^2$.
The liquid crystalline state is more compressible and the parameter a_1 may
be by up to an order of magnitude lower than a_r corresponding to the rigid
crystalline state.In Fig.9 the ener-
gy profile of the two states is depicted. The energy barrier is given by
the energy level at the intersection
of the twoenergy well profiles. At the
equilibrium phase transution temperature the free energy minima of the
two states are on the same level and
the free energy difference between
the two states is zero:

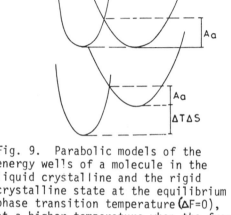

$$\Delta F= \Delta H - T\Delta S =0 \qquad (7$$

When we raise or lower the temperature, the free energy of the rigid
crystalline or of the liquid crystalline phase respectively is raised
with respect to the other phase, for
constant ΔS, by $\Delta F = \Delta T\Delta S$.
At the equilibrium transition tempe-
rature when the free energy differences between the two phases $F = 0$
the activation energies for the
transition between the two phases
are equal: $A_{r\rightarrow 1}= A_{1\rightarrow r}$

$$A_{r\rightarrow 1}=A_{1-r}=a_r r_r^2=a_1 r_1^2=a_1 (r-r_r)^2 \qquad (8$$

$$r_r + r_1 =r= \delta_1 - \delta_r \cong 1.7A \qquad (9$$

Fig. 9. Parabolic models of the
energy wells of a molecule in the
liquid crystalline and the rigid
crystalline state at the equilibrium
phase transition temperature $(\Delta F=0)$,
at a higher temperature when the free
energy of the rigid state is higher
by $\Delta S\Delta T$ and vice versa at a lower
temperature.

When δ_1 and δ_r are the average distances between centers of molecules
in the liquid crystalline and rigid
crystalline phases respectively.
Hence,

$$r_r= r a_1^{\frac{1}{2}}(a_r^{\frac{1}{2}}-a_1^{\frac{1}{2}})/(a_r -a_1) \qquad (10$$

and the transition activation energy per a couple of molecules is

$$A= r^2 a_r a_1 (a_r^{\frac{1}{2}}-a_1^{\frac{1}{2}})/(a_r -a_1)^2 \qquad (11$$

But the transition may involve a cooperative unit of nucleation of n
molecules which will make the effective activation energy

$$A_{ef.}=(n-1)A \qquad (12$$

Increasing the temperatures of the rigid phase by T above the equilibrium
transition temperature, the liquid crystalline phase becomes the stable
one, but the rigid crystalline phase can still persist as a metastable
state if the potential barrier between the two states is high enough.
The energy of the rigid crystalline phase becomes higher than the liquid
crystalline one by $\Delta T \Delta S$

$$\Delta T \Delta S= a_r r_r^2= a_1 (r-r_r)^2 \qquad (13$$

$$r_r = \frac{a_1 r}{a_r - a_1}\left\{\left[1 + \frac{a_r - a_1}{a_1}\left(1 - \frac{\Delta T \Delta S}{a_1 r^2}\right)\right]^{\frac{1}{2}} - 1\right\} \tag{14}$$

$\Delta T \Delta S / a_1 r^2$ varies between 0 (at $\Delta T = 0$) and 1 when the transition becomes diffusion controlled. This state is reached at not very large values of ΔT and therefore the values of the parameters a_1 and even of a_r are below $1KT/A^2$. The activation energy for the rigid to liquid crystalline transition $A_{r \rightarrow l} = a_r r_r^2$ when $T > T_{eq}$ becomes:

$$A_{r \rightarrow l} = \frac{a_r a_1^2 r^2}{(a_r - a_1)^2}\left\{2 + \frac{a_r - a_1}{a_1}(1 - \frac{\Delta T \Delta S}{a_1 r^2}) - 2[1 + \frac{a_r - a_1}{a_1}(1 - \frac{\Delta T \Delta S}{a_1 r^2})]^{\frac{1}{2}}\right\} \tag{15}$$

Upon decreasing the temperature to T T_{eq}, the liquid crystalline state changes into the rigid crystalline one at a rate depending on the activation energy $A_{l \rightarrow r} = a_1 r_1^2$

$$r_r = \frac{a_r r}{a_r - a_1}\left\{1 - (1 - \frac{a_r - a_1}{a_r}(1 - \frac{\Delta T \Delta S}{a_1 r^2}))^{\frac{1}{2}}\right\} \tag{16}$$

and the activation energy $A_{l \ r}$ becomes.

$$A_{l \rightarrow r} = \frac{a_1 a_r^2 r^2}{(a_r - a_1)^2}\left\{2 + \frac{a_r - a_1}{a_1}(1 - \frac{\Delta T \Delta S}{a_1 r^2}) - 2[1 + \frac{a_r - a_1}{a_r}(1 - \frac{\Delta T \Delta S}{a_r r^2})]^{\frac{1}{2}}\right\} \tag{17}$$

At $\Delta T \Delta S - a_1 r^2$ the potential barrier is completely abolished. Oleoyl cerebroside shows a phase transition hysteresis of the type illustrated in Fig. 8. The measured transition enthalpy in the case of oleoyl cerebroside is about 10 kCal/mole or about 16 RT units. The energy change between the two phases - $\Delta S \Delta T$, corresponding to the change of temperature from T_{eq} which is probably close to 47°C, to 29°C, where the transition half time from the rigid to the liquid crystalline phase is less than 0.5 min, amounts to about 0.8 kT per molecule. If this would be close to the temperature when the barrier is completely abolished $a_1 r^2 = 0.8$ kT, it would be around 0.27 kT/A^2 (r = 1.7 A). This would yield according to equation 8 a maximal activation energy of about 0.134 kT at T_{eq}. Such a low value cannot explain the slow transition rates and the hysterisis phenomena.

Phase transition is known to be a cooperative phenomenom with a cooperative unit size varying between several tens and several hundreds of molecules (Hinz and Strutevan, 1973) around the equilibrium. In a rate process the controlling unit is a stable nucleus (which reaches the point of no return and transits into the other phase) (Klein, 1982). It is not obvious what determines the size of the nucleus. It is certainly smaller than the cooperative unit at the equilibrium temperature but its size decreases probably when lowering the temperature for the liquid crystalline to rigid crystalline transition. No nucleation may be needed when increasing the temperature to drive the opposite process and the equilibrium transition temperature is probably identical with T_m of the oleoyl cerebroside measured during the heating scan or very close to it. The activation energy of a rigid nucleus formation or of liquidification of a rigid cooperative unit is the sum of the activation energies of all the molecules involved.

The diffusion controlled rate constant of the phase transition of a couple of molecules of lipids should be of the order of 10^7 sec^{-1} since the rotational time constant τ_{rot} of a lipid molecule is of the order of 10^{-8} sec and only during one out of Z rotations will a lipid molecule find a neighbouring molecule at the right orientation for phase transition. $1/Z$ is the fraction of the molecular perimeter which can give a rigid interaction product with a neighbouring molecule. The time required to form a stable n-membered nucleus τ_{ncl} will then become $\tau_{nucl} = Z^{(n-1)}\tau_{rot}$. Even if $Z = 2$ or less. This may become for $n = 20$ a very large time constant. Indeed the observed transition rate constant of phospholipidcholines, where no hysteresis phenomena are observed and there is no reason to assume high activation energies, are of the orders of $0.1 - 10$ sec^{-1} (Yager and Peticolas, 1982).

In the case of cerebrosides with predominantly saturated long chan acyl groups (Bunov, 1979; Estep et al, 1980; Freire et al, 1980) metastable rigid crystalline phase exists besides the stable one. During fast cooling the liquid crystalline phase transforms predominantly into the metastable rigid crystalline one. This converts then very slowly at low temperatures and its rate increases fast with temperatures when approaching the liquid crystalline-metastable rigid transition temperature. The tridimensional diagram representing the potential profiles at different temperatures given in Fig. 10 represents this behaviour. The interpretation of the behaviour represented in Fig. 10 is similar to that given for the two state transition. Here the size of the nuclei and of the cooperative units may vary with the temperature and with the three possible states from which the transition proceeds. The formation of the metastable rigid crystalline state from the liquid crystalline one occurs at a higher temperature than the stable state. The transition between the two rigid crystalline states requires nucleation in either direction.

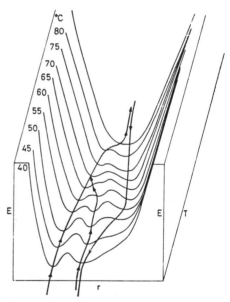

Fig. 10. Three dimensional diagram of potential barrier profiles between one liquid crystalline and two rigid phases at different temperatures.

Conclusion

Facilitated transport is possible only in liquid crystalline lipid membranes. In the phase transition region, when even only a small part of the lipid bilayer is in the fluid state, facilitated transport is possible, since the facilitator tends to perturb and to liquefy the lipids around. The liquefaction is a fast process since it does not require any cooperativity and single lipid molecules can do it independently and at random. On the other hand resealing imperfections created at the liquid-rigid boundary lines formed in course of the phase transition may be a slow process, since it requires cooperative adjustment of several boundary lipid molecules.

References

Benz, J., Duzgunes, N. and Nir, S. (1983) Biochemistry $\underline{22}$:3320-3330

Blok, M.C., van Deenen, L.M., and de Gier, J. (1976) Biochim. Biophys. Acta $\underline{433}$:1-12

Bunov, M.R. (1979) Biochim. Biophys. Acta $\underline{74}$:542-546

Ciani, S., Laprade, R., Eisenman, G., and Szabo, G. (1973) J. Memb. Biol. $\underline{11}$:255-292

Davis, P.J., Coolbear, K.P., and Keough, K.M.W. (1980) Can. J. Biochem. $\underline{54}$:851-859

Deamer, D.D. and Bangham, A.D. (1976) Biochim. Biophys. Acta $\underline{443}$:629-634

Eisenman, G., Kranse, S., and Ciani, S. (1975) Ann. NY Acad. Sci $\underline{264}$:34-60

Elamrani, K., and Blume, A. (1984) Biochemistry $\underline{22}$:3305-3311

Estep, T.N., Calhoun, W.I., Barneholz, Y., Bilton, R., Shipley, G.G., and Thompson, T.E. (1980) Biochemistry $\underline{19}$:20-24

Everett, D.H. (1978) Some biological implications of hysteresis in ions on macromolecular and biological systems, Ed. D.H. Everett and B. Vincent, Colston Papers no. 29, pp. 129-200

Freire, E., Bach, D., Correa-Freire, M., Miller, I.R., and Barenholz, Y., (1980) Biochemistry $\underline{19}$:3662-3665

Gutknecht, J.(1983) Biochim. Biophys. Acta $\underline{735}$:185-188

Hinz, H.J., and Sturtevan, J.M. (1972) J. Biol. Chem. $\underline{247}$:6071-6075

Hladky, S.B. (1979) The carrier mechanism in current topics. In: Membranes and Transport, $\underline{12}$:53-164. Eds: F. Bonner and A. Kleinzeller. Academic Press

Klein, R.A. (1982) Quart. Rev. of Biophys. $\underline{15}$:667-757

Lauger, P. and Stark, G. (1970) Biochim. Biophys. Acta $\underline{211}$:458-466

Lauger, P., Benz, R., Start, G., Bamberg, E., Jordan, P.C., Fahr, A., and Brock, W. (1981) Quart. Rev. of Biophys. $\underline{14}$:513-598

Lelkes, P.I. and Tandeter, H.V. (1982) Biochim. Biophys. Acta $\underline{716}$:410-418

Markin, V.S. and Liberman, Y.A. (1973) Dohl. Akad. Nauk. SSSR $\underline{201}$:975-978

Papahadjopoulos, D., Jacobson, K., Nir, S., and Isac, T. (1973) Biochim. Biophys. Acta $\underline{311}$:330-348

Ralston, E., Hiemland, L.M., Klausner, R.D., Weinstein, J.N., and Blumenthal, R. (1981) Biochim. Biophys. Acta $\underline{649}$:133-141

Sequaris, J.M. and Miller, I.R. (1984) Bioelectrochem. and Bioenerg. In press.

Smith, R.M. and Martell, A.F. (1976) Critical Stability Constants. Plenum Press

Urban, B.W., Hladky, S.B., and Haydon, D.A. (1980) Biochim. Biophys. Acta 602:331-354

Wilshut, J. and Papahadjopoulos, D. (1979) Nature 282:690-692

Yager, P. and Peticolas, W.L. (1982) Biochim. Biophys. Acta 668:775

SINGLE CHANNEL CONDUCTANCE CHANGES OF THE DESETHANOLAMINE-

GRAMICIDIN THROUGH pH VARIATIONS

R. Reinhardt[1], K. Janko and E. Bamberg[2]

Fakultät für Biologie, Universität Konstanz,D-7750 Konstanz
1) II. Physiologisches Institut, Universität des Saarlandes
   D-6650 Homburg (FRG)
2) Max-Planck-Institut für Biophysik
   D-6600 Frankfurt/Main (FRG)

INTRODUCTION

The surface potential of biological membranes seems to have influence
on the ion transport across the membranes (Frankenhaeuser and Hodgkin,
1957). Changing it also changes the voltage drop across the membrane and
therefore alters the field sensed by the transport protein. This will
influence the voltage dependent gating of membrane channels (for survey
see Hille, 1985). In literature a difference between specific charges at
the channel entrance and a charge distribution in the membrane surface
mainly created by the lipids is made. The theoretical treatment based on
electrostatics by Debye and Hückel, Gouy, Chapman and Stern provide the
experimentalist with useful equations to compare with their data. Since
it is known that the biological membranes consist of a lipid bilayer into
which proteins are embedded (Singer, 1972), the artificial bilayer - in-
vented by P. Müller and coworkers in 1963 (Müller et al., 1963) provides
the model system for testing the predictions based on a physical concept
applied to biology (for review see McLaughlin, 1977). This article shows
the influence of a discrete negative charge on the ion transport through
a channel formed by a gramicidin derivative in an artificial bilayer. But
before this a survey of the gramicidin in bilayers will be given followed
by a short description of other bilayer experiments that have been designed to study the influence of charges on the ion transport through the
gramicidin channel.

THE GRAMICIDIN A

Valine gramicidin A is a linear pentadecapeptide with the structure
HCO-NH-L-Val-Gly-L-Ala-D-Leu-L-Ala-D-Val-L-Val-D-Val-L-Tryp-D-Leu-L-Trp-D-
Leu-L-Trp-D-Leu-L-Trp-CO-NH CH_2 CH_2 OH (Sarges and Witkop, 1965). It increases the permeability for alkali ions of biological membranes (Pressman, 1965) and of artificial membranes (Liberman and Topaly, 1965). Today
there is a general agreement on the structure of the gramicidin channel,
which consists of a helical dimer formed by a head to head (formyl end
to formyl end) association (Urry, 1971; Ramachandran and Chandrasekaran,
1972). The channel is stabilized by intra- and intermolecular hydrogen
bonds. The central ion pathway through the π^6(L,D)-helix has a diameter
of about 0,4 nm. The total length of a dimer is about 3 nm which is the

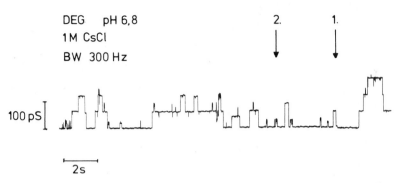

Fig. 1: Typical record of current fluctuations of the DEG at pH 6.8
with 1M CsCl. The arrows indicate examples of conductances be-
longing to the higher peak (1) and lower peak (2) in the histo-
grame.

lower limit of the thickness of a lipid bilayer. Different dimeric structures of the gramicidin A may exist in organic solvents (Veatch and Blout,
1974), but virtually all experiments with gramicidin A incorporated into
lipid bilayers are consistent with Urry's model of a head to head associated dimer.

That gramicidin forms ion channels in lipid bilayers has been first
demonstrated by Hladky and Haydon in 1972 (Hladky and Haydon, 1972). They
observed stepwise current changes in an artificial bilayer after the addition of very small amounts of gramicidin to one side of the bilayer. A
typical record of such current fluctuations from the gramicidin analog
DEG is shown in figure 1. The smallest step is taken as the conductance
of one channel, which is 90 pS for gramicidin A at 1M CsCl on both sides
of the bilayer. The high turnover rate of ions together with only a weak
dependence of the conductance on membrane thickness has provided the evidence for a channel as the transport mechanism rather than a carrier. The
distribution of the lifetimes of the channel shows one exponential with
a mean lifetime of 1,5 s in bilayers of GMO in hexadecane. The current
voltage relationship is nearly linear up to 100 mV.

Experiments with different ions have shown that gramicidin A is
selective for monovalent cations with the permeability sequence $Li^+ < Na^+ <$
$K^+ < Cs^+ < Rb^+$ (Mayers and Haydon, 1972; Eisenmann et al., 1978). Divalent
cations like Ca^{2+} and Ba^{2+} reduce the single channel current if they are
present in the concentration range of 0,1 to 1M (Bamberg and Läuger, 1977).
It is suggested that the divalent cations bind to a site at or near the
channel mouth and thereby reduce the transport rate of the channel.

Today the gramicidin is the best known ion channel. Although it
does not resemble very closely the channels known from cell membranes
it gives an excellent view to the basic principles of ion transport.
In the last five years several reviews appeared about gramicidin looking
at the different aspects and the more interested reader may be refered to
these (Läuger, 1980; Finkelstein and Anderson, 1981; Andersen, 1984;
Hladky and Haydon, 1984).

HO—CH$_2$ CH$_2$NH—CO—L—Trp —   gramicidin A

$^{\ominus}$OOC—L—Trp—   desethanolaminegramicidin

Fig. 2: Comparison between gramicidin A and the desethanolamine
gramicidin.

Two ways are possible to investigate what influences the ion transport through the gramicidin channel. One can change the gramicidin itself
or modify the surrounding of the channel which means using different lipids to make the bilayer. With membranes made of dioleoyl-phosphatidyl-
choline which has a polar headgroup, the single channel conductance has
been found to be reduced about 50% compared to the value in GMO membranes
(Zingsheim and Neher, 1974; Bamberg and Läuger, 1974). It is not yet
understood whether the change in the headgroup of the lipids changes the
single channel conductance through an influence on the ions or through
a conformational change of the gramicidin at the channel entrance.

Using phosphatidylserine, Apell et al. (1979) studied the influence
of negatively charged lipids on the ion transport of gramicidin. They
found an increase of the single channel conductance which could be explained qualitatively with the Gouy Chapmann theory.

The knowledge of the structure of the gramicidin channel opens the
way for chemical modifications and the investigation of the structure
function relationship. Numerous analogs and derivatives have been synthesized to study this subject. Using the single channel method (Hladky
and Haydon, 1972) as well as electrical relaxation experiments (Bamberg
and Läuger, 1973) and noise analysis (Zingsheim and Neher, 1974; Kolb et
al.,1975) the ion transporting properties of these analogs have been
characterized (Bamberg et al., 1976; Tredgold et al., 1977; Ovchinnikov
and Ivanov, 1977; Bamberg et al., 1978; Morrow et al., 1979; Heitz et al.,
1982). Because the main subject of this article is the discription of the
effect of a negative charge at the channel month, experiments with similar
modifications that have already been published shall be mentioned first.

In contrast to a charge distribution in the membrane a gramicidin
derivative has been synthesized containing three carboxyl groups at the
channel entrance, the O-pyromellityl gramicidin (OPG)(Apell et al., 1977).
Again the single channel conductance at low ionic strength is much higher
than for the neutral channel. But the comparison of the charged channel
in neutral lipids with the neutral channel in charged lipids shows that
the electrostatic effect on conductance is much stronger for the latter
one. When the effect of a negatively charged channel entrance on ion
transport is studied the OPG has the disadvantage of having the pk's
of the three carboxyl groups between 1 and 2.5, which is too low to be
titrated in a bilayer experiment. The comparison has to be made between
gramicidin A and the OPG which are two different molecules. To overcome
this problem another gramicidin derivative has been synthesized. It contains one free carboxyl group at the channel entrance. In the remainder
of this article the single channel analysis of this compound will be described.

93

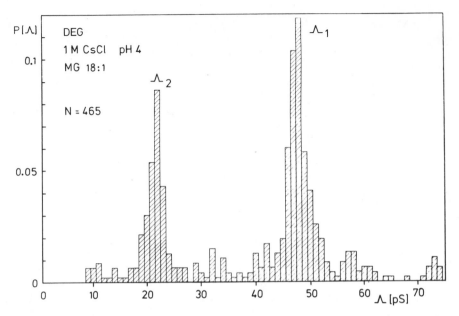

Fig. 3: Histogram of the single channel conductances at pH 4 with the
two conductance states. The ion concentration was 1M CsCl and
the total number of events was 465 (see text).

THE DESETHANOLAMINE GRAMICIDIN

Cleaving the ethanolamine end of the gramicidin A gives a peptide with
a free carboxyl group, the desethanolamine gramicidin (DEG) (figure 2)
(Reinhardt et al., in prep.). According to Urry's model the ethanolamine
is the end that faces the electrolyte. Therefore it should be possible to
titrate the carboxyl group through the pH of the bulk phases and to look
for changes of the channel properties. Regarding the results with the OPG
an increased single channel conductance for the deprotonated molecule can
be expected. If Λ° is the single channel conductance of the protonated
form and Λ^- the single channel conductance of the deprotonated form, than
the single channel conductance can be expressed with the equation

$$\Lambda = \alpha\Lambda^- + (1-\alpha)\Lambda^\circ, /0 \leq \alpha \leq 1 \qquad (1)$$

where α represents the fraction of time the channel spends in the deprotonated form. The relationship between the appearent pk of the carboxyl
group and the quantity α is given by the Henderson-Hasselbalch equation

$$pH = pk + \log \frac{\alpha}{1-\alpha} \qquad (2).$$

To demonstrate the titration of the carboxyl group a single channel analysis has been performed.

Planar lipid bilayers were raised in an orifice separting two compartments of a teflon chamber. The bilayers were made of glycerolmonooleate
(GMO) in hexadecane in all cases and the single channel currents were
measured with silver/silver-chlorid electrodes across the bilayer. The
data were stored on a tape recorder for further analysis.

Adding small amounts of the DEG (10^{-13}M) to one side of the bilayer
results in stepwise current changes as shown in figure 2. The distribution
of the single channel conductances shows two clearly distinct peaks

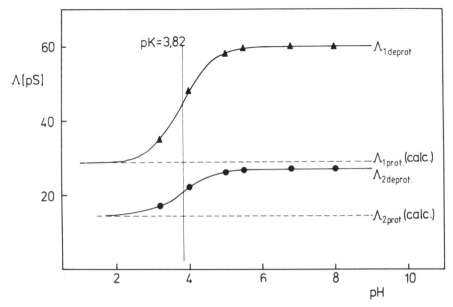

Fig. 4: Dependence of the single channel conductance of the DEG on pH
at 1M CsCl and 50 mV for both conductances Λ_1 and Λ_2. The drawn
curves are fits according to equation (1) and (2).

(figure 3). This is in contrast to gramicidin A that shows only one peak
under identical conditions. Extensive purification of the DEG excludes
chemical heterogenity as an explanation for the second peak. Since both
peaks vary with pH in the same manner – as will be shown below – the
possibility that one of the two peaks belongs to the protonated and the
other to the deprotonated form of the channel is excluded. A possible explanation arise with the assumption that the DEG can assume two different
conformations in the region near the carboxyl terminus, where the bulky
tryptophane residues are located.

The result of the variation of the pH in the bulk phase from 3.2 up
to 8 is shown in figure 4. The data points are the maximum values of a
histogram of the single channel conductances at a fixed pH. Λ_1 denotes
the values of the peak with the higher conductance and Λ_2 are the values
of the peak with the lower conductance in the histogram. The curve is
the fit-curve with equation (1) and (2). It shows that the values of both
peaks can be fitted with the same pk of 3.82. This is a reasonable value
for a terminal carboxylgroup of a peptide; for instance N-acetylalanine
has a pk-value of 3.72 (King and King, 1956). The data points at low pH
have been corrected for the hydrogen conductance.

Now that the accessibility of the carboxyl group through the electrolyte pH has been demonstrated, the effect of the negative charge on ion
transport can be investigated with the same channel. In the following
figures only the higher conductance peak is shown. The lower one appears
with less probability and behaves in a parallel manner through all variations.

For gramicidin A the single channel conductance increases with ion
concentration up to a saturation value which is reached at a salt concentration higher than 1M. For the alkali ions Rb^+ and Cs^+ and much less for
K^+ the single channel conductance decreases again if the ion concentration

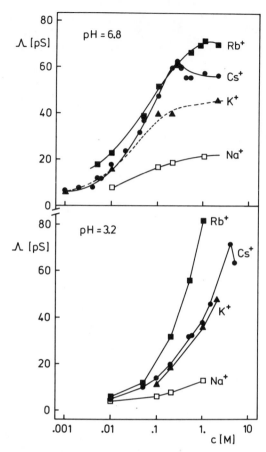

Fig. 5: Dependence of the single channel
        conductance of the DEG on ion con-
        centration for the alkali ions Na[+],
        K[+], Rb[+] and Cs[+] at pH 3,2 and 6,8.
        The curves are data connecting
        lines without theoretical meaning.

is further increased. This is probably due to ion-ion repulsion because
at high ionic strength there is an increased probability that both binding sites of the gramicidin channel are occupied (Urban et al., 1980).
The dependence of the DEG channel conductance on ion concentration in the
protonated form (at pH 3.2) and the deprotonated form (at pH 6.8) is
shown in figure 5. Comparison of the protonated form with gramicidin a
shows qualitatively the same behaviour. Only the conductance for Rb[+] is
higher relative to Cs[+]. But looking at the deprotonated channel it sa-
turates at much lower ion concentrations and the conductance at high
ionic strength is lower than the conductance for the protonated channel.
This is expected because the negative charge at the channel entrance
accumulates the positive ions and the saturation concentration is reached
at much lower bulk concentrations.

    If the different saturation concentrations for the single channel
conductances at the two pH values are caused by the presence or absence
of the negative charge at the channel entrance, it should be possible to
bring the two curves to coincidence with a recalculation of the ion con-

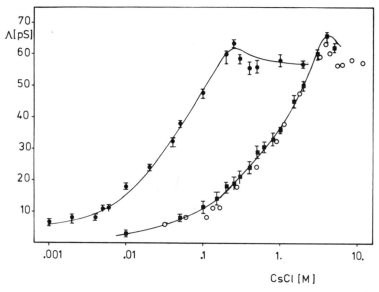

Fig. 6: Single channel conductance of the DEG for the Cs^+-
        ion at pH 6,8 (o) and pH 3,2 (■). The open circles
        are the recalculated values of the conductances
        at pH 6,8 according to equations (3) to (5). For
        detailes see text. The curves are data connecting
        lines with no theoretical meaning.

centrations using the Debey Hückel theory (Apell et al., 1977). The
potential ϕ at a radius a around a point charge is

$$\phi(a) = \frac{z \, e_0}{4\pi\varepsilon_0\varepsilon} \frac{1}{a(1-\kappa a)} \; ; \; \kappa = \frac{1}{l_D} \tag{3}$$

The ion concentration at a is then

$$c(a) = c_\infty \exp \frac{\phi(a)e_0}{kT} \tag{4}$$

Assuming that the single channel conductance is proportional to the
ion concentration the ratio of Λ^- and Λ^0 is

$$q = \frac{\Lambda^-}{\Lambda^0} = \exp \frac{-\phi(a)e_0}{kT} \tag{5}$$

From a fit with equation (5) a capture radius a=2Å is obtained. With
a given a the recalculation of the concentration with equations (3)
and (4) is possible. The result for Cs^+ is shown in figure 6.

The ion selectively for the gramicidin A channel shows the sequence
$Li^+ < Na^+ < K^+ < Cs^+ < Rb^+$ which is consistent with a decreasing ionic hydration
energy for higher permeability (Mayers, Haydon, 1972; Eisenmann et al.,
1977). Entering the channel, the ion has to strip off part of its hydration shell. The carbonyls along the channel interior serve as a substi-
tute for the water.

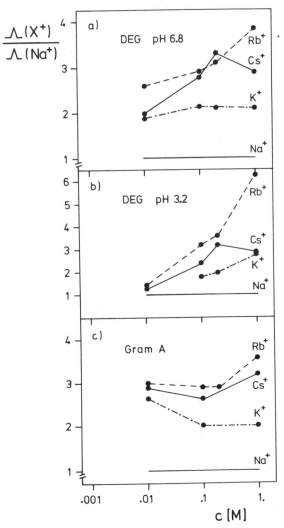

Fig. 7: Ion selectivity ratios relative to Na$^+$
from single channel conductances for DEG
at pH 6.8 and 3.2 and the gramicidin A as
a function of ion concentration. The
values for gramicidin A are taken from
Neher et al., (1978).

The selectivity for the DEG can be seen in figure 7, where the
ratios of the single channel conductances of ion X and the corresponding
conductance for Na$^+$ are plotted. The sequence in the protonated state
is the same as for gramicidin A. A difference appears in the deprotonated
state where the sequence is changed to Na$^+$<K$^+$<Rb$^+$<Cs$^+$ around 0.2 M salt
concentration. So changing the pH at this ionic strength from 3.2 to 6.8
causes a change in selectivity between Cs$^+$ and Rb$^+$.

For the gramicidin A a binding site for divalent cations near the
channel entrance is proposed, which is no binding site for the permeating ion (Bamberg and Läuger, 1977). The presence of 500 mM CaCl$_2$ causes
a reduction of the single channel conductance of about 30%. Bamberg and
Läuger (1977) were able to show that this is consistent with the idea

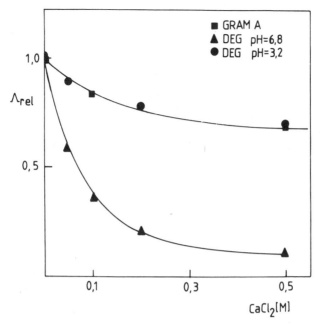

Fig. 8: Influence of calcium on the single channel conductance of the DEG at pH 3,2 and 6,8 at 100 mM
CsCl and its comparison with gramicidin A; V=50 mV.
The conductances have been normalized to the gramicidin A conductance without Calcium which has been
set to 1.

that the transit time of the permeating ion is much longer than the
lifetime of the blocked or unblocked state of the channel, so that the
permeating ion "sees" an entrance barrier of averaged hight.

The conductances of both states of the DEG are also strongly reduced
in the presence of Ca^{2+}. The effect is much more pronounced in the deprotonated state where the single channel conductance is half blocked in
the presence of 60 mM $CaCl_2$ (figure 8). For the protonated state a calcium concentration of 500 mM is needed to reduce the single channel conductance 30%. This is the same amount of calcium that is necessary to
get a 30% reduction of the single channel conductance of gramicidin A.

For electrostatic reasons it can be assumed that the calcium interacts with the carboxyl group, which means it binds near the channel entrance as proposed for gramicidin A. Therefore the same model can be
applied so that a fast association dissociation reaction of the calcium
at the carboxyl group creates an increased entrance barrier for the
permeating ion. A higher affinity of H^+ to the carboxyl group reduces
the effect of the Ca^{2+} at lower pH and the blocking effect is much less.

At a cesium concentration of 100 mM the reduction of the single
channel conductance of the DEG in the deprotonated state by 60 mM calcium
is similar to the reduction from changing the pH from 6.8 to 3.2 (data
not shown). With a further increase of the calcium conentration a reduc-

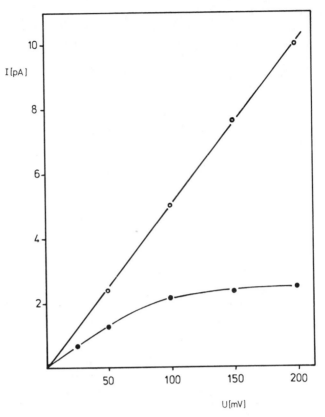

Fig. 9: Current-voltage relationship of the DEG at pH 6,8
and 200 mM CsCl (open circles) and in the presence
of 100 mM CaCl₂

tion of 90% in the deprotonated state is reached at 500 mM. This is the
maximum block that can be obtained with calcium which is much more than
possible with changing the pH. This might be due to the two positive
charges of the calcium.

The I(V)-relationship of the DEG at pH 6,8 is also effected by
calcium. Without calcium it is linear up to 200 mV, but with calcium a
strong saturation is observed (figure 9). A stabilization of the calcium
binding by the voltage seems to play a role.

ACKNOWLEDGMENT

Supported by the Deutsche Forschungsgemeinschaft, SFB, 138.

REFERENCES

Anderson, O.S., 1984, Gramicidin channels, Ann. Rev. Physiol. 46: 531-548
Apell, H.-J., Bamberg, E., Alpes, H., Läuger, P., 1977, Formation of ion
        channels by a negatively charged analog of gramicidin A.
        J. Membrane Biol. 31: 171-188
Apell, H.-J., Bamberg, E., Läuger, P., 1979, Effects of surface charge on
        the conductance of the gramicidin channel. Biochem. Biophys.
        Acta 552: 369-389
Bamberg, E. and Läuger, P., 1973, Channel formation kinetics of gramicidin

A in lipid bilayer membranes. J. Membrane Biol. 11: 177-194

Bamberg, E., and Läuger, P., 1974, Temperature-dependent properties of gramicidin A channels. Biochim. Biophys. Acta 367: 127-133

Bamberg, E.,Noda, K., Gross, E. and Läuger, P., 1976, Single channel parameters of gramicidin A, B and C, BBA 419: 223-228

Bamberg, E. and Läuger, P., 1977, Blocking of gramicidin channel by divalent cations. J. Membrane Biol. 35: 351-375

Bamberg, E., Apell, H.-J., Alpes, H., Gross, E., Morell, J.L., Harbaugh, J.F., Janko, K. and Läuger, P., 1978, Ion channel formed by chemical analogs of gramicidin A. Fed. Proc. 37: 2633-2638

Eisenmann, G., Sandblom, J. and Neher, E., 1978, Interactions in cation permeation through the gramicidin channel Cs, Rb, K, Na, Li, Tl, H and effects of anion binding. Biophys. J. 22: 307-340

Finkelstein, A. and Andersen, O.S., 1981, The gramicidin A channel: A review of its permeability characteristics with special reference to the single-file aspects of transport. J. Membrane Biol. 59: 155-171

Frankenhaeuser, B. and Hodgkin, A.L., 1957, The action of calcium on the electrical properties of squid axons. J. Physiol. (Lond) 137: 218-244

Heitz, F., Spack, G. and Trudelle, Y., 1982, Single channels of 9,11,13,15 destryptonphyl-phenylalanyl-gramicidin A, Biophys. J. 40: 87-89

Hille, B., 1984, Ionic channels of excitable membranes, Sinauer Associates Inc. Sunderland, Massachusetts

Hladky, S.B. and Haydon, D.A., 1984, Ion movements in gramicidin channels in: current topics in membrane and transport, Vol. 21, 327-372

King, E.J. and King, G.W., 1956, The thermodynamics of amino acids. II. The Ionization constants of some N-Acyl Amino Acids. J. Am. Chem. Soc. 78: 1089

Kolb, H.-A., Läuger, P. and Bamberg, E., 1975, Correlation analysis of electrical noise in lipid bilayer membranes: kinetics of gramicidin A channels. J. Membrane Biol. 20: 133-154

Läuger, P., 1980, Kinetic properties of ion carriers and channels, J. Membrane Biol. 57: 163-178

Liberman, E.A. and Topaly, V.P., 1968, Selective transport of ions through bimolecular phospholipid membranes, BBA 163: 125-136

McLaughlin, S., 1977, Electrostatic potentials at membrane-solution interfaces, in: "Current Topics in Membranes and Transport" 9: 71-144

Morrow, J.S., Veatch, W.R. and Stryer, L., 1979, Transmembrane channel activity of gramicidin A analogs. Effects of modification and deletion of the aminoterminal residue, J. Mol. Biol. 132: 733-738

Müller, P., Rudin, D.O., Tien, H.T. and Wescott, W.C., 1963, Methods for the formation of single bimolecular lipid membranes in aqueous solution, J. Phys. Chem. 67: 534-535

Myers, V.B. and Haydon, D.A., 1972, Ion transfer across lipid membranes in the presence of gramicidin A. II. The ion selectivity. BBA 274: 313-322

Neher, E., Sandblom, J. and Eisenmann, G., 1978, Ionic selectivity, saturation and in gramicidin A channels. J. Membrane Biol. 40: 97-116

Ovchinnikov, A., YU, and Ivanov, V.T., 1977, Recent developments in the structure-functional studies of peptide ionophores. in: Biochemistry of Membrane Transport, FEBS Symposium No. 42, Semenza, G. and Carafoli, E., Eds, Springer Verlag, Berlin, Heidelberg, New York, 123-146

Pressman, B.C., 1965, Induced active transport of ions in mitochondria PNAS, 63, 1076-1083

Ramachandran, G.N. and Chandrasekaran, R., 1972, Studies on Dipeptide conformations and on peptides with sequences of alternating L and D Redidues with special reference to Antibiotic and Ion transport peptides. Progr. Peptide Res. 2: 195-215

Reinhardt, R., Janko, K. and Bamberg, E., in Prep.

Sarges, R. and Witkop, B., 1965, Gramicidin. VIIl. The structure of Va-
line- and Isoleucine-Gramicidin C. Biochemistry 4: 2491-2494

Singer, S.J. and Nicolson, G.J., 1972, The fluid mosaic model of the
structure of cell membranes. Science 175: 720-731

Tredgold, R.H., Hole, P.N., Sproule, R.C. and Elgamal, M., 1977, Single
channel characteristics of some synthetic gramicidins. Biochim.
Biophys. Acta 471: 189-194

Urban, B.W. and Hladky, S.B., 1979, Ion transport in the simplest single
file pore. BBA 554: 410-429

Urban, B.W., Hladky, S. and Haydon, D.A., 1980, Ion movements in gramicidin pores. An Exemple of single-file transport, BBA 602: 331-354

Urry, D.W., 1971, The Gramicidin A transmembrane channel: A proposed (L,D)-
Helix, PNAS 68: 672-676

Veatch, W.R., Fossel, E.T. and Blout, E.R., The conformation of gramicidin
A, Biochem. 13: 5249-5256

Zingsheim, H.P. and Neher, E., 1974, The equivalence of fluctuation analysis and chemical relaxation measurements: A kinetic study of
ion pore formation in thin lipid membranes. Biophys. Chem. 2:
197-207

MEMBRANE POTENTIAL OF SQUID AXONS

Shinpei Ohki

Department of Biophysical Sciences
School of Medicine
State University of New York at Buffalo
Buffalo, New York 14214 U.S.A.

ABSTRACT

The resting membrane potentials of squid axons were measured as a function of the nature of the ionic species as well
as the ionic strength of the extracellular and intracellular
media. The observed membrane potentials were analyzed using
two equations: a) the Goldman-Hodgkin-Katz equation, and b)
the surface/diffusion potential equation. The overall experimental results were found to be explained better by the sur-
face/diffusion potential equation. The surface charge densities
of the outer and inner surfaces (on the surface responsible
for the ion transfer) of the squid axons were found to be
approximately $-e/180 \text{ Å}^2$ and $-e/1300 \text{ Å}^2$, respectively. Using
the obtained surface charge densitities and the membrane potential equation involving divalent cations, the relative
permeabilities (with respect to K^+) of divalent cations across
the squid axon membrane were obtained to be 0.033 for Ca^{2+},
0.031 for Sr^{2+} and 0.022 for Mg^{2+} from the experimental data.

INTRODUCTION

For many years, it has been known that many biological cells maintain
different distribution in ionic concentrations and an electrical potential
difference between their intracellular and extracellular phases. This
electrical potential difference, called the resting membrane potential,
was thought to be due to the Donnan equilibrium potential caused by the
semi-permeable nature of most biological cells with respect to ions.
Bernstein (1902) suggested that the testing membrane potential was determined by the ratio of potassium ion concentrations inside and outside the
cell. The validity of the theory of the membrane potential for biological cells was then supported by others (Boyle & Conway, 1941; Hodgkin &
Horowicz, 1959).

However, the accurate measurements of the resting membrane potential
(Ling & Gerard, 1949) and the ion flux measurements (Keynes & Lewis, 1951)
across cell membranes revealed that the Donnan equilibrium potential theory
was not sufficient to explain the observed membrane potential.

The closest approximate formula describing the observed resting membrane potential of squid axon was presented by Hodgkin and Katz (1949) about three decades ago, which was based solely on ion diffusion potential (Planck, 1890; Goldman, 1943).

Since then the equation has been widely used to explain the membrane potential of many biological cells. However, it was reported that the equation was not adequate to account for changes in resting membrane potential, when the internal salt solution of the squid axon was altered in a certain manner (Tasaki, et al., 1963, 1964). On the other hand, others (Baker, et al., 1962, 1964; Narahaski, 1963) attempted to explain such an observed membrane potential by retaining the Goldman-Hodgkin-Katz equation and introducing modified ionic permeability factors. In order to explain a shift in the inactivation curve during the nerve activity which occurred when the intracellular KCl was diluted with isotonic sucrose, Chandler, et al., (1965) introduced a layer of fixed negative charges on the inside of the axon membrane, the magnitude of which was about one electronic charge per 700 Å^2. Recently, Ohki and Aono (1979) have measured the membrane potential of squid axons with variation of the internal and external ionic species and concentrations. The results are also found to be difficult to explain by the Goldman-Hodgkin-Katz equation and it is suggested that the consideration of the surface potential of the membrane in addition to the membrane diffusion potential may resolve this difficulty.

In order to examine this latter possibility for the membrane potential theory, we have measured the membrane potential of squid axon membranes as a function of the nature of the ionic species and ionic strength of the extracellular as well as intracellular media, and have analyzed the observed membrane potentials using both the Goldman (1943)-Hodgkin and Katz (1949) equation and the surface/diffusion potential equation (Ohki, 1979).

The membrane potential equation involving polyvalent cations which are permeable across the membrane in the bathing solution is also obtained. Using this equation and the experimental data of membrane potential with variation of extracellular divalent cation concentrations and the obtained surface charge densities for the axon membrane, an attempt is made to obtain the relative permeabilities of divalent cations across the squid axon membrane.

MATERIALS AND METHODS

Materials. Experiments were performed using giant axons of squids (Loligo pealeii) at the Marine Biological Laboratory, Woods Hole, Mass. during the summers of 1982 and 1984. The giant axons having 400 - 450 μm in diameter were isolated and cleaned under a dissecting microscope by removing as much of the surrounding connective tissues and small nerve fibers as possible. The length of each axon between the two cut ends was at least 5 cm. The isolated axons were tied with thread at each end and used immediately. All chemicals used were of analytical reagent grade and were obtained from Baker Chemical or Fisher Chemical Companies.

Perfusion of extracellular solution. The external solutions used were either natural sea water, available at Woods Hole (Cavanaugh, 1956), or artifical solutions whose ionic components are given in Table 1. The pH's of the solutions were adjusted to 8.0 with 2 mM Tris-buffer and hydrochloric acid.

TABLE 1

Ionic Compositions of Various Extracellular Solutions (number in mM)

A) $[Na]_o$ substitution

Na	Choline	K	Ca	Cl
488	0	10	20	~540
244	244	10	20	~540
122	366	10	20	~540
61	428	10	20	~540
20	468	10	20	~540

B) $[Cl]_o$ substitution

Na	Glutamate	K	Ca	Cl
488	0	10	20	~540
488	244	10	20	~295
488	366	10	20	~184
488	427	10	20	~113
488	488	10	20	~52

C) $[NaCl]_o$ dilution

Na	Sucrose	K	Ca	Cl
488	0	10	20	~540
244	373	10	20	~294
122	560	10	20	~172
61	727	10	20	~102
20	747	10	20	~70

D) $[Ca^{2+}]_o$ alteration

Na	Sucrose	K	Ca	Cl
0	762	0	20	~52
0	729	0	33	~78
0	648	0	66	~144
0	324	0	132	~276
0	162	0	264	~540

E) $[Mg^{2+}]_o$ alteration

Na	Sucrose	K	Mg	Cl
0	762	0	20	~52
0	729	0	33	~78
0	648	0	66	~144
0	324	0	132	~276
0	162	0	264	~540

F) $[Sr^{2+}]_o$ alteration

Na	Sucrose	K	Sr	Cl
0	762	0	20	~52
0	729	0	33	~78
0	648	0	66	~144
0	124	0	132	~276
0	162	0	264	~540

An isolated axon was first placed in a lucite chamber containing sea water. Firstly, its resting and action potentials were checked to ensure that the axon was healthy. During each experiment, the external solution was continuously flushed around the axon. The flow rate of the extracellular solutions was about 30 ml/min. The effective volume of the cell compartment in which the axon was immersed was ~1 cm^3. The device used permitted a complete change of the external solution around the axon within 15 sec.

Perfusion of intracellular solution. The method of perfusing the axoplasm was similar to Tasaki, et al., (1962) and was as follows: a glass cannula of about 180 μm in outside diameter was inserted into one end of the axon and penetrated through the axon to the other end of the axon, and then the cannula was withdrawn back to the original insertion

TABLE 2

Ionic Compositions of Various Intracellular Solutions (numbers in mM).

(A) Internal salt dilution

K	Glutamate	Cl	F	PO_4
~400	250	50	50	30
~200	125	25	25	15
~100	62	12.5	12.5	7.5
~50	31	6.0	6.0	3.7

point in the axon to serve as an inlet for the perfusing solution. As the
inlet cannula was withdrawn, another glass cannula of about 300 μm in outer
diameter was inserted into the other end of the axon and it was used as
the outlet for the perfusing solution. The inlet cannula was connected
through a thin polyethylene tubing to a reservoir for the internal perfusing fluid. The flow rate of the internal solution was about 40 μl/min.
The internal perfusion solution used in the cases of the extracellular
solution change consisted of 0.25 M potassium glutamate, 0.05 M potassium
fluoride, 0.05 M potassium chloride, and 30 mM potassium phosphate buffer,
pH 7.3.

Measurements of membrane potential. The standard electrophysiological
technique (intracellular potential recording via an agar-agar salt bridge
glass capillary electrode introduced into the axon longitudinally with
respect to the extracellular potential via a reference calomel electrode)
was employed for the measurements of cell membrane potential. The output
signal of the internal electrode was fed to an oscilloscope (via a pre-
amplifier) in order to detect a transient change in transmembrane potential
as well as to a chart recorder to record the resting membrane potential.
Nerve activity was induced by stimulating the nerve extracellularly using
a pair of platinum electrodes and was monitored from time to time during the
course of the experiments. Experimental data were collected for those axons
in which the resting and action potentials of axons were restored to the
same values when the external solutions were changed back to the normal
physiological solution (sea water) after various solution changes were made.
A schematic diagram of the experimental cell arrangement is shown in Fig. 1.
Before inserting the internal electrode into the axon, its tip potential
was corrected against the reference electrode by immersing them both in
the internal perfusion solution. Each value of the observed membrane resting
potential shown in the following figures was the arithmetic average of data
using at least four different axons. All experiments were performed at
room temperature (20°C).

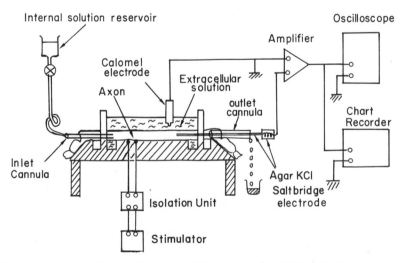

Fig. 1. Schematical diagram of the experimental set-up.

EXPERIMENTAL RESULTS

The first series of experiments were the examination of the effect of substitution of extracellular Na^+ with other impermeable monovalent cations on membrane potential. Figure 2 shows the observed membrane potential (indicated by o) of the squid axons as a function of the extracellular Na^+ concentration, where various amounts of Na^+ were substituted with corresponding amounts of choline$^+$ in order to maintain constant ionic strength, while at the same time the other extracellular ion concentrations were kept constant ($[K^+]_o$ = 10 mM, $[Ca^{2+}]_o$ = 20 mM, $[Cl^-]_o$ = 540 mM (see the detail in Table 1)). For the experiments to examine the effect of the extracellular salt solution on membrane potential, all axons were perfused with the same internal solution (0.25 M potassium glutamamate, 0.05 M potassium fluoride, 0.05 M potassium chloride, and 30 mM potassium phosphate buffer, pH 7.3) at a constant flow rate ($\sim$ 40 μl/min.) in order to maintain constant internal salt contents and concentrations. The membrane potential hyperpolarized slightly as the extracellular Na^+ concentration was reduced and seemed to show saturation at an extremely low Na^+ concentration.

In the second series of experiments, the membrane potential was measured as a function of the extracellular Cl^- which was substituted with various concentrations of glutamate. At the same time the concentrations of the other ions were kept constant ($[Na^+]_o$ = 488 mM, $[K^+]_o$ = 10 mM,

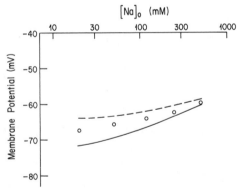

Fig. 2. Membrane potential of squid axon versus extracellular Na^+ concentrations (Table 1-A). Various amounts of Na^+ were substituted with the corresponding amounts of choline$^+$. The ordinate refers to the intracellular potential (ϕ_i) with respect to the extracellular potential (ϕ_o) as zero. o: experimental data; ----: calculated values by the G-H-K equation; and ———: calculated values by the S-D potential equation (the surface charge density $-e/180$ $Å^2$ was used). The same relative ionic permeabilities ($P_K:P_{Na}:P_{Cl}:P_{choline}:P_{glutamate}:P_F:P_{PO_4}$ = 1:0.04:0.45:0:0:0:0) were used for both cases.

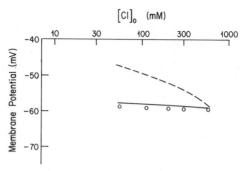

Fig. 3. Membrane potential versus extracellular Cl^- concentration.
Various amounts of the extracellular Cl^- were replaced
with glutamate (Table 1-B). o: experimental data;
----: calculated values by eq. (1); ——: calculated
values by eq. (2) ($\sigma = -e/180$ Å^2). The same relative
ionic permeabilities ($P_K:P_{Na}:P_{Cl}:P_{glutamate}:P_F:P_{PO_4} =$
$1:0.04:0.45:0:0:0$) were used for both cases.

$[Ca^{2+}]_o = 20$ mM). The observed membrane potentials were unchanged or
slightly depolarized (a few mV for a ten-fold change in Cl^- concentration)
as the concentration of Cl^- was reduced (Fig. 3).

In the third series of experiments, the membrane potential for various dilutions of the external NaCl was measured, while the medium was
kept isotonic with the sucrose. The other extracellular salt concentrations
were kept constant ($[K^+]_o = 10$ mM, $[Ca^{2+}]_o = 20$ mM). The membrane potentials were hyperpolarized more with dilution of the extracellular
NaCl. The membrane potential tended to saturate as the concentration of
NaCl was diluted to a low value (less than 20 mM). The experimental data
are shown in Fig. 4.

The fourth series of experiments was done to examine the changes in
membrane potential with respect to changes in the extracellular divalent
cation concentration. The divalent cations used were Ca^{2+}, Mg^{2+}, and Sr^{2+}.
The ionic components and concentrations of the solutions are given in
Table 1-D,E,F. In these experiments, the extracellular solution contained
chloride salts of divalent cations, sucrose, and 2 mM Tris, but no Na^+
and K^+. The total osmolarity of the extracellular solution was kept constant with sucrose as divalent cation concentrations were changed. As the
divalent cation concentration was increased, the membrane potential depolarized. The membrane potential change was about 13 mV for a 3 fold
change in divalent cation concentration for all three divalent cations,
and the potential change was approximately linear with respect to the
logarithm of divalent cation concentration (Fig. 5). The latter two ions

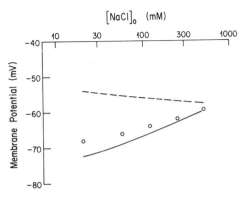

Fig. 4. Membrane potential with respect to various dilutions of extracellular NaCl by isotonic sucrose, while $[K^+]_o$ = 10 mM and $[Ca^{2+}]_o$ = 20 mM were kept constant (Table 1-C). o: experimental value; ----: calculated value by eq. (1); ———: calculated values by eq. (2) (σ = $-e/180$ Å^2). The same relative ionic permeabilities ($P_K:P_{Na}:P_{Cl}:P_{sucrose}:P_F:P_{PO_4}$ = 1:0.04:0.45:0:0:0) were used.

(Sr^{2+}(o) and Ca^{2+}(x)) resulted in the similar membrane potential with respect to variation of divalent cation concentrations. The Mg^{2+} case (shown by a symbol of Δ in Fig. 5) gave greater resting potentials than those in the Sr^{2+} and Ca^{2+} cases at a given concentration of divalent cation. The observed membrane potentials shown in Figs. 2, 3, 4, and 5 were identical within the experimental error to those (Ohki, 1985) obtained under a slightly different internal perfusion condition.

In the fifth series of experiments, the effect of dilution of the intracellular salt concentration of membrane potential was examined (fig. 6). The intracellular perfusion salt solution (250 mM K-glutamate/ 50 mM KF/50 mM KCl/30 mM phosphate buffer) was diluted with glucose so as to maintain an iso-osmolarity of the intracellular solution, while the extracellular perfusion salt solution was kept constant (488 mM NaCl, 30 mM CaCl$_2$, 10 mM KCl). When the concentration of K^+ was reduced, the change in membrane potential with respect to K^+ concentration change became larger but the dependence of the membrane potential change on the intracellular K^+ concentration was not as large as that expected from the Nernst equation. For the change in the intracellular K^+ concentration (by dilution) from 400 mM to 50 mM, the resting potential increased about 16 mV.

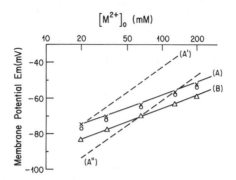

Fig. 5. Membrane potential versus various external divalent cation concentrations (Table 1-D,E,F). x: observed membrane potential for the
CaCl$_2$ case; Δ: for MgCl$_2$; O: for SrCl$_2$; ----: calculated membrane
potential by use of eq. (1); ——: calculated membrane potential
by use of eq. (2) (σ = -e/180 Å^2). The relative ionic permeabilities
used are $P_K:P_{Na}:P_{Cl}:P_{Ca}$ = 1:0.04:0.45:0.26 for the dotted line A'
and $P_K:P_{Na}:P_{Cl}:P_{Ca}$ = 1:0.04:0.45:0.14 for the dotted line A" and
$P_K:P_{Na}:P_{Cl}:P_{Ca}(P_{Mg})$ = 1:0.04:0.45:0.035(0.023) for the solid line
A(solid line B).

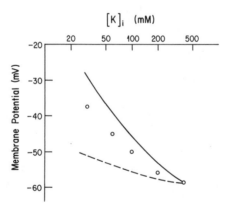

Fig. 6. Membrane potential with respect to various dilutions of intracellular salt solution by isotonic glucose, while the extracellular
solution (488 mM Na$^+$/10 mM K$^+$/20 mM Ca^{2+}/∿540 mM Cl$^-$/pH 8.0) was
maintained constant. The components and concentrations of the intracellular perfusion solution are given in Table 2. o: experimental
data; ----: calculated values by the G-H-K equation (eq. (1)); and
——: calculated values by the S-D potential equation (eq. (2)).
The surface charge densities used were -e/180 Å^2 for the external
and -e/1300 Å^2 for the internal surfaces, respectively. The same
relative ionic permeabilities ($P_K:P_{Na}:P_{Cl}:P_{glutamate}:P_{PO_4}$ = 1:0.04:
0.45:0:0) were used for both cases.

ANALYSIS AND DISCUSSION

The observed membrane potentials were analyzed with the use of two membrane potential equations: 1) the Goldman-Hodgkin and Katz (G-H-K) equation and 2) the membrane potential theory including their surface potentials (Ohki, 1979) using the set of relative ion permeabilities obtained by Hodgkin and Katz (1949); $(P_K:P_{Na}:P_{Cl} = 1:0.04:0.45)$. The validity of the use of these relative ionic permeabilities will be discussed later.

The analysis of membrane potential by the G-H-K equation is straightforward. The membrane potential E_m is given by

$$E_m = \phi_i - \phi_o = E_D = \frac{RT}{F} \ln \frac{P_{Na}[Na^+]_o + P_K[K^+]_o + P_{Cl}[Cl^-]_i}{P_{Na}[Na^+]_i + P_K[K^+]_i + P_{Cl}[Cl^-]_o}, \qquad (1)$$

where ϕ_i and ϕ_o are the electrical potentials of the internal and external bulk solutions, respectively, $P_K:P_{Na}:P_{Cl} = 1:0.04:0.45$ and [M] refers to the activity of each ion in the bulk solutions. These [M] values were obtained from the activity coefficients (Moore, 1962), the extracellular ionic concentrations given in Table 1, and the intracellular ionic concentrations given in Table 2. The subscripts o and i of the ionic activity refer to the outside and inside of the axon, respectively. This potential theory gives a potential profile across a membrane which is approximately depicted in Fig. 7A. There is no consideration of the fixed charge effect at the membrane surface.

Analysis of the membrane potential using the second theory (the surface/diffusion potential (S-D) equation) is as follows: The membrane potential is expressed (Ohki, 1979) by:

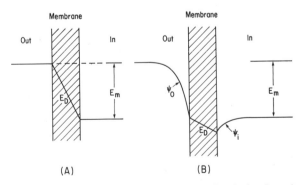

(A)   (B)

Fig. 7.   Schematic profiles of membrane electrical potential (A) Diffusion potential dominated (Goldman-Hodgkin-Katz equation) (B) Surface/diffusion potential for a membrane having negatively charged surfaces.

$$E_m = \phi_i - \phi_o = \psi_o(0) + E_D' - \psi_i(0)$$

(2)

$$= \psi_o(0) + \frac{RT}{F} \ln \frac{P_{Na}[Na]_o^S + P_K[K]_o^S + P_{Cl}[Cl]_i^S}{P_{Na}[Na]_i^S + P_K[K]_i^S + P_{Cl}[Cl]_o^S} - \psi_i(0),$$

where the relative ionic permeabilities used were the same as those (P_K:P_{Na}: P_{Cl} = 1:0.04:0.45) of Hodgkin and Katz. However, $[M]^S$ was the activity of each ion at the membrane outer surface, which can be expressed by

$$[M_j]_o^S = [M_j]_o \exp\left(\frac{-Z_j e\psi_o(0)}{kT}\right)$$

and

(3)

$$[M_j]_i^S = [M_j]_i \exp\left(\frac{-Z_j e\psi_i(0)}{kT}\right)$$

where $[M_j]_o$ and $[M_j]_i$ are the activities of the jth ionic species in the bulk extracellular and intracellular solutions, respectively, and $\psi_o(0)$ and $\psi_i(0)$ are surface potentials at the outer and inner surfaces of the axon membrane. This potential theory gives a potential profile across a membrane as depicted in Fig. 7-B. The signs and magnitudes of surface potentials are varied depending on the sign and magnitude of the surface fixed charges and surrounding ionic conditions.

The following procedure was taken to determine the surface charge densities on both sides of the membrane. First, we assumed the internal surface potential to be zero since the surface charge density of the internal surface of squid axon membranes was reported to be relatively low (Chandler, et al., 1965, Rojas, et al., 1970). Then, the outer surface potential $\psi(0)$ was calculated by using the Grahame equation (see Appendix), using the known salt concentrations of the extra- & intracellular solutions & various surface charge densities of axon membrane, assuming that ions do not bind to the fixed surface charge sites of the membrane. The most suitable value among the surface charge densities was obtained so as to give the best correspondence between the calculated membrane potential and the results of the first three series of experiments (Figs. 2, 3, and 4). Then, by using the obtained outer surface charge density, σ_o, the internal surface charge density, σ_i, was calculated from equation (2) and the Grahame equation, knowing internal and external salt concentrations and varying the internal surface charge densities. The value was chosen to give the best correlation between the theoretical membrane potentials and the experimentally obtained membrane potentials in the case where the potassium concentration of the internal perfusion solution was diluted (Fig. 6). Then, using the obtained internal surface charge density, the external surface charge density was determined as described above by the itaration method so as to fit the observed membrane potential. The values among the surface charge densities which give the most suitable correlation between the calculated membrane potentials and the experimental results (Figs. 2, 3, 4, and 6), were $-e/180$ Å^2 for the outer surface charge density, σ_o, and $-e/1300$ Å^2 for the inner surface charge density, σ_i, respectively.

In analysis of the data of the first three series of experiments (Figs. 2, 3 and 4), and for the fifth series of experiments (Fig. 6) (the membrane potentials for the case of the internal salt dilution), the diffusion potential E_D' used was that expressed as the Goldman-Hodgkin-Katz equation involving monovalent ions only, since the contribution of divalent cations (20 mM Ca^{2+}) to the diffusion potential should be relatively small compared to those of monovalent cations; except for the cases of low Na^+ concentration in the third series of experiments. In the fourth series of experiments, however, the extracellular solution contained predominantly divalent cations and monovalent anions, and practically no monovalent cations. Therefore, we should use the diffusion potential equation involving not only monovalent ions but also polyvalent cations for this potential analysis. In the case where only the divalent cation and the monovalent anion are in the extracellular phase and the monovalent cation in the intracellular phase, which are all permeable across the membrane (this corresponds to the case of the fourth series of experiments), the diffusion potential is expressed by

$$E_D = \frac{RT}{F} \ln \left\{ -0.5 + (0.25 + \frac{4[M^{2+}]_o P^{2+}}{P^+[M^+]_i + P^-[M^-]_o})^{1/2} \right\} \tag{4}$$

where $[M^{2+}]_o$, $[M^+]_i$, $[M^-]_o$ are the activities of divalent cation, monovalent cation and monovalent anion in the bulk solution, respectively, and the suffixed i and o refer to the intracellular and extracellular phases, respectively. P^+, P^- and P^{2+} are the relative permeability coefficients of the monovalent cation, monovalent anion and divalent cation, respectively.

In such a case, the G-H-K equation and surface/diffusion equation (S-D equation) are expressed by

$$E_m = \phi_i - \phi_o = \frac{RT}{F} \ln \left\{ -0.5 + (0.25 + \frac{4[M^{2+}]_o P^{2+}}{P^+[M^+]_i + P^-[M^-]_o})^{1/2} \right\} \tag{5}$$

and

$$E_m = \phi_i - \phi_o = \psi_o + \frac{RT}{F} \ln \left\{ -0.5 + (0.25 + \frac{4[M^{2+}]_o^s P^{2+}}{P^+[M^+]_i^s + P^-[M^-]_o^s})^{1/2} \right\} - \psi_i \tag{6}$$

respectively, were $[M]^s$ are the activity of the M-th ionic species at the membrane surface:
$$[M_i]^s = [M_j] \exp\left\{ \frac{-Z_j e \psi(0)}{kT} \right\}.$$

Using the obtained values for the surface charge densities and these two equations (5) or (6), the membrane potentials obtained in the fourth series of experiments were analyzed. From the best fit of theoretical values to the experimental data, the relative permeability of divalent cations to the axon membrane with respect to K^+ was obtained by using the two equations.

As seen in Fig. 2, both membrane potential theories gave relatively good fitting results which are within the same degree of deviation to the experimental data. However, the results of the second series of experiments (Fig. 3) fit better with the values calculated with the S-D equation than with the G-H-K equation. Similarly, the calculated membrane potential using the S-D equation seemed to be in better agreement with the results of the third series of experiments (Fig. 4) than those obtained by the G-H-K equation, although the deviations (to the further hyperpolarized direction) of the calculated values from the experimental values was slightly greater than those in the previous two cases. These deviations seen in the latter case may be due to the neglection of the contribution of divalent cation permeation (diffusion potential) across the membrane. The consideration of divalent cation permeation would shift the (calculated) membrane potential more in the direction of depolarization.

The G-H-K equation resulted in small changes in membrane potential with respect to dilution of the internal salt solution which did not correspond well with the experimental data of the fifth series of experiments (Fig. 6). Although the membrane potentials calculated by use of the S-D potential equation showed a rather good correlation with the experimental data, there were some degrees of discrepancy between the calculated values and the experimental results (Fig. 6).

In the fourth series of experiments, the analysis using the G-H-K equation gave concentration dependent permeability coefficients of divalent cations (e.g. P_{Ca} = 0.26 at 20 mM Ca^{2+} and P_{Ca} = 0.14 at 132 mM Ca^{2+}; P_{Mg} = 0.17 at 20 mM Mg^{2+} and P_{Mg} = 0.10 at 132 mM Mg^{2+}): the use of a constant permeability in the equation does produce a much larger slope of the membrane potential versus divalent cation concentration curve compared to that of the observed membrane potential versus the divalent cation concentration (Fig. 5(A)' and (A)"). On the other hand, the S-D equation gave a constant (practically no concentration dependence) permeability coefficient for each divalent cation for which the theoretical values fit the experimental data well. The obtained relative permeability coefficients are 0.033 for Ca^{2+}, 0.031 for Sr^{2+} and 0.022 for Mg^{2+}. The value of 0.033 for Ca^{2+} is comparable to that estimated from experiments (Tasaki, 1968). On the other hand, those values (P_{Ca} = 0.25, P_{Mg} = 0.17) of the relative permeability coefficients obtained by use of the G-H-K equation appear to be too large to correspond to the experimental data. This analysis clearly indicates that the surface/diffusion potential theory is better suited for the analysis of these potential data than the G-H-K potential theory and that the membrane surface potential does contribute to the total membrane potential at certain ionic environments.

Fig. 8 shows membrane potential profiles of the squid axon membrane at various ionic environmental conditions. The use of the relative ionic permeability values obtained by Hodgkin and Katz (1949) in the present analysis of membrane potential with the S-D equation might be thought to be the one-sided treatment, while demonstrating the unsatisfactoriness of the use of G-H-K equation for membrane potential analysis. However, this can be defended as follows: As the ionic strength in the bathing solution, especially those of divalent cations, is increased, the magnitude of surface potential becomes small. In the normal physiological solution (Hodgkin, 1964), the outer and inner surface potentials of squid axon would be relatively small ($\sim$ 30 mV for the outer surface potential and 8 mV for the inner surface potential as seen in Fig. 8-A) due to the presence of high divalent ion concentrations in the extracellular solution, and due to a low surface charge density at the inner surface, respectively, as well as the effect of surface potential on the overall relative ionic permeabilities is small. In addition, if ions, especially divalent ions, can bind to the surface charge sites of the membrane, the magnitude of the surface potential will be reduced significantly as shown elsewhere (Ohki & Ohshima,

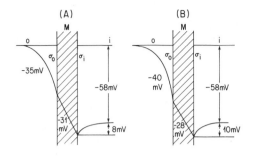

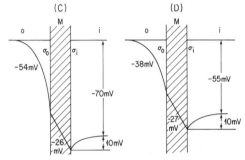

Fig. 8. Electrical potential profiles across squid axon membranes for various ionic environments using eq. (2)'. The assumed values used for surface charge densities were $\sigma = -e/180\ \text{Å}^2$ for the outer membrane surface and $\sigma' = 0$ for the inner surface.

(A) The normal physiological environment (Hodgkin, 1964)

Outside: $[K^+]_o = 10$ mM, $[Na^+]_o = 460$ mM, $[Cl^-]_o = 540$ mM, $[Ca^{2+}]_o = 10$ mM and $[Mg^{2+}]_o = 53$ mM.

Inside: $[K^+]_i = 400$ mM, $[Na^+]_i = 50$ mM, $[Cl^-]_i = 100$ mM, $[Ca^{2+}]_i = 0.4$ mM, and $[Mg^{2+}]_i = 10$ mM.

(B) Outside: $[K^+]_o = 10$ mM, $[Na^+]_o = 488$ mM, $[Ca^{2+}]_o = 20$ mM, $[Cl^-]_o = 540$ mM.

Inside: $[K^+]_i = 400$ mM, $[\text{glutamate}^-]_i = 250$ mM, $[Cl^-]_i = 50$ mM, $[F^-]_i = 50$ mM, $[PO_4]_i = 30$ mM.

(C) Outside: $[Ca^{2+}]_o = 33$ mM, $[Cl^-]_o = 78$ mM, $[\text{sucrose}]_o = 729$ mM.

Inside: the same as in (B)

(D) Outside: $[Ca^{2+}]_o = 132$ mM, $[Cl^-]_o = 276$ mM, $[\text{sucrose}]_o = 324$ mM.

Inside: the same as in (B)

1985). Therefore, the permeability obtained under this condition may be approximately correct values for the squid axon membrane which corresponds to a situation of no surface charges. This may be the reason why the relative permeabilities of ions estimated by Hodgkin and Katz (1949) are approximately proper values for squid axon membranes. But, the essential point of difference between the two membrane potential theories (G-H-K and S-D) is that the G-H-K potential theory does not include the fixed charge effect on the membrane potential.

From the above analysis, we conclude that the membrane potential theory, which incorporates the surface potential of the membrane (the S-D equation), is better suited to characterize the membrane potential of the squid axon, especially in the cases where ionic strength and ionic species are altered. This is because the G-H-K equation does not adequately describe the fixed charge effect of the membrane. Although the partition coefficient for each ion in the equation (eq. (1)) seems to take care of the effect of the solution-membrane boundary which may include the charge effect, it is a constant for each ion on both sides of the membrane surface, and the asymmetrical effect of membrane surface charges, which should be acting upon most biological membranes, are not taken into account for the effective surface concentrations of ions. It is concluded here that the external surface of squid axon membranes possesses a relatively high surface charge density ($-e/180 \, \text{Å}^2$). This does not necessarily mean, however, that the surface charges are distributed uniformly over the entire membrane outer surface. It is possible that the obtained surface charge density may represent those located only at the specific surface responsible for the ionic passages of the production of resting membrane potential (see Fig. 9). Our analysis does not exclude the latter possibility. In fact, it has been reported that the electrophoresis study of the squid and lobster axon membranes gives a very small surface charge density ($-1.2 \, e/10^5 \, \text{Å}^2$ for squid axon and $-2.6 \, e/10^5 \, \text{Å}^2$ for lobster axon)(Segal, 1968), whereas the membrane outer surface charge density estimated from the ionic conductance

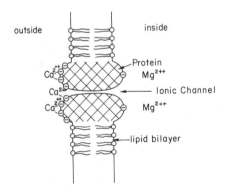

Fig. 9. Schematic diagram of an ionic channel in the lipid bilayer of biological membranes.

measurements (Gilbert & Ehrenstein, 1969) is a relatively large value ($-e/120$ $\mathring{A}^2$), which is more or less comparable to that obtained in the present study. The electrophoresis measurements give us an overall value of surface charge density on the outer membrane surface, while the charge density deduced from ionic current (action current) measurements may reflect the surface charge density at the ionic channel region. The obtained surface charge density of the inner surface in this study is compared reasonably well to that obtained for the squid axon membrane by Chandler, et al., (1965).

The present analysis of the observed membrane potential is not yet completely satisfactory for the squid axon system, probably because in our experiments, as well as in theoretical treatments used, there are a number of points to be improved in order to obtain a more accurate and valid analysis of the membrane potential (i.e., analysis by the use of proper ionic permeabilities, the effect of ion bindings to the surface charged sites (Ohki & Kurland, 1982; Ohki & Ohshima, 1985), the contribution of electrogenic potential (De Weer, 1975), etc.). It should be noted that the choices of proper values for the above factors (ionic permeability (Mullin, 1979) and ion binding) are indeed very difficult to be solved. However, our treatment seems to be in the correct direction to analyze the membrane potential of charged membranes, reflecting more closely the case of many actual bioloigcal cell systems (Lakshminarayanaiah, 1977; Sherbet, 1978). In our analysis, there is a reasonable agreement between the theoretical and experimental results.

APPENDIX

The Gouy-Chapman double layer potential (which we term "Surface Potential") at the planar membrane-electrolyte interface is expressed (Delahay, 1965) by

$$\sigma = (1/273) \left\{ \sum_i C_i \left(\exp\left[-\frac{e z_i \psi(0)}{kT} \right] -1 \right) \right\}^{1/2} \text{(Grahame equation)} \quad \text{(A)}$$

where σ is the surface charge density (electronic charge per $\mathring{A}^2$), C_i the molar concentration of the i-th ionic species in the bulk phase; z_i the valency of the i-th ion; $\psi(0)$ the surface potential at the membrane interface; k the Boltzman constant, and T the temperature, 20°C.

Using equation (A) and knowing all ionic concentrations in the bulk phase, and assuming the value of the surface charge density of the membrane, the surface potential was calculated by a University computer (CDC 6400) at SUNY/Buffalo.

ACKNOWLEDGEMENTS

This author would like to express his thanks to Dr. Ichiji Tasaki (National Institutes of Health) for the use of his laboratory facility at MBL in performing the experiments.

This work was partially supported by a grant from the U. S. National Institutes of Health (GM24840).

REFERENCES

Baker, P.F., Hodgkin, A.L. & Shaw, T.I., 1962, J. Physiol. (Lond), 164:355.
Baker, P.F., Hodgkin, A.L. & Meves, H., 1964, J. Physiol. (Lond), 170:541.
Bernstein, J., 1902, Pflügers Arch., 92:521.
Boyle, P.J. & Conway, E.J., 1941, J. Physiol., 100:1.
Cavanaugh, D.M. (ed.), 1956, Formula and Method V, MBL, Woods Hole, Mass., p. 52.
Chandler, W.K., Hodgkin, A.L. & Meves, H., 1965, J. Physiol., 180:821.
Delahay, P., 1965, in "Double Layer and Electrode Kinetics", Interscience Publ., New York.
De Weer, P., 1975, MTP International Review of Science, Physiology series one, Vol. 3, C. C. Hunt, ed., Butterworths, London.
Gilbert, D.L. & Ehrenstein, G., 1969, Biophys. J., 9:447.
Goldman, D.E., 1943, J. Gen. Physiol. 27:37.
Hodgkin, A.L. & Horowicz, P., 1959, J. Physiol., 145:405.
Hodgkin, A.L., 1964, in "The Conduction of the Nervous Impulse", Charles C. Thomas, publisher, Springfield, Illinois, p. 28.
Hodgkin, A.L. & Katz, B., 1949, J. Physiol. (Lond), 108:37.
Keynes, R.D. & Lewis, P.R., 1951, J. Physiol., 113:73.
Lakshminarayanaiah, N., 1977, Bull. Math. Bio., 39:643.
Ling, G.N. & Gerard, R.W., 1949, J. Cell. Comp. Physiol., 34:382.
Moore, W.J., 1962, Physical Chemistry, Prentice-Hall, Inc., N.J., p. 351.
Mullin, L.J., 1979, in "Membrane Transport in Biology, Vol. II", ed. Giebisch, G., Tosteson, D.C. & Ussing, H.H., Springer-Verlag, New York, Chapter 5.
Narahashi, T., 1963, J. Physiol., 169:91.
Ohki, S., 1976, in "Progress in Surface and Membrane Sciences, Vol. 10", ed. Cadenhead, D.A. & Danielli, J.F., Academic Press, N.Y., p. 117.
Ohki, S., 1979, Physics Letters 75A:149.
Ohki, S., 1981, Physiological Chemistry and Physics, 13:195.
Ohki, S., 1985, Bioelectrochem. Bioenerg., in press.
Ohki, S., & Kurland, R., 1981, Biochim. Biophys. Acta, 645:170.
Ohki, S., & Aono, O., 1979, Jap. J. Physiol., 29:373.
Ohki, S. & Ohshima, H., 1985, the paper of "Donnan Potential and Surface Potential, and the Effect of Ion Binding", in this book.
Planck, M., 1890, Ann. Physik., 40:561.
Rojas, E., Atwater, I. & Bezanilla, F., 1970, in "Permeability and Function of Biological Membrane" ed. by L. Bolis, North-Holland Publishing Co., Amsterdam-London, p. 273.
Segal, J.R., 1968, Biophys. J., 8:470.
Sherbet, G.V., 1978, in "The Biophysical Characterization of the Cell Surface", Academic Press, London and New York.
Tasaki, I. & Takenaka, T., 1963, Proc. Natl. Acad. Sci. USA, 50:619.
Tasaki, I. & Takenaka, T., 1964, in "The Cellular Functions of Membrane Transport", ed. by J. F. Hoffman, Prentice-Hall, Inc., Englewood Cliffs, N.J., p. 95.
Tasaki, I., Watanabe, A. & Takenaka, T., 1962, Proc. Natl. Acad. Sci. USA 48:1177.
Tasaki, I., 1958, in "Nerve Excitation", Charles C. Thomas, Publisher, Springfield, Ill.

ELECTRICAL DOUBLE LAYERS IN ION TRANSPORT AND EXCITATION

Martin Blank

Biological Sciences Division
Office of Naval Research
Arlington, VA 22217

INTRODUCTION

The dimensions of biological cells and biopolymers are in the range of classic colloids, a region where surfaces play an important role in determining the properties of a system. When the biological surfaces are charged, the effects on the properties are usually more pronounced, and can be calculated in terms of the resulting electrical double layers.

In this paper, we shall consider the mechanism of a classic biological problem, nerve excitation, from the point of view of surface chemistry. To understand ion transport processes across excitable membranes, we have found it necessary to include ionic processes in the electrical double layer regions at the membrane surfaces (Blank & Britten, 1978; Blank et al., 1982). In addition, the opening and closing of voltage gated ion channels in membranes can be related to the aggregation-disaggregation reactions of oligomeric proteins and their variation with surface charge (Blank, 1982). Both approaches enable us to describe the unusual ion fluxes during nerve excitation in terms of the physical chemistry of charged surfaces, and provide an insight into the mechanisms of channel selectivity and channel inactivation.

THE SURFACE COMPARTMENT MODEL (SCM)

Natural membranes are complex structures, but physiologists studying ion transport usually consider them a single homogeneous barrier and use equations that involve the intra- and extra-cellular solutions only. This has frequently led to elaborate schemes for ion transport that make little or no sense in terms of physical chemistry. Such a theoretical approach is obviously oversimplified, and in the last few years we have been able to make it more realistic by including ionic processes at membrane interfaces. This extension of the basic model has greatly simplified the description of the unusual ion transport processes during excitation.

Although electrical double layer theory was developed shortly after the turn of the century, it was not until 1935 that Teorell applied it to membrane processes and, introduced the idea that the membrane potential includes the effects of two phase boundary potentials at the surfaces in addition to an electro-diffusion potential across the membrane

(Teorell, 1935.) However, additional surface properties were not considered. In the electrical double layer regions at membrane-solution interfaces, the potential, charge, capacitance, and ionic concentrations are all different from bulk values and should play an important role in ion transport.

As a first approach to studying the influence of various surface properties on ion transport, we have devised the Surface Compartment Model (SCM), which approximates the electrical double layers at the surfaces of natural membranes as compartments (Blank & Britten, 1978; Blank & Kavanaugh, 1982.) Specifically, the SCM includes two surface potentials and two surface capacitances in addition to the dielectric capacitance of the membrane. It also considers ion concentration changes and cation binding and release processes in the surface layers. The set of differential equations for this system can be considered in four groups:

1) Six for the conservation of mass for the two cations and one anion at the two surfaces;
2) Four for the kinetics of cation binding at the two surfaces;
3) Two for the conservation of charge at the two membrane surfaces; and
4) Three for the conservation of charge during current flow (including gating currents) across the membrane and the electrical double layers.

Conservation equations are fundamental, while the equations describing the kinetics of ion binding are based on chemical kinetics principles, with a bimolecular forward rate and a monomolecular reverse rate. In all of the equations, the ion fluxes are given by Nernst-Planck expressions, where the driving forces are the electrochemical potential differences. (Provision

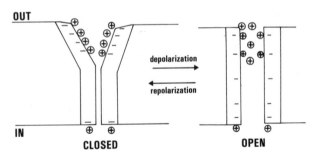

Fig. 1. A model of a voltage gated oligomeric channel having an asymmetric distribution of charge. When the channel is polarized (left hand side), it is closed because of the low charge density on the inside face. When the channel is depolarized (right hand side), the rapid displacement of charge within the oligomer causes the charge density on the inside face to increase and the channel to open. The channel will close upon membrane repolarization or as a result of the combination of cations with the surface charges and a lowering of the surface charge density. (Reproduced in modified form, with permission, from Bioelectrochemistry and Bioenergetics 9, 615-624, 1982.)

is also made for gating currents that arise from the redistribution of surface charges following changes in membrane polarization.)

A VOLTAGE SENSITIVE ION CHANNEL

We have tested the SCM under conditions that reflect the properties of excitable membranes, i.e., with a voltage sensitive permeability that is characteristic of ion channels in axons. Although any molecular mechanism that leads to a voltage dependent permeability would give similar results, we have developed a rationale for the opening and closing of channels, based on the observed charge-sensitive dissociation equilibria of oligomeric proteins. Oligomer aggregation reactions are usually described as "entropy driven" (Lauffer, 1975), because they proceed spontaneously with positive change of enthalpy. For the net free energy change to be negative, there must also be a large positive entropy change accompanying this reaction, probably arising from the release of adsorbed water molecules when the protein oligomers associate. This description is compatible with the energetics of aggregation when it occurs, but it does not indicate why it occurs at particular values of pH, i.e., surface charge. An explanation that includes the pH dependence appears to arise naturally when considering the aggregation process in terms of surface energy, i.e., where the loss of adsorbed waters upon oligomer interaction is equivalent to the loss of interfacial area. From the point of view of surface chemistry, disaggregation is a process similar to the formation of an additional protein water interface.

Consider the dissociation of hemoglobin (Blank, 1980.) Ordinarily, dissociation involves an increase in surface area, which is energetically unfavorable. However, as the hemoglobin tetramer becomes charged, positively or negatively, there is a compensatory decrease in surface energy when the charge is spread over a greater area, and a greater tendency for the molecule to dissociate into two dimers. This model of the variation of the hemoglobin dissociation with surface charge can also account for the Bohr effect during oxygenation (Blank, 1975), the Hill coefficient as a surface excess (Blank, 1976) and the unusually high viscosity of concentrated solutions (Blank, 1984).

We can use this model for ion channels since they are aggregated oligomeric proteins or the associated helices of a single protein. In the squid axon, the membrane has a much higher negative charge on the outside than on the inside, so we can consider the subunits dissociated (i.e., open) on the outer surface, where the charge is high, and associated (i.e., closed) on the inner surface, where the charge is low (Figure 1). Upon depolarization there is a shift of some negative charge from the outer to the inner surface and this causes the channel to open (Blank, 1982). Qualitatively, the properties of this voltage sensitive ion channel are compatible with observed behavior in terms of:

1) the steady state distribution of charge;
2) the magnitude and direction of charge flow during depolarization;
3) the range of surface charge where opening (dissociation) occurs;
4) the cation selectivity of the open channel; and
5) the cation binding.

THE VOLTAGE CLAMP CURRENTS

Using the SCM for voltage clamp, together with an equation for a channel whose ion permeability increases upon depolarization as in the model channel, we obtain the membrane currents shown in Figure 2 (Blank, 1983). After the initial gating current in the channel, there is an inward

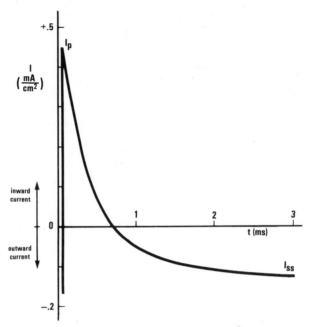

Fig. 2. The ionic current, I (in mA/cm^2), as a
function of time, t(in ms), for a voltage gated channel under a depolarizing
voltage clamp. The peak inward current,
Ip, and the steady outward current, Iss,
are designated to be indicative of the
sodium and potassium currents, respectively. (Reproduced in modified form,
with permission, from Bioelectrochemistry
and Bioenergetics 10, 451-465, 1983.)

(positive) ionic current, followed by an outward (negative) ionic current,
as observed in the voltage clamp of squid axons.

    The numerical values of the properties used in the calculations are
taken from studies of the squid axon, and the few values that have to be
assumed (i.e., the ion desorption rate constant, the surface capacitance,
the ionic mobility in the double layer) are based on physical systems.
Except for differences in the permeability of the resting membrane to sodium
and potassium ions, the two cations are assumed to have identical mobility
in the double layer region, as well as ion binding and release rate
constants, so the apparent specificity in the system arises primarily from
the asymmetry of the resting concentration gradients.

    When the permeability to potassium is set equal to zero, the current
resembles the curves obtained when introducing non-permeating tetraethyl-
ammonium ions (TEA) in place of potassium. When the permeability to sodium
is set equal to zero, the current resembles the curves obtained when sodium
ions are replaced by non-permeating choline (or the specific toxin tetro-
dotoxin is used). These results suggest that the voltage clamp current can
be resolved into an early inward component due to sodium ion and a later
outward component due to potassium, even though both ions contribute to the
normal current at all times. (The individual ion fluxes have approximately
the same kinetics, both reaching a peak within 100 usec and then decaying to
a steady state level. The peak current is higher for the sodium, so the

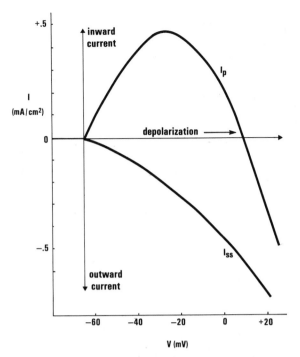

Fig. 3. A plot of the early peak inward current,
Ip, and the later steady state outward
current, Iss, both in mA/cm^2, as functions of the clamp voltage, V (in mV).
(Reproduced in modified form, with permission, from Bioelectrochemistry and
Bioenergetics, 10, 451-465, 1983.)

initial current is dominated by the sodium ion. However, the sodium flux
decays faster, so the later current is dominated by the potassium ion.)

The membrane current can be characterized in terms of the peak current
(for the early inward current) and the current at 3 msec (for the steady
state current). Figure 3, which shows these currents for different clamp
voltages, is similar to the curves published for squid axon. These currents
are calculated for a single channel, but similar curves can be obtained for
a mixture of channels that show greater apparent selectivity, as discussed
below.

CHANNEL SPECIFICITY AND INACTIVATION

The speed of the gating current in the voltage dependent channel
appears to have a particularly strong effect on the apparent specificity of
the ionic current. For example, a more outward membrane current at all
voltages appears if the ion channel opens slowly. (Such a process can be
due to either a decrease in the speed of the gating current or a delay in
the coupling of the gating current to the channel opening.) The solid
lines in Figure 4 show how the typical voltage clamp current, (dashed line),
is displaced when the rate constant controlling the gating current is increased by a factor of 5 or decreased to 0.3 of the original value. The
result in both cases is greater apparent specificity. It appears that a
"sodium channel" could be characterized by rapid gating currents and a

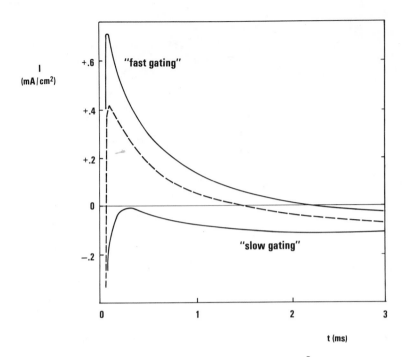

Fig. 4. A plot of the ionic current, I (in mA/cm^2, vs. the time, t (in ms), for different rates of the gating current. The dashed line is for conditions similar to those of Figure 2, while the solid lines are for faster and slower gating processes, as labeled. The apparent specificity of the channel changes dramatically with an order of magnitude change of gating speed, as explained in the text. (Reproduced in modified form, with permission, from Bioelectrochemistry and Bioenergetics, 13, 93-101, 1984.)

"potassium channel" by slower gating currents, with an order of magnitude difference between them. A difference in the speed and magnitude of the gating currents shown in the SCM, between the sodium and potassium channels is in line with measurements on ion channels in excitable membranes. In the squid axon, the gating current in the sodium channel is about 20 times faster than in the potassium channel. It should be stressed that the currents in the SCM are due to all cations in a system, and the ion specificity arises only from the kinetics.

The process of channel inactivation, i.e., the closing or decrease in permeability of a channel with time, also appears to be related to the gating current. In Figure 5, we see that increases in the order of magnitude of the gating conductance bring about or increase the inactivation. This too is in line with the observed inactivation of fast gating "sodium channels", but not of slow gating "potassium channels".

The ion selectivity of a channel is usually thought of in terms of a tight filter or bottleneck that discriminates largely on the basis of ion size, although factors such as ionic hydration and polarizability influence the process. This way of thinking leads to the paradox of a constricted pathway with a relatively large ionic conductance. The high rate of ion flow through the ion selective channels of excitable membranes, which is

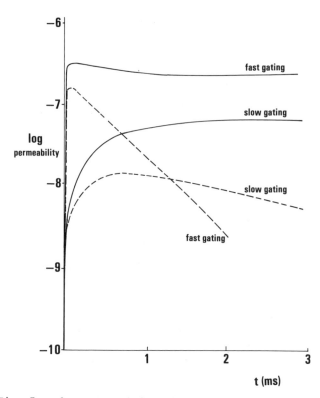

Fig. 5. The permeability (on a logarthmic scale) of the ion channel as a function of time, t (in ms), for fast gating and slow gating channels. The difference in gating speed is two orders of magnitude and the two curves at each gating speed correspond to the depolarizing clamp voltages -20 mV and +20 mV, for the lower and higher curves, respectively. (Reproduced in modified form, with permission, from Bioelectrochemistry and Bioenergetics 10, 451-465, 1983.)

not much lower than the rate in free solution, is incompatible with a bottle-neck type of selective channel. Also, the selectivity is unusual in that lithium passes through the fast channel almost as rapidly as sodium. Furthermore, the potassium channel of the squid axon allows the larger thallium ion to pass through at almost twice the rate. It is apparent that ion selectivity is much more complex than filtration, and may not even involve filtration. In the SCM, the selectivity arises largely from the kinetics of channel opening and is influenced by various electrical double layer properties.

Another system where there is suggestive evidence for a kinetic basis of ion selectivity is from bacteriorhodopsin (BR). When light of an appropriate frequency activates the photoreceptor, an outwardly directed proton movement occurs (Spudich and Bogomolni, 1983.) A related protein structure, halorhodopsin (HR), present in many of the same membranes, shows a chloride ion movement into the cell after illumination. HR shows slower photocycle kinetics than BR. A third membrane protein, slow rhodopsin (SR), has still

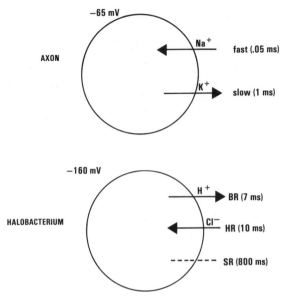

Fig. 6. A summmary of the rate constants of the
stimulating processes in axon and halobacterium membranes, and the resulting
directions of ion flow across the membranes. For the axon, the rate constants are for the gating currents in
electrical stimulation, while in halobacterium the rate constants are for
the photocycle periods. Although each
arrow relates to a different membrane
protein, there appears to be a relationship between the rate constant and the
direction of the flux.

slower kinetics and is the basis of the phototactic response, but causes
none of these ion movements. The three proteins have the same molecular
weight and are related in structure, so it appears that changes in the rate
constants of the pigment photocycles may affect ionic movement and the
selectivity of the channel. These ideas are summarized in Figure 6.

THE LIGAND GATED CHANNEL

The ligand gated acetylcholine receptor, which binds the neurotrans-
mitter acetylcholine, can be considered in terms of the same channel opening
mechanism. In the electric organ of the electric eel, the receptor has a
molecular weight of about 270,000 daltons, and is composed of five protein
chains that form a long channel through the membrane. When two acetylcho-
lines bind to the outer surface of two of the chains, the channel opens.

Under normal conditions, the acetylcholine receptor is negatively
charged and binds about 60 calcium ions. When acetylcholine binds to the

receptor, about 4-6 of those calcium ions are released (Chang & Neumann, 1976.) A displacement of calcium ions from the receptor by acetylcholine ions could lead to an increase in the total negative charge on the protein, and trigger the partial disaggregation (i.e. opening) of the receptor protein. Biopolymer aggregation-disaggregation phenomena appear to be charge mediated and the change in surface charge due to ion displacement could be the basis for opening and closing of oligomeric channels for ligand gated as well as the voltage gated case illustrated in Figure 1.

MIXED ION FLOWS IN A SINGLE CHANNEL

If ion selectivity derives from the kinetic constants of channel processes as demonstrated by the SCM, then it should be possible to observe a mix of ion currents in the same channel, e.g., where an initially inward "sodium" current is followed by an outward "potassium" current, as in Figure 1. This may be happening in an Aplysia axon (Wachtel & Kandel, 1967) where there is an excitatory followed by an inhibitory post-synaptic signal, and both are mediated by acetylcholine. The usual explanation given is that there are two different types of cholinergic receptors in the same membrane. However, according to the SCM, a single channel with intermediate kinetic constants could give that type of signal. Other evidence supports this idea. High frequency stimulation, which causes the release of large quantities of transmitter and has the same effect as a large depolarization in the voltage stimulated SCM, leads to inward inhibiting currents only. At low frequencies, like small depolarizations, one observes excitatory followed by inhibitory currents. (In Figure 3, at large depolarizations we see that both I_p and I_{ss} are outwardly directed, while for small depolarizations there is an inward I_p and an outward I_{ss}.) Both effects would be expected in a channel with mixed ion flows. The fact that the excitatory signal is always followed by the inhibitory one and never seen alone also suggests that a single channel is involved.

SUMMARY

In summary, it appears that excitable membranes which contain fast gating and slow gating ion channels will show sodium and potassium selectivity, respectively. The selectivity is apparent and arises from the kinetics of ion transport between electrical double layer regions on the two surfaces of membranes when the channels open. This theoretical approach provides an alternative to the "bottleneck theory" of channel selectivity and is more compatible with the observed high conductance and unusually high permeabilities to other ions.

The introduction of surface compartments as approximations to electrical double layer regions has led to new insights into mechanisms of membrane function in general, and may explain the function of ligand gated channels as well as the existence of channels with both excitatory and inhibitory ion flows. The SCM has also drawn our attention to physical properties like surface capacitances that have not generally been considered as relevant to membrane behavior, and that may help to explain the actions of different chemical agents.

Finally, it is interesting to note that the conclusions of the SCM approach are in many ways more compatible with the original hypothesis to explain nerve excitation on the basis of physical chemistry, due to Bernstein (1902), than with more recent ideas. Bernstein postulated a breakdown in selective permeability, and according to the SCM, a transiently open channel presents a relatively non-selective, open pathway for ion diffusion. The asymmetric concentration gradients and the kinetic parameters cause the apparent selectivity.

ACKNOWLEDGEMENT

This work was done at Columbia University and supported by contract N00014-83-K-0043 from the Office of Naval Research.

REFERENCES

Bernstein, J. (1902). Untersuchungen zur Thermodynamik der bioelecktrischen Strome. Pflugers Arch. 92, 521-562.

Blank, M. (1975). A Model for Calculating the Bohr Effect in Hemoglobin Equilibria. J. Theoret. Biol. 51, 127-134.

Blank, M. (1976). Hemoglobin Reactions as Interfacial Phenomena. J. Electrochem. Soc. 123, 1653-1656.

Blank, M. (1980). A Surface Free Energy Model for Protein Structure in Solution: Hemoglobin Equilibria. Colloids and Surfaces 1, 139-149.

Blank, M. (1982). The Surface Compartment Model-Role of Surface Charge in Membrane Permeability Changes. Bioelectrochem. Bioenerg. 9, 615-624.

Blank, M. (1983). The Surface Compartment Model with a Voltage Sensitive Channel. Bioelectrochem. Bioenerg. 10, 451-465.

Blank, M. (1984). Molecular Association and the Viscosity of Hemoglobin Solutions. J. Theoret. Biol. 108, 55-64.

Blank, M., and J. S. Britten (1978). The Surface Compartment Model of the Steady State Excitable Membrane. Bioelectrochem. Bioenerg. 5, 528-540.

Blank, M., and W. P. Kavanaugh, (1982). The Surface Compartment Model During Transients. Bioelectrochem. Bioenerg. 9, 427-438.

Blank, M., and W. P. Kavanaugh, and G. Cerf (1982). The Surface Compartment Model-Voltage Clamp. Bioelectrochem. Bioenerg. 9, 439-458.

Chang, H. W., and E. Neumann (1976). Dynamic Properties of Isolated Acetylcholine Receptor Proteins: Release of Calcium Ions Caused by Acetylcholine Binding. Proc. Nat. Acad. Sci. USA 73, 3364-3368.

Lauffer, M. A. (1975). "Entropy Driven Processes in Biology: Polymerization of Tobacco Mosaic Virus Protein and Similar Reactions". Springer Verlag, New York.

Spudich, J. L., and R. A. Bogomolni (1983). Spectroscopic Discrimination of the Three Rhodopsin-Like Pigments in Halobacterium Membranes. Biophys. J. 43, 243-246.

Teorell, T. (1935). An Attempt to Formulate a Quantitative Theory of Membrane Permeability. Pro. Soc. Exp. Biol. Med. 33, 282-285.

Wachtell, H., and E. R. Kandel (1967). A Direct Synaptic Connection Mediating Both Excitation and Inhibition. Science 158, 1206-1209.

ELECTRICAL DOUBLE LAYERS IN PIGMENT-CONTAINING BIOMEMBRANES

Felix T. Hong and T. L. Okajima

Department of Physiology
Wayne State University School of Medicine
Detroit, Michigan 48201

ABSTRACT

Displacement photocurrents are unique to pigment-containing biomembranes. They belong to a special class of bioelectric signals that are generated by rapid charge displacements in the membranes. Interpretation of displacement photocurrents is not straightforward, and direct application of classical electrophysiological methodology is not adequate. However, a more comprehensive understanding of displacement photocurrents can be achieved by applying the Gouy-Chapman diffuse double layer theory to two possible prototype mechanisms of light-induced charge displacements. Macroscopically, a chemical capacitance in addition to the ordinary membrane capacitance must be invoked in the analysis of experimental data. By doing so, some apparent discrepancies of experimental observations can be readily resolved. In this paper, we use the bacteriorhodopsin membrane system to illustrate how the concept of chemical capacitance can assist experimentalists to achieve a meaningful decomposition of a multi-component displacement photocurrent signal. The general applicability of such an approach in other pigment-containing membrane systems of greater complexity is suggested.

INTRODUCTION

Displacement currents belong to a special class of bioelectric signals that are distinguished by extremely fast risetimes. The best known example is the gating current in squid axons (Armstrong and Bezanilla, 1974). A less well understood example is the early receptor potential (ERP) in visual photoreceptor membranes discovered by Brown and Murakami (1964) about two decades ago. This fast photosignal was identified as a capacitative current, which is a manifestation of charge displacements of the membrane-bound pigment rhodopsin (Hagins and Rüppel, 1971). The development of a technique of forming artificial bilayer lipid membranes by Mueller, Rudin, Tien, and Wescott (1962) opened up possibilities for studying these fast photosignals under rigorously controlled experimental conditions. Tien (1968) was the first to apply this technique to study a reconstituted chloroplast bilayer lipid membrane and reported a photovoltage. Subsequently, displacement photocurrents were found in many different artificial systems in which the pigments are membrane-bound (for reviews see Tien, 1974; Hong, 1976, 1977, 1980; Mauzerall, 1979; Skulachev, 1982; Trissl, 1982; Bamberg et al., 1984a; Keszthelyi, 1984).

There have been some controversies concerning the measurement and the interpretation of displacement photocurrents (Trissl, 1981; Huebner et al., 1984). In the study of electrokinetic behaviors in biomembranes, two methods are generally used. One of them is the current clamp method, in which the access impedance of the measuring device is considerably larger than the source impedance of the membrane system being measured and in which the membrane voltage is reported. The other method is the voltage clamp method, in which the access impedance of the measuring device is negligible as compared to the source impedance of the membrane system and in which the membrane current is reported. In principle, these two methods provide identical information content. Use of these two methods has contributed to the wealth of our current knowledge and insight about the mechanisms of excitability and ion transport across biomembranes. Transplantation of these approaches to the study of displacement photocurrents is not straightforward, however. This is because of a drastic reduction of the source impedance in pigment-containing biomembranes over the time range where the photosignal can be resolved (Hong, 1976). Based on this realization, Hong and Mauzerall (1974, 1976) have proposed a tunable voltage clamp measurement method and have applied this method to a Mg porphyrin containing lipid bilayer which was coupled to an aqueous redox gradient. They further proposed a concept of chemical capacitance in the interpretation of their tunable voltage clamp data. They described in detail a method to obtain a pseudo-first order relaxation time constant of an interfacial photoreaction involving the photoactive pigment in the membrane phase and the aqueous redox reagents in one of the two aqueous phases. The second order rate constant was then obtained by varying the electron donor concentration (Hong, 1976).

Until recently, the analytical method based on the concept of chemical capacitance has never been applied to more complex pigment-containing membrane systems. Therefore, the general applicability of this approach remains to be demonstrated. In contrast to this approach, Trissl (1981) objected to the concept of chemical capacitance. He argued that the chemical capacitance is actually the ordinary membrane capacitance in disguise. The apparent existence of two capacitances arises from partial illumination of the membrane with a focused laser beam: one capacitance from the illuminated region and another from the non-illuminated region. Furthermore, Trissl (1981) maintained that the open-circuit method is superior to the voltage clamp method in the measurements of displacement photocurrents. In contrast, Hong (1980) cautioned about interpretation of open-circuit data, since the apparent kinetics always contains an overwhelming contribution from the membrane RC relaxation and thus appears distorted (cf: p. 293 in Trissl et al., 1984 and p. 125, Trissl, 1985). Okajima and Hong (1985) also demonstrated that the relaxation time course of the fast bacteriorhodopsin photosignal is virtually unaltered when the fraction of the illuminated area is varied progressively from 6 % to 100 %, whereas the time course is altered by even a few per cent change of the access impedance. Thus, the chemical capacitance is not the ordinary membrane capacitance in disguise as a consequence of partial illumination. In this paper, we shall use data obtained in our laboratory from bacteriorhodopsin model membranes to illustrate the usefulness of the concept of chemical capacitance. We shall demonstrate that some discrepancies between data reported by various groups of investigators can be readily reconciled by applying the Gouy-Champ theory of diffuse double layers.

FAST PHOTOELECTRIC SIGNALS IN BACTERIORHODOPSIN MODEL MEMBRANES

Among the reported displacement photocurrents in biomembranes, one that closely mimics the time course and behavior of the ERP appears in reconstituted bacteriorhodopsin membranes. It was discovered by Montal and his co-workers (Trissl and Montal, 1977; Hong and Montal, 1979). Bacterio-

Table 1. Relaxation Time Constants of Bacteriorhodopsin Photosignals

Source	τ_1	τ_2	τ_3	τ_4
Fahr et al. (1981)	1.3 µs	17 µs	0.06 ms	0.9 ms
Keszthelyi and Ormos (1980)	4.4 µs	81 µs	2.5 ms	8 ms
Rayfield (1983)[a]		57 µs	1.06 ms	13 ms
Trissl (1983a)[ab]		115 µs	4.5 ms	640 ms
Drachev et al. (1981)	<0.2 µs	200 µs	2 ms	1000 ms

[a]τ_1 not reported
[b]τ_4 derived from Fig. 1d in Trissl (1983a)

rhodopsin is the major protein component in the purple membrane fraction of
Halobacterium halobium (reviewed by Stoeckenius and Bogomolni, 1982). Bacteriorhodopsin resembles rhodopsin chemically, but its function is to translocate protons from the intracellular space to the extracellular space upon
illumination. The resultant proton electrochemical gradient is utilized in
ATP synthesis in the red membrane fraction of Halobacteria. The bacteriorhodopsin membrane exhibits a displacement photocurrent with at least two
components. Hong and Montal (1979) named the fast component B1 because of
its temperature-resistance which is similar to the R1 component of the ERP.
The slow component B2 can be reversibly suppressed by low temperature (0°C),
again similar to the ERP R2 component. However, up to four components were
reported by subsequent investigators, but the relaxation time constants
reported by various groups do not agree with one another (Table 1). The
diverse differences of the reported kinetic parameters of presumably identical molecular systems can hardly be explained by differences in temperature,
nor can they be readily explained by differences in the methods of forming
model bacteriorhodopsin membranes. As we shall explain later, the discrepancy arises mainly from differences in a hidden (measurement) parameter,
namely the access impedance. The peril of interpreting open-circuit data
without a detailed analysis is further exemplified by the elusive pH dependence expected of an interfacial proton transfer reaction which exists both
in rhodopsin and in bacteriorhodopsin membranes, but which has not been evident in open circuit data (see reviews by Hong, 1978, 1980).

PHYSICAL NATURE OF A CHEMICAL CAPACITANCE

The clue that led Hong and Mauzerall (1974) to propose the concept of
chemical capacitance is the presence of two exponential time constants for a
single pseudo-first order reaction and the presence of a signal waveform that
is characteristic of ac-coupling of the input light pulse. Since the ordinary membrane capacitance appears to be in parallel with the photoemf, Hong
and Mauzerall (1974) were forced to postulate an additional series capacitance. However, a similar need to invoke an additional series capacitance
also arose in the interpretation of a photovoltage recorded in a reconstituted bacteriorhodopsin membrane, reported by Drachev et al. (1976). These
investigators developed an ingenious method to incorporate bacteriorhodopsin
into a planar lipid bilayer by first incorporating the pigment molecules into

phospholipid vesicles and then letting the vesicles fuse with the non-pigmented planar lipid bilayer. The appearance of a signal waveform characteristic of ac-coupling upon addition of proton ionophore CCCP (2,4,6-tri-chlorocarbonylcyanide phenylhydrazone) to the aqueous phase prompted them to postulate a membrane structure shown in Fig. 1. The attached vesicle provides an additional capacitance needed to explain the waveform. However, Hong (1980) pointed out that similar behavior appeared in the earlier work of Drachev et al. (1974), in which the bacteriorhodopsin membranes were formed by a different method. Instead, Hong (1977, 1980) suggested that the peculiar behavior in CCCP-doped membrane is a manifestation of the existence of a chemical capacitance and hence of a displacement photocurrent. The effect of CCCP is mediated through alteration of the membrane discharging time constant (see Hong, 1980 for detail).

In order to gain a deeper insight concerning the physical nature of the chemical capacitance and its distinction from an ordinary membrane capacitance, one must analyze the electrical behavior of a pigment-containing membrane based on realistic molecular models. Attempts to correlate the displacement photocurrent with the known photochemistry of rhodopsin or bacteriorhodopsin have been made since the discovery of ERP and the ERP-like signal in bacteriorhodopsin membranes (e.g., see reviews by Hong, 1978, 1980). The widely accepted conclusion is that the charge displacement in question is a transient oriented dipole formation that accompanies the light-induced conformational changes in the membrane-bound pigments (to be referred to as the oriented dipole (OD) mechanism). There is, however, a

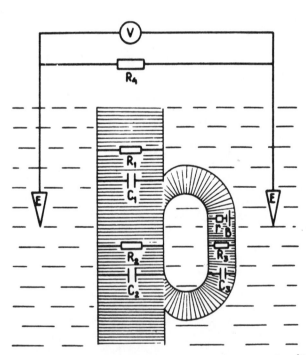

Fig. 1. Membrane structure proposed by Drachev et al. (1976) to explain the effect of CCCP on the photovoltage recorded from a reconstituted bacteriorhodopsin membrane. Bacteriorhodopsin molecules are incorporated in the phospholipid vesicles which are incompletely fused with the planar lipid bilayer membrane. (Reproduced from Drachev et al., 1976).

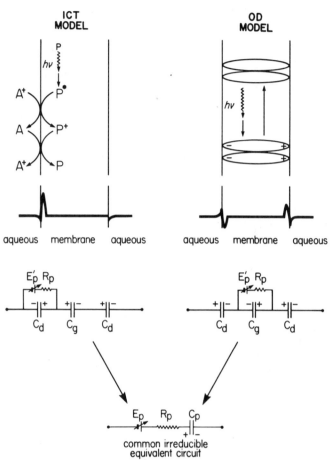

Fig. 2. Diagrams showing the difference between an interfacial
charge transfer (ICT) mechanism and an oriented dipole (OD)
mechanism in the generation of a displacement photocurrent.
The upper half of the diagram shows the two models and
their charge distribution profiles across the membrane
during the generation of the photocurrent. In the ICT
model, P is the ground-state photopigment, and P^* is one of
the intermediates. P^+ is a positively charged intermediate, a product of an interfacial proton transfer reac-
tion: $P^* + A^+ \rightleftarrows P^+ + A$, where A^+ and A are proton donor and
acceptor, respectively. The OD model is self-explanatory.
However, the diagram does not imply the actual shape of the
pigment molecule and the location of the pigment in the
membrane. In both diagrams, photoreactions are indicated
by wavy arrows, while dark thermal reactions are indicated
by smooth arrows. The lower half of the diagram shows the
respective equivalent circuit of the two models as a result
of interpreting the Gouy-Chapman calculation. E_p' is the
photoemf and R_p is the internal resistance associated with
it. Notice that the equivalent circuits of the two models
can be reduced to the same irreducible equivalent circuit.
The C_p is a combination of three fundamental capacitances:
a geometric capacitance, C_g, and two double layer capacitances, C_d. (Reproduced from Hong, 1978).

possible alternative mechanism with interfacial charge transfers (to be referred to as the interfacial charge transfer (ICT) mechanism). This latter mechanism is responsible for the displacement photocurrent reported by Hong and Mauzerall (1974) in the Mg porphyrin/aqueous redox membrane system mentioned earlier. This mechanism is compatible with the known photochemistry, namely, an interfacial electron transfer reaction (Mauzerall and Hong, 1975). Besides, there is no conceivable oriented dipole formation in the membrane, since the porphyrin molecules have been shown to be mobile in the membrane phase and are presumably symmetrically oriented with respect to the two aqueous phases. Hong (1978) subsequently applied a kinetic analysis based on the Gouy-Chapman diffuse double layer theory to these two models and derived two separate microscopic equivalent circuit models for the two distinct molecular models (Fig. 2). The basic strategy is the recognition of the fact that the ionic relaxation in the aqueous phase is so rapid that equilibrium prevails in the aqueous phase during the rise and decay of the displacement photocurrent, which typically has microsecond relaxation times. In contrast, the ionic relaxation of photo-generated charges in the membrane phase is so slow that they can be regarded as fixed surface charges.

These basic premises permit the application of an electrostatic theory such as the Gouy-Chapman theory in the aqueous phase and also justify the constant field condition in the membrane phase. Thus, an oriented dipole mechanism will lead to formation of two sheets of surface charges of opposite polarity at the two membrane-water interfaces. In the case of the ICT mechanism, a single sheet of surface charge will form at the membrane-water interface where a charge transfer reaction occurs. Solving these boundary value problems leads to the potential profile, as well as the space charge density profile, across the membrane. Re-interpretation of the linearized charge-potential relationship in terms of geometric (dielectric) capacitance

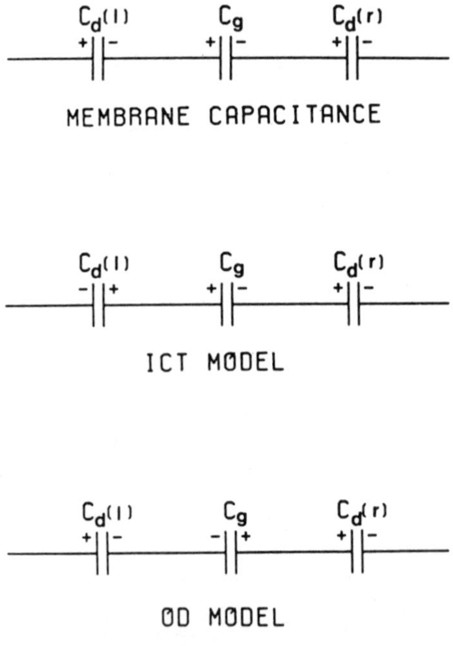

$C_d(l)$ C_g $C_d(r)$
+||- +||- +||-

MEMBRANE CAPACITANCE

$C_d(l)$ C_g $C_d(r)$
-||+ +||- +||-

ICT MODEL

$C_d(l)$ C_g $C_d(r)$
+||- -||+ +||-

OD MODEL

Fig. 3. Charge distribution pattern for the ICT model, the OD model and a charged ordinary membrane capacitance.

and double layer capacitance leads to the two distinct microscopic equivalent circuits (Fig. 2). Note that the charge distribution patterns on the capacitances are consistent with the surface charge pattern as well as the polarity of charges in the double layer regions. These charge distributions cannot be duplicated by an ordinary membrane capacitance. The charge distribution patterns of the OD model as well as the ICT model are shown in Fig. 3 along with that of a charged ordinary membrane capacitance. All three models are constructed from the same geometric capacitance and two double layer capacitances in the aqueous phases. Simple geometric reasoning indicates that there are three, and only three, distinct types of charge distribution patterns involving the same three capacitances. Type I pattern (seen in a charged membrane capacitance) is characterized by the same polarity in all three elements: (+,-),(+,-),(+,-) or (-,+),(-,+),(-,+). Type II pattern (seen in the ICT model) is characterized by one single reversal of polarity: (-,+),(+,-),(+,-) or (+,-),(-,+),(-,+). Type III pattern (seen in the OD model) is characterized by two consecutive reversals of polarity: (+,-),(-,+),(+,-) or (-,+),(+,-),(-,+). This analysis indicates that the capacitance that one encounters in the generation of displacement photocurrents cannot be accounted for by the ordinary membrane capacitance and must be included in the analysis of kinetic data obtained experimentally.

In attempting to determine the molecular mechanism of the ERP and the ERP-like photosignal in bacteriorhodopsin membrane, one cannot rely on the difference between the equivalent circuits of the two mechanisms. This is because both circuits are equivalent to the same irreducible equivalent circuit that was proposed by Hong and Mauzerall (1974) on an empirical basis. Further distinction of the two mechanisms must be based on chemical manipulation. The diagram in Fig. 2 suggests one possible way of discriminating between the ICT mechanism and the OD mechanism. The pair of separated charges in the OD model resides within the same molecule and is statistically correlated with each other throughout the entire time course of the photosignal relaxation. The relaxation of the charge separation is most likely a first order reaction. In contrast, one of the pair of separated charges in the ICT model resides in the aqueous double layer. The pair of separated charges become rapidly de-correlated as the ionic clouds relax in subnanoseconds. The relaxation is therefore a second order or a pseudo-first order reaction. Thus, the photosignal relaxation may depend on the concentration of the charge acceptor or donor in a pseudo-first order manner. The ERP R2 component has been correlated with the metarhodopsin I to metarhodopsin II reaction (Cone and Pak, 1971), and the bacteriorhodopsin B2 component has been correlated with the formation and decay of M_{412} (e.g., Keszthelyi and Ormos, 1980; Fahr et al., 1981; Drachev et al., 1984a). Both reactions involve interfacial proton transfer processes. But the measured pH dependence has been less than spectacular in both cases (Fig. 4). The R2 component in a reconstituted rhodopsin membrane was shown to be virtually pH independent from pH 5 to 8 (Fig. 4a; Trissl, 1979). Similar behavior was reported by Drachev et al. (1981) for the ERP-like photosignal in bacteriorhodopsin membranes for pH higher than 5 (Fig. 4b). In Fig. 4, both the R2 and the B2 amplitude were shown to decline to zero at lower pH, but at the same time the B1 and the R1 (R_+ in Fig. 4a) component increase in amplitude from pH 2 to 0. Since these data were obtained by means of open circuit methods, and the decomposition into components was presumably carried out without taking into account the effect of the access impedance and the presence of a chemical capacitance, the significance of the apparent lack of pH effect must be viewed with due caution. Qualitatively, this apparent lack of a significant pH effect can be understood on the basis of interaction of the chemical capacitance and the ordinary membrane capacitance (Hong, 1980). Thus, the undistorted relaxation kinetics can be recorded only when true short circuit conditions are maintained. The distortion is most severe when the photosignal is measured under open-circuit conditions. However, it is important to make sure that this conclusion is also correct at a quantitative level.

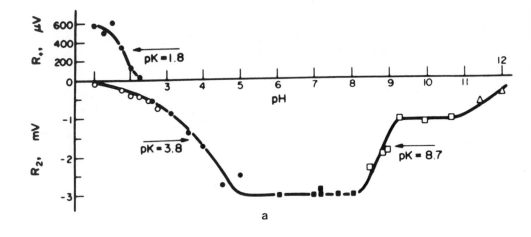

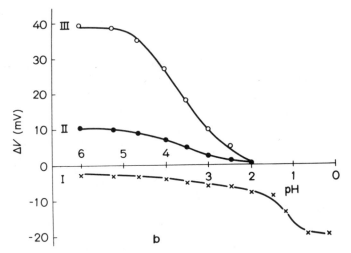

Fig. 4. pH dependence of fast photoelectric signals from rhodopsin and bacteriorhodopsin model membranes. (a) The photosignal from rhodopsin model membranes was decomposed into two components of opposite polarity, R_+ (i.e., R1) and R2. The peak amplitudes of each component are plotted as the ordinate. The membranes were formed by the Trissl-Montal method in 1 M NaCl. (Reproduced from Trissl, 1979). (b) The photosignal from bacteriorhodopsin model membranes was decomposed into three components. Component I corresponds to the B1 component as well as the time constant τ_1 in Table 1. Components II and III correspond to the time constants, τ_2 and τ_3, respectively, in Table 1. The membranes were formed by incorporating bacteriorhodopsin into collodion films according to a method developed by Drachev et al. (1978). The bathing solution contained 0.1 M NaCl. See text for further explanation. (Reproduced from Drachev et al., 1981).

INTERPRETATION OF DISPLACEMENT PHOTOCURRENTS IN BACTERIORHODOPSIN MODEL MEM-
BRANES BASED ON THE CONCEPT OF CHEMICAL CAPACITANCE

Using a method originally developed by Trissl and Montal (1977) to form
bacteriorhodopsin membranes on a thin Teflon film support, we measured
photocurrents over a wide range of pH under conditions of reduced access
impedance. The B2 peak is gradually reduced in amplitude as pH is lowered,
while the B1 peak is concurrently increased (Okajima and Hong, 1985). This
behavior is reminiscent of what was demonstrated by Drachev et al. (1981).
However, the pH effect can be observed throughout the entire pH range.
Furthermore, the B2 component is reduced in amplitude if water is replaced by
D_2O. This strongly suggests that the B2 component is generated by an interfacial proton transfer mechanism.

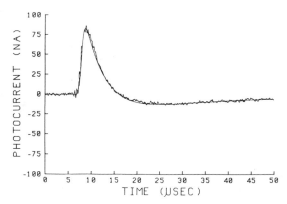

Fig. 5. Equivalent circuit analysis of the B1 component. The membrane was formed by the multi-lamellar coating technique on
a Teflon film. The temperature was 25°C. The bathing
electrolyte solution contained 3 M KCl and 0.05 M L-histidine in D_2O at pH 7. The measurement was made at an access
impedance of 40 K and an instrumental time constant of
0.355 μs. The computed (smooth) curve is superimposed on
the measured (noisy) curve. The root-mean-square value of
noise is 1.0 nA, whereas the root-mean-square deviation of
the computed curve from the experimental curve is 2.1 nA.
This analysis yields the relaxation time constant:
τ_p = 12.7 μs.

In order to test the equivalent circuit model experimentally, we must
first decompose the photosignals into the B1 and B2 components and then
analyze them separately. We do not think the peak amplitudes can be taken
as the respective amplitudes of these components, because we think that both
components consist of two apparent exponential decays. The strategy is to

Table 2. Kinetic Data of the B1 Component (from Okajima and Hong, 1985)[a]

Parameter	H_2O	D_2O
τ_p at 25°C and pH 2 (μs)	12.3 ± 0.7 (n = 8)	12.6 ± 1.4 (n = 6)
Activation Energy (kcal/mole)	2.54 ± 0.24 (n = 20)	2.45 ± 0.19 (n = 16)

[a]no difference between the H_2O and D_2O data at the 5 % level of significance

isolate the B1 component first. The B2 component can then be obtained by subtraction. Attempts to isolate the B1 component by lowering the temperature failed because we found a small temperature dependence of the B1 component, rendering the method of subtraction unreliable (Okajima and Hong, 1985). This is because the relaxation time course of the B1 component at room temperature differs from that observed at low temperature. Furthermore, we think the photosignal contains a residual B2 component over all temperature ranges and all pH ranges. This latter view was inferred from our inability to fit the data with the equivalent circuit model, and therefore cannot to be taken as conclusive without an additional proof.

Fortunately, we developed a slightly different method of forming the model membrane by direct deposition of purple membranes on the Teflon film (Okajima and Hong, 1985). This latter method offers photosignals with very little B2 content. Between pH 0 and 2, the photosignals fit the prediction of the equivalent circuit model. Beyond pH 3, the deviation of the model from the experimental data is small but is nevertheless significant and reproducible. In addition to changes in relaxation kinetics, the amplitude also declines at higher pH. We think that the deviation is the consequence of a residual B2 component which become detectable beyond pH 3. However, this residual B2 contribution can be eliminated by replacing water with D_2O. Thus, we were able to fit the photosignal from pH 0 to 10, provided that water is replaced by D_2O. A signal so obtained in D_2O at 25°C and pH 7 is shown in Fig. 5. The agreement between the calculated and the measured time course is within the noise level. This analysis allows us to obtain a first order relaxation constant of 12.7 μs for the B1 component. The pH dependent variation in amplitude can be explained on the basis of pH dependent absorbance changes of the bacteriorhodopsin. For example, the B1 amplitude at pH 10 is about 40 % of that observed at pH 1. However, after correction of pH dependent absorbance changes, the amplitude at pH 10 agrees with that at pH 1 to within 3 %. Thus, as shown in Fig. 6, the B1 photosignal so obtained is virtually unaltered by pH changes both in kinetics (not shown but documented in Okajima and Hong, 1985) and in amplitude over the wide range where experiments were carried out (from pH 1 to 10). Furthermore, this signal behaves in an Arrhenius fashion as a single component in agreement with the equivalent circuit over a temperature range from 5°C to 35°C (Okajima and Hong, 1985). For these reasons and for the lack of either better models or explanation, we tentatively regard this signal as a pure B1 component. In keeping with the notion that D_2O substitution suppresses the B2 component but leaves the B1 component unaltered, we also found the first order relaxation time constant (τ_p) of the B1 component, as well as its activation energy, to be virtually unaltered by D_2O substitution (Table 2).

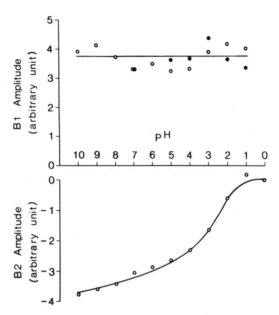

Fig. 6. pH dependence of the B1 and B2 amplitudes. The B1 amplitudes were taken from two bacteriorhodopsin membranes (open
circles and filled circles) formed by the multi-lamellar
coating method at 25°C in 3 M KCl and 0.05 M L-histidine in
D_2O. These two sets of amplitude data were normalized with
respect to each other by superimposing the data points at
pH 7. The measurements were made at an access impedance of
40 K and an instrumental time constant of 0.457 μs (open
circles) or 0.355 μs (filled circles). The B2 signals were
derived from photosignals recorded in a bacteriorhodopsin
membrane formed by the Trissl-Montal method at 25°C in 3 M
KCl and 0.05 M L-histidine in water. The measurements were
made at an access impedance of 40 K and an instrumental
time constant of 33 μs. In all measurements, the photoresponses are within the range of linear light dependence.
The light intensity was held constant during the measurements of a given membrane. All signal amplitudes shown
here have been corrected for the pH dependent absorbance
change of bacteriorhodopsin. The photoresponses are in the
range of linear light dependence. The two separate amplitude units are arbitrary. The relaxation of the B1 component is independent of pH in the range from 0 to 10. In
contrast, variation of pH affects the relaxation kinetics
of the B2 component. The details of analysis are documented elsewhere (Okajima and Hong, 1985).

Accurate separation of the B2 component free of the B1 contribution is
not currently feasible owing to the uncertainty of the normalization factor
relating photosignals obtained by means of the two slightly different membrane forming methods. However, for a semi-quantitative estimate of the pH
dependence of the B2 component, we can do the following. Though not strictly
true, we treat the photosignal obtained by means of Trissl-Montal method at
pH 0 as a pure B1 signal. Subtraction of this signal from the photosignal
obtained at different pH yields the approximate isolated B2 component. The
B2 amplitude as a function of pH so obtained is also shown in Fig. 6. Although a quantitative demonstration of the B2 relaxation as a second order
interfacial proton transfer reaction is not yet available, the crude data
showing the pH dependence of the B2 amplitude in Fig. 6 are closer to what
one expects from an interfacial proton transfer reaction. In reconciliation
with the data of Drachev et al. (1981), we interpret Fig. 4b as the result of
an incomplete separation of the two components in the decomposition. This
can be made clear by referring to Fig. 7. Basically, each component has two
exponential terms except in data obtained under true voltage clamp conditions. Since these two components have opposite polarity, there is considerable overlap of the rises and the decays of the two components. The
suppression of the B2 component at lower pH would lead to an increase of
the positive (B1) peak of the photosignal, thus giving an impression of an
apparent increase of the B1 amplitude. Guided by the concept of chemical
capacitance, we were able to avoid this pitfall. Without a detailed model,
a method of decomposition could turn out to be arbitrary and misleading.

We believe that the four exponential time constants reported in the
literature represent the four apparent time constants of the two components

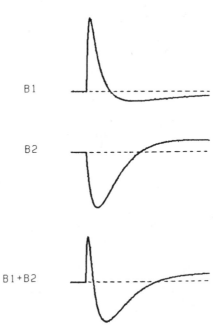

Fig. 7. Schematic diagram showing the decomposition of the photosignal into its components. Note that both the B1 and B2
components decay with two different exponential time
constants. See text for explanation. (Reproduced from
Hong and Montal, 1979).

140

as shown here. The disparity of the reported time constants shown in Table 1 can be explained as a consequence of differences in the access impedance. The data of Fahr et al. (1981) were obtained under conditions close to, but not quite equal to, short-circuit conditions. The data of Drachev et al. (1981) were obtained under open circuit conditions. The latter investigators correctly identified τ_4 as the passive membrane RC relaxation time constant.

As for the underlying molecular mechanism, it is clear that the B2 component is generated by an interfacial proton transfer mechanism. Specifically, the rise of the B2 signal corresponds to a proton uptake by bacteriorhodopsin from the intracellular space and the decay corresponds to the release of protons into the same intracellular space. That is, the B2 decay corresponds to a <u>back reaction</u>. So far we have no evidence of any B1 dependence on the concentration of an aqueous component. For lack of any such dependence, we tentatively conclude that the B1 component is generated by an oriented dipole mechanism. It is probably related to internal charge movement during light-induced proton translocation (cf: p. 2059 in Smith et al., 1984).

A REFINED MOLECULAR MODEL THAT INCLUDES A BOUNDARY LAYER

Trissl (1981) was correct in pointing out that the calculated value of the chemical capacitance as defined by the ICT equivalent circuit shown in Fig. 2 is much bigger than the experimental value. However, as indicated earlier, the discrepancy cannot be attributed to partial illumination or to an ordinary membrane capacitance in disguise. It is instructive to examine possible alternative explanations. First of all, one would not expect the value of chemical capacitance calculated on the basis of the Gouy-Chapman theory to agree with the experimental value, since it is well known that simplifying assumptions in that theory render the molecular picture unrealistic. For example, the smeared charge assumption is certainly not valid in the Mg porphyrin membrane system studied by Hong and Mauzerall (1974). In that system, the transient surface charge density at the peak of photocurrent response is estimated to be about 100 charges/μm^2, which is sufficiently low to reveal the discrete charge effect. Besides, the linearization of the Poisson-Boltzmann equations invoked in the derivation of the equivalent circuit is also not fully justified because the estimated transient surface potential in the Mg porphyrin membrane is about 50 mV (Hong, 1978) and this is not sufficiently small relative to 2RT/F ($\simeq$ 50 mV at room temperature). Furthermore, the assumption of a homogeneous aqueous phase with a unique value of the dielectric constant and of the interface as a mathematical plane are also oversimplifications (Losev and Mauzerall, 1983). But one of the most important factors that may account for a substantial portion of the discrepancy is the finite size of the pigment ions in the membrane phase (Mauzerall, 1979).

That the surface charge assumption may be an oversimplification can be seen by considering the fact that the porphyrin molecule has a diameter of about 10 Å and that electron transfer from excited porphyrin molecules can extend to a center-to-center distance of 20 Å (Carapellucci and Mauzerall, 1975; Ballard and Mauzerall, 1980), yet the membrane thickness is of the order of 60 - 100 Å. Thus, the light-induced charges may be localized <u>inside</u> the membrane dielectric and, therefore, the value of C_p could be considerably smaller. This can be made clear by considering a refined ICT model in which the transient fixed charge layer is now located some finite distance from the membrane-water interface as depicted in Fig. 8a. This process of charge separation leads to a boundary potential (Andersen et al., 1978) instead of a surface potential. Appropriate modification of the model would require that the element $C_d(1)$ in Fig. 3 be replaced by a composite capacitance, C_x'', which itself is a combination in series of the left double layer capacitance,

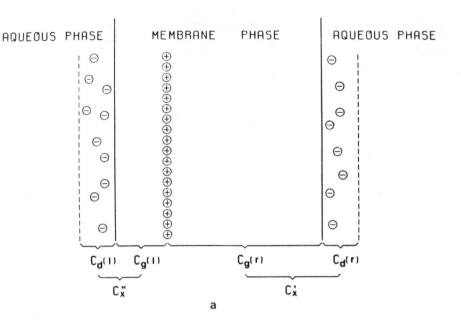

$C_d(l)$ $C_g(l)$ $C_g(r)$ $C_d(r)$

C_x'' C_x'

a

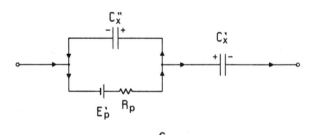

b

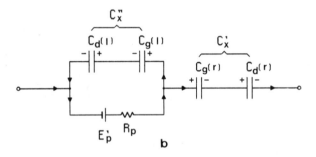

c

Fig. 8. Refined ICT model with a boundary layer (a) and its equivalent circuits (b and c). In this molecular model, the
fixed boundary charges generated by photoreactions are
deep-seated inside the membrane phase. The geometric
capacitance due to the boundary layer is designated $C_g(l)$,
and that due to the remaining thickness of the membrane is
designated $C_g(r)$. Elements C_x' and C_x'' are composite
capacitances with two different dielectrics. The remaining
symbols have the same meaning as in Figs. 2 and 3. The
arrows indicate the direction of the photocurrent under a
short circuit condition.

$C_d(1)$, and a geometric capacitance, $C_g(1)$, that is attributed to the boundary layer (Fig. 8b). The original C_g in the ICT model is now replaced by $C_g(r)$, which is the geometric capacitance attributed to the remaining thickness of the membrane. This is a situation similar to the case of absorption of tetraphenylborate ions to bilayer lipid membranes (Andersen et al., 1978). This is also suggested to be the case for the interfacial proton transfer accompanying the R2 component of the ERP (Cafiso and Hubbell, 1980). In the light of the above-mentioned findings from Mauzerall's laboratory, the boundary layer in the Mg porphyrin membrane could be as much as 10 Å thick. Thus, inclusion of a boundary layer capacitance, $C_g(1)$, in the composite capacitance, $C_x"$, may drastically reduce the value of $C_x"$, since $C_g(1)$ and $C_d(1)$ are in series. As a consequence, the theoretical value of C_p diminishes accordingly. While this refined model is more realistic than the previously proposed ICT model, it can be seen that many of the points of the foregoing discussion concerning the nature of C_p remain valid. Notably, the required charge distribution pattern remains type II (with the capacitance $C_d(1)$ now replaced by a composite capacitance $C_x"$), and is therefore in agreement with the proposed molecular picture (Fig. 8c). It is also obvious that similar refinement can be applied to the molecular model for the OD mechanism, since the separated pair of charges may not span the entire thickness of the membrane.

CONCLUSIONS AND PERSPECTIVES

In this paper, we have demonstrated the applicability of the concept of chemical capacitance in analyzing displacement photocurrents recorded from bacteriorhodopsin model membranes. Displacement photocurrents are unique to biomembranes with membrane-bound pigments. Because of the localization of the pigment molecules, light-induced charge injection must occur either near the membrane-water interface (ICT mechanism) or inside the membrane phase (OD mechanism). In contrast, the charging process of an ordinary membrane capacitance always involves an emf source which is placed completely outside the membrane. The difference in localization of the emf accounts for the difference in the charge distribution patterns shown in Fig. 3 as well as for the difference in the effect of these capacitances in shaping the time course of externally measured photosignals. Notably, the chemical capacitance (due to either an OD mechanism or an ICT mechanism) is always placed in series with the photoemf, and thus endows the measured photosignal a waveform that is characteristic of ac-coupling. In fact, the chemical capacitance is a built-in high pass filter. As a consequence, the source impedance of the membrane can be drastically reduced at the high frequency range of the photochemical relaxation. As the source impedance drops precipitously, the previously negligible access impedance in a short-circuit measurement set-up may become no longer negligible. Thus, it is possible that an attempted short circuit measurement using current amplifiers designed for low frequency applications may turn out to be inadvertently an open circuit measurement. As pointed out by Hong (1980; p. 222), the commonly observed first derivative relationship between the measured photocurrent and the measured photovoltage (Trissl, 1979, 1980, 1981) is tell-tale evidence indicating that the intended short circuit measurements were actually made under near open circuit conditions.

Although the analytical method based on the concept of chemical capacitance has so far been applied to only two pigment-containing biomembrane systems, it is likely that the method may also be applicable to membranes reconstituted from halorhodopsin (Bamberg et al., 1984b, 1984c, 1984d), from rhodopsin (Chapron, 1980; Drachev et al., 1980; Bayramashvili et al., 1984), or even from more complex systems such as chloroplasts (Trissl and Gräber, 1980a, 1980b; Gräber and Trissl, 1981; Trissl et al., 1982; Trissl and Kunze, 1985) and bacterial photosynthetic reaction centers (Schönfeld et al., 1979;

Packham et al., 1982; Trissl, 1983b; Drachev et al., 1984b). As demonstrated previously (Hong, 1980), the concept of chemical capacitance provides a unifying approach for comprehending diverse phenomenology in many different photobiological systems. The validity of the method certainly requires an independent proof. Further refinements of the molecular models and theoretical computations are required to relate the experimentally measured value of C_p to fundamental molecular constants. However, despite all the simplification and all the flaws associated with the Gouy-Chapman theory, the chemical capacitance as an experimental parameter turns out to be remarkably self-consistent in the analysis of photokinetic data from pigment-containing biomembranes. The concept also serves as a valuable guide for the optimization of experimental protocols. Conversely, the present experimental system may also be useful for quantitative testing of a refined theory about the electrical double layers.

ACKNOWLEDGEMENTS

The authors thank Richard Needleman, James Sedensky and Robert Shepard for critical reading of the manuscript. F.T.H. acknowledges the generous help and fruitful interaction with the following individuals during the experimental work which forms the basis of this paper: Victor Chen, George Feher, David Mauzerall and Mauricio Montal. This work has been supported by the following research grants: NIH grants GM-25144, EY-03334, EY-04068 and RR-05384, American Heart Association of Michigan grant No. 34, and a Wayne State University Neuroscience Small Grant.

REFERENCES

Andersen, O. S., Feldberg, S., Nakadomari, H., Levy, S., and McLaughlin, S., 1978, Electrostatic interactions among hydrophobic ions in lipid bilayer membranes, Biophys. J., 21:35-70.

Armstrong, C. M., and Bezanilla, F., 1974, Charge movement associated with the opening and closing of the activation gates of the Na channels, J. Gen. Physiol., 63:533-552.

Ballard, S. G., and Mauzerall, D. C., 1980, Photochemical ionogenesis in solutions of zinc octaethyl porphyrin, J. Chem. Phys., 72:933-947.

Bamberg, E., Fahr, A., and Szabó, G., 1984a, Photoelectric properties of the light-driven proton pump bacteriorhodopsin, in: "Electrogenic Transport: Fundamental Principles and Physiological Implications, Soc. Gen. Physiol. Ser., Vol. 38," M. P. Blaustein and M. Lieberman, eds., Raven Press, New York, 381-394.

Bamberg, E., Hegemann, P., and Oesterhelt, D., 1984b, Reconstitution of the light-driven electrogenic ion pump halorhodopsin in black lipid membranes, Biochim. Biophys. Acta, 773:53-60.

Bamberg, E., Hegemann, P., and Oesterhelt, D., 1984c, The chromoprotein of halorhodopsin is the light-driven electrogenic chloride pump in Halobacterium halobium, Biochem. (Wash.), 23:6216-6221.

Bamberg, E., Hegemann, P., and Oesterhelt, D., 1984d, Reconstitution of halo-rhodopsin in black lipid membranes, in: "Information and Energy Transduction in Biological Membranes," C. L. Bolis, E. J. M. Helmreich, and H. Passow, eds., Alan R. Liss, Inc., New York, 73-79.

Bayramashvili, D. I., Drachev, A. L., Drachev, L. A., Kaulen, A. D., Kudelin, A. B., Martynov, V. I., and Skulachev, V. P., 1984, Proteinase-treated photoreceptor discs: photoelectric activity of the partially-digested rhodopsin and membrane orientation, Eur. J. Biochem., 142:583-590.

Brown, K. T., and Murakami, M., 1964, A new receptor potential of the monkey retina with no detectable latency, Nature (London), 201:626-628.

Cafiso, D. S., and Hubbell, W. L., 1980, Light-induced interfacial potentials in photoreceptor membranes, Biophys. J., 30:243-264.

Carapellucci, P. A., and Mauzerall, D., 1975, Photosynthesis and porphyrin excited state redox reactions, Ann. N. Y. Acad. Sci., 244:214-238.

Chapron, Y., 1980, Rhodopsin-induced transient photopotentials in retinal and vesicle membranes, Photobiochem. Photobiophys., 1:297-304.

Cone, R. A., and Pak, W. L., 1971, The early receptor potential, in: "Handbook of Sensory Physiology, Vol. I. Principles of Receptor Physiology," W. R. Loewenstein, ed., Springer-Verlag, Berlin, 345-365.

Drachev, L. A., Kaulen, A. D., Ostroumov, S. A., and Skulachev, V. P., 1974, Electrogenesis by bacteriorhodopsin incorporated in a planar phospholipid membrane, FEBS Lett., 39:43-45.

Drachev, L. A., Frolov, V. N., Kaulen, A. D., Liberman, E. A., Ostroumov, S. A., Plakunova, V. G., Semenov, A. Yu., and Skulachev, V. P., 1976, Reconstitution of biological molecular generators of electric current: bacteriorhodopsin, J. Biol. Chem., 251:7059-7065.

Drachev, L. A., Kaulen, A. D., and Skulachev, V. P., 1978, Time resolution of the intermediate steps in the bacteriorhodopsin-linked electrogenesis, FEBS Lett., 87:161-167.

Drachev, L. A., Kalamkarov, G. R., Kaulen, A. D., Ostrovsky, M. A., and Skulachev, V. P., 1980, Animal rhodopsin as a photogenerator of an electric potential that increases photoreceptor membrane permeability, FEBS Lett., 119:125-131.

Drachev, L. A., Kaulen, A. D., Khitrina, L. V., and Skulachev, V. P., 1981, Fast stages of photoelectric processes in biological membranes. I. bacteriorhodopsin, Eur. J. Biochem., 117:461-470.

Drachev, L. A., Kaulen, A. D., and Skulachev, V. P., 1984a, Correlation of photochemical cycle, H^+ release and uptake, and electric events in bacteriorhodopsin, FEBS Lett., 178:331-335.

Drachev, L. A., Dracheva, S. M., Samuilov, V. D., Semenov, A. Yu., and Skulachev, V. P., 1984b, Photoelectric effects in bacterial chromatophores: comparison on spectral and direct electrometric methods, Biochim. Biophys. Acta, 767:257-262.

Fahr, A., Läuger, P., and Bamberg, E., 1981, Photocurrent kinetics of purple-membrane sheets bound to planar bilayer membranes, J. Membrane Biol., 60:51-62.

Gräber, P., and Trissl, H.-W., 1981, On the rise time and polarity of the photovoltage generated by light gradients in chloroplast suspensions, FEBS Lett., 123:95-99.

Hagins, W. A., and Rüppel, H., 1971, Fast photoelectric effects and the properties of vertebrate photoreceptors as electric cables, Fed. Proc., 30:64-68.

Hong, F. T., 1976, Charge transfer across pigmented bilayer lipid membrane and its interfaces, Photochem. Photobiol., 24:155-189.

Hong, F. T., 1977, Photoelectric and magneto-orientation effects in pigmented biological membranes, J. Colloid Interface Sci., 58:471-497.

Hong, F. T., 1978, Mechanisms of generation of the early receptor potential revisited, Bioelectrochem. Bioenerg., 5:425-455.

Hong, F. T., 1980, Displacement photocurrents in pigment-containing biomembranes: artificial and natural systems, in: "Bioelectrochemistry: Ions, Surfaces, Membranes, ACS Advances in Chemistry Ser., Vol. 188," M. Blank, ed., American Chemical Society, Washington, D.C., 211-237.

Hong, F. T., and Mauzerall, D., 1974, Interfacial photoreactions and chemical capacitance in lipid bilayers, Proc. Natl. Acad. Sci. USA, 71:1564-1568.

Hong, F. T., and Mauzerall, D., 1976, Tunable voltage clamp method: application to photoelectric effects in pigmented bilayer lipid membranes, J. Electrochem. Soc., 123:1317-1324.

Hong, F. T., and Montal, M., 1979, Bacteriorhodopsin in model membranes: a new component of the displacement photocurrent in the microsecond time scale, Biophys. J., 25:465-472.

Huebner, J. S., Arrieta, R. T., Arrieta, I. C., and Pachori, P. M., 1984, Photo-electric effects in bilayer membranes; electrometers and voltage clamps compared, <u>Photochem. Photobiol</u>., 39:191-198.

Keszthelyi, L., 1984, Intramolecular charge shifts during the photoreaction cycle of bacteriorhodopsin, <u>in:</u> "Information and Energy Transduction in Biological Membranes," C. L. Bolis, E. J. M. Helmreich, and H. Passow, eds., Alan R. Liss, Inc., New York, 51-71.

Keszthelyi, L., and Ormos, P., 1980, Electric signals associated with the photocycle of bacteriorhodopsin, <u>FEBS Lett</u>., 109:189-193.

Losev, A., and Mauzerall, D., 1983, Photoelectron transfer between a charged derivative of chlorophyll and ferricyanide at the lipid bilayer-water interface, <u>Photochem. Photobiol</u>., 38:355-361.

Mauzerall, D,, 1979, Photoinduced electron transfer at the water-lipid bilayer interface, <u>in</u>: "Light-Induced Charge Separation in Biology and Chemistry," H. Gerischer and J. J. Katz, eds., Verlag Chemie GmbH, Weinheim, 241-257.

Mauzerall, D., and Hong, F. T., 1975, Photochemistry of porphyrins in membranes and photosynthesis, <u>in</u>: "Porphyrins and Metalloporphyrins," K. M. Smith, ed., Elsevier, Amsterdam, 701-725.

Mueller, P., Rudin, D. O., Tien, H. T., and Wescott, W. C., 1962, Reconstitution of excitable cell membrane structure <u>in</u> <u>vitro</u>, <u>Circulation</u>, 26:1167-1171.

Okajima, T. L., and Hong, F. T., 1985, Kinetic analysis of displacement photocurrents elicited in two types of bacteriorhodopsin model membranes, manuscript submitted to Biophysical Journal.

Packham, N. K., Dutton, P. L., and P. Mueller, 1982, Photoelectric currents across planar bilayer membranes containing bacterial reaction centers: response under conditions of single electron turnover, <u>Biophys. J.</u>, 37:465-473.

Rayfield, G. W., 1983, Events in proton pumping by bacteriorhodopsin, <u>Biophys. J.</u>, 41:109-117.

Schönfeld, M., Montal, M., and G. Feher, 1979, Functional reconstitution of photosynthetic reaction centers in planar lipid bilayers, <u>Proc. Natl. Acad. Sci. USA</u>, 76:6351-6355.

Skulachev, V. P., 1982, A single turnover study of photoelectric current-generating proteins, <u>Methods Enzymol</u>., 88:35-45.

Smith, S. O., Myers, A. B., Pardoen, J. A., Winkel, C., Mulder, P. P. J., Lugtenburg, J., and Mathies, R., 1984, Determination of retinal Schiff base configuration in bacteriorhodopsin, <u>Proc. Natl. Acad. Sci. USA</u>, 81:2055-2059.

Stoeckenius, W., and Bogomolni, R. A., 1982, Bacteriorhodopsin and related pigments of <u>Halobacteria</u>, <u>Ann. Rev. Biochem</u>., 51:587-616.

Tien, H. T., 1968, Light-induced phenomena in black lipid membranes constituted from photosynthetic pigments, <u>Nature (London)</u>, 219:272-274.

Tien, H. T., 1974, "Bilayer Lipid Membranes (BLM): theory and practice," Marcel Dekker, New York, 245-321.

Trissl, H.-W., 1979, Light-induced conformational changes in cattle rhodopsin as probed by measurements of the interface potential, <u>Photochem. Photobiol</u>., 29:579-588.

Trissl, H.-W., 1980, I. Novel capacitative electrode with a wide frequency range for measurements of flash-induced changes of interface potential at the oil-water interface, <u>Biochim. Biophys. Acta</u>, 595:82-95.

Trissl, H.-W., 1981, The concept of chemical capacitance: a critique, <u>Biophys. J.</u>, 33:233-242.

Trissl, H.-W., 1982, Electrical responses to light: fast photovoltages of rhodopsin-containing membrane systems and their correlation with the spectral intermediates, <u>Methods Enzymol</u>., 81:431-439.

Trissl, H.-W., 1983a, Charge displacements in purple membranes adsorbed to a heptane/water interface: evidence for a primary charge separation in bacteriorhodopsin, <u>Biochim. Biophys. Acta</u>, 723:327-331.

Trissl, H.-W., 1983b, Spatial correlation between primary redox components in reaction centers of <u>Rhodopseudomonas sphaeroides</u> measured by two electrical methods in the nanosecond range, <u>Proc. Natl. Acad. Sci. USA</u>,

80:7173-7177.

Trissl, H.-W., 1985, I. Primary electrogenic processes in bacteriorhodopsin probed by photoelectric measurements with capacitative metal electrodes, Biochim. Biophys. Acta, 806:124-135.

Trissl, H.-W., and Gräber, P., 1980a, II. Electrical measurements in the nanosecond range of the charge separation from chloroplasts spread at a heptane-water interface: application of a novel capacitative electrode, Biochim. Biophys. Acta, 595:96-108.

Trissl, H.-W., and Gräber, P., 1980b, Properties of chloroplasts spread at the heptane/water interface: measurements of the photosynthetic charge separation in the nanosecond range, Bioelectrochem. Bioenerg., 7:167-186.

Trissl, H.-W., and Kunze, U., 1985, II. Primary electrogenic reactions in chloroplasts probed by picosecond flash-induced dielectric polarization, Biochim. Biophys. Acta, 806:136-144.

Trissl, H.-W., and Montal, M., 1977, Electrical demonstration of rapid light-induced conformational changes in bacteriorhodopsin, Nature (London), 266:655-657.

Trissl, H.-W., Kunze, U., and Junge, W., 1982, Extremely fast photoelectric signals from suspensions of broken chloroplasts and of isolated chromatophores, Biochim. Biophys. Acta, 682:364-377.

Trissl, H.-W., Der, A., Ormos, P., and Keszthelyi, L., 1984, Influence of stray capacitance and sample resistance on the kinetics of fast photovoltages from oriented purple membranes, Biochim. Biophys. Acta, 765:288-294.

ELECTRONIC PROPERTIES OF ELECTROACTIVE BILAYER LIPID MEMBRANES

H. Ti Tien, Jan Kutnik,[*] Pawel Krysinski,[+] and Z.K. Lojewska[*]

Membrane Biophysics Lab, Department of Physiology
Michigan State University
East Lansing, MI 48824 (USA)

Abstract.   The lipid bilayer postulated as the basic structural
matrix of biological membranes is widely accepted.   Experiments in
the early 1960s have made direct studies of lipid bilayer possible.
At present, the bilayer lipid membrane (BLM) upon suitable modification serves as a unique model for biological membranes.   This paper,
after   a   minireview,   describes   our   recent   experiments   with   BLMs
containing   TCNQ   (7,7',8,8'-tetracyano-p-quinodimethane)   or   TTF
(tetrathiafulvalene).   These doped BLMs have been investigated by a
voltammetric technique which has shown that suitably modified bilayer
lipid membranes can act as an electronic conductor partaking in redox
reactions at membrane/solution interfaces.

INTRODUCTION

Under the electron microscope, all the biomembranes thus far examined
are on the order of 100 Å in thickness and are generally interpreted as
consisting of a lipid bilayer of the Gorter-Grendel-Davson-Danielli-
Robertson type with sorbed protein or nonlipids (Robertson, 1981; Koryta,
1982).   Biomembranes are complex structures involved in diverse functions
including transport, nerve impulse propagation, antigen-antibody reaction,
energy transduction, ATP synthesis, and sensory reception (Blank, 1970,
1980).   Until recently, the study of biomembranes was constrained to
intact organelles or membrane fragments since methods for the isolation
and reconstitution of simple membrane systems were not available.   Under
those conditions, the   specific functions of biomembranes could not be

---

[*] On leave from the Institute of Physics, Maria Curie-Sklodowska
University, Lublin, Poland.

[+] On leave from the Department of Chemistry, University Warsaw, Poland.

149

easily delineated and completely understood at the molecular level. The availability of artificially constituted bilayer lipid membranes of planar configuration (Mueller, Rudin, Tien & Wescott, 1963; Tien, 1974) and of vesicular configuration (Antolini et al., 1982; Bader et al., 1985), as models of biomembranes, have played an important role in providing insights into molecular processes which are pertinent to the much more complicated natural membranes. These artificially constituted planar lipid bilayers, now generally termed bilayer lipid membranes (or BLM), have since proved ideal for investigations of the electrical, mechanical, photoelectrical, and a host of other properties of the generally postulated lipid bilayer model of biomembranes. The past and current status of work on BLMs and liposomes has been reviewed (McLaughlin & Eisenberg, 1975; Ohki, 1976; Hong, 1980; Blumenthal & Klauser, 1982; Tien, 1985).

Among the outstanding problems of biological membranes, ion selectivity and electrical "excitability" are two of the most investigated topics. The latter is closely related to and most probably predicated upon the former. However, in spite of the increasingly detailed descriptions of ion permeability and membrane structure that have been offered, the exact mechanisms by which ions and other species translocate biological membranes are not known. It is well known, however, that certain types of biological membranes (such as the plasma membrane of red blood cells and squid axons) possess the ability to discriminate between Na^+ and K^+ ions. When results of this kind can be demonstrated with the BLM, a systematic investigation is of value in that we can at long last search for physical mechanisms underlying this important process. With a BLM together with its modification the structure is much better defined and simpler, and their composition and surrounding aqueous environment can be varied in a controlled manner. It is presently conceivable that a number of working hypotheses can be put to critical tests. Many authors (Blank, 1980; Martonosi, 1982; Miller, 1983) have pointed out that the mechanisms of ion permeabilities are central to the understanding of bioelectric potentials and their associated phenomena. In electrophysiology, for example, when an axon is stimulated by a brief current pulse, an electrical transient phenomenon known as "action potential" is frequently observed. Earlier, it had been demonstrated that BLM, when treated with a certain proteinaceous material (known as Excitability Inducing Material or EIM), can be made electrically "excitable" in response to an imposed threshold voltage (Scott, 1977). The current-voltage curves were highly non-linear and exhibited the proper N-shaped negative resistance behavior. Such a negative resistance phenomenon has also been observed in a plant algae (Valonia) and in squid axon. It seems probable that, from physico-chemical points of view, these structures can behave as redox systems. For a 70 Å BLM, electrons could "tunnel" through in a way analogous to a solid state tunneling device in which the observed membrane current (J_m) through the BLM is the sum of ionic current (J_i) and electron tunneling current (J_e), that is $J_m = J_i + J_e$ for BLM bearing fixed charge groups, J_i is rectified and is directly proportional to the applied potential. Further, the temperature has the opposite effect on J_i and J_e, in which J_i increases with the temperature. Since BLM possesses high electrical resistance (10^8 ohms cm^2) and can also be made ion-selective, the system offers a practical means to test whether oxidation and reduction actually take place at the BLM-solution interface. Thus, the theoretical current-voltage characteristic as a result of electron tunneling between two redox systems through BLM should be examined in more detail (Milazzo, 1983).

In areas of the bioelectrochemistry of membranes, the proposed experiments are centered on reconstitution (for recent reviews, see Blumenthal & Klauser, 1982; Tien, 1985). At present, a great deal of experimental investigation is being carried out on chloroplasts, mitochondria and H. halobium (Barber, 1979; Bolton & Hall, 1979; Hauska & Orlich, 1980). In the case of chloroplast, for instance, light-induced electron transport is associated with an influx of hydrogen ions. It has also been found that if acidified chloroplasts are rapidly transferred to a medium containing ADP and inorganic phosphate at pH 9, a substantial amount of ATP is produced. In the case of mitochondria, Mitchell and Moyle have shown that ATP hydrolysis is accompanied by an acidification of the bathing solution. Hydrogen ions are pumped out of the mitochondria during ATP hydrolysis and re-equilibrate slowly. These findings and other work are interpreted in terms of the chemiosmotic theory of Mitchell (1979). According to this theory a pH gradient must exist across the membrane (cristae or thylakoid or H. halobium) in order that the separation of charge due to redox reaction can bring about a change in the so-called "hydro-dehydration" reaction with an ATPase which activates the ATP, ADP, P_i and water reaction. Further, the Mitchell theory purports to explain a number of phenomena associated with an electron-transport system, membrane potential, charge separation, photo-/oxidative phosphory-lation and ion exchange.

From a biophysical point of view, the use of whole mitochondria (or chloroplasts or halobacteria) to test a "molecular" theory does not seem satisfying, even if it is possible. The translocation of ions such as protons across a complex system, such as a cristae or thylakoid membrane, is in itself ill-defined. The interactions between various fluxes (ion and water movements, electron and hole transport, etc.) and their conjugate forces (osmotic pressure and electrical potential, etc.) are far too complex to be amenable to a simple analysis. We must conclude that the chemiosmotic hypothesis as it stands today remains as an excellent hypothesis. Experimental evaluation of the hypothesis using a simpler membrane system is therefore in order (Tien, 1968; Bockris, 1979; Antolini, Gliozzi & Gorio, 1982; Tien, 1985).

This paper describes a voltammetric study of BLM containing TCNQ (7,7',8,8'-tetracyano-p-quinodimethane) or TTF (tetrathiafulvalene) as model systems for electron conducting biological membranes. TCNQ and TTF were chosen for their well known function as electron acceptor and donor, respectively (Perlstein, 1977; Bryce & Murphy, 1984). From the resulting voltammograms, information about thermodynamic and kinetic parameters of electron translocation and ensuing redox reactions may be gleaned, thereby providing insights into the mechanism of electron transfer in the system under investigation.

EXPERIMENTAL

All reagents and special organic compounds such as 7,7',8,8'-tetracyanoquinodimethane (TCNQ) and tetrathiafulvalene (TTF) were obtained commercially and were of the highest quality available. Ths BLMs were formed using microsyringe technique according to the published methods (Tien, 1984). The BLM forming solution was made as follows: a 10% natural lecithin in standard oxidized cholesterol solution was first prepared. This was then diluted with an equal volume mixture of n-decane and n-butanol (1:1). The other BLM forming solution used was a mixture of phosphatidylethanolamine (PE) (3.3%) and phosphatidylserine (PS) (1.3%) in n-decane. The bathing solution usually contained 0.1 M KCl and 0.01 M sodium acetate buffer (pH 5.5). Unless otherwise noted, the outer solution also contained 0.01 M $K_3Fe(CN)_6$ and $K_4Fe(CN)_6$. To endow the BLM with desired properties, the membrane forming solution was saturated with TCNQ or TTF. These solutions were considered as containing 100% of the

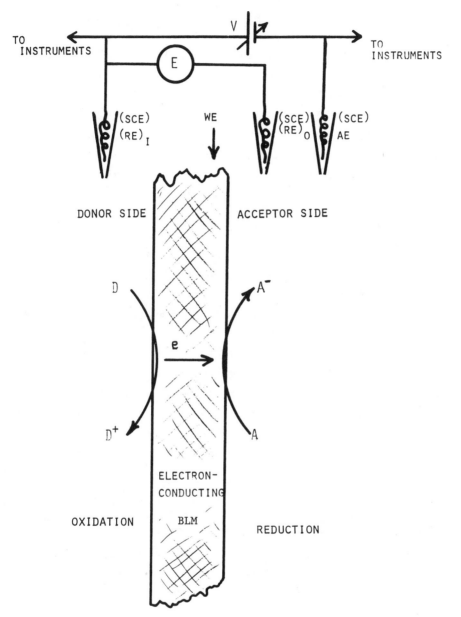

Fig. 1. Equivalent circuit diagram for electrical measurements of electron-conducting bilayer lipid membranes (BLM). V = voltage source. SCE = saturated calomel electrode. $(RE)_I$, $(RE)_O$ and AE (auxiliary electrode) denote SCE in the outer and inner chamber, respectively. E = electrometer, A = electron acceptor, D = electron donor, A^- and D^+ are the reduced A and oxidized D, respectively.

preceded mentioned compounds. For concentration dependence studies the saturated solutions of TCNQ or TTF used for BLM formation were diluted with the stock lipid solution to obtain solutions of lower concentration.

The BLM was formed in the aperture (1.5 mm) in the wall of a 10 ml Teflon cup, which was held in a plexiglass (Lucite) block having a second chamber of equal volume. After membrane formation, small amounts of 0.5 M ascorbic acid (AscA) in a 0.1 M KCl were added to the Teflon cup (donor side) whereas equal amounts of 0.5 M $K_3Fe(CN)_6$ in 0.1 M KCl were added to the other compartment (acceptor side) at the same time. The membrane was studied by use of voltammetric technique. In the conventional practice of cyclic voltammetry (CV), a stationary working electrode in quiescent solu-tion is used with a reference electrode (usually a saturated calomel electrode, SCE) and an auxiliary electrode. Between it and the working electrode a repetitive potential of triangular waveform is applied. In our experiments, the following configuration was used: one saturated calomel electrode (SCE) was placed in the Teflon cup and the other one or two others were placed outside (Fig. 1). An IBM EC/225 Voltammetric Analyzer for most studies of BLMs containing TTF was used, whereas, Keithley 417 High Speed Picoammeter and PARC 175 Universal Programmer were used for most studies of TCNQ containing BLMs. Current-voltage curves (voltammo-grams) were recorded on X-Y recorder (and/or oscilloscope) according to the convention that positive potential coordinate was directed to the left and cathodic current coordinate was directed upward. The applied voltage changed linearly in the range of 200 mV - 400 mV with scan rate 2.5 mV/s to 400 mV/s. Using a three position switch, I-V characteristics were monitored in either two-electrode or three-electrode mode. Open circuit voltage (V_{oc}) was measured in a two-electrode system using the high impedance electrometer (Keithley, Model 610 BR). Membrane capacitance (C_m) was registered by Low Level Capacitance Meter (ICE/Electronics - Model I-6). Open circuit voltage and capacitance time courses were recorded using a two-channel recorder (Linseis).

RESULTS

Two kinds of experiments with modified BLMs were carried out. In the first series of experiments, mainly exploratory, TCNQ-containing BLMs were investigated in the presence of a variety of redox couples such as AscA/ dehydro-AscA, iodine/iodide, cupric/cuprous, stannic/stannous, cysteine/

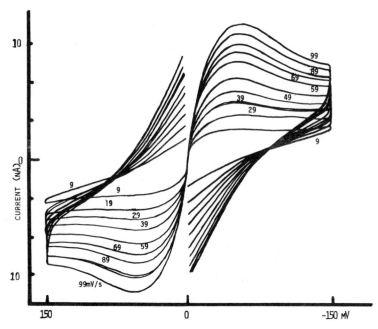

Fig. 2. Voltammograms of a TCNQ-BLM in 2 M KCl. The membrane was scanned every 9.9 sec at the various indicated mV/s rates.

cystine, chromic/chromous, hexaminecobalt(III)/hexaminecobalt(II) chloride, nitrate/nitrite and ferri-ferrocyanide. Interesting voltammograms however, even in the absence of added redox couples (strictly speaking, oxygen, an excellent electron acceptor, was always present), were obtained with TCNQ-BLMs as shown in Fig. 2. The scan rates varied from 9 mV/s to 99 mV/s. Figs. 3 and 4 present voltammograms of a number of aforementioned redox couples. Usually, there was no potential difference and further, no I-V curve asymmetry was noted when control tests were run on pure lipid membrane for the redox systems mentioned above. Remarkable I-V curve asymmetry and large V_{oc} were observed when AscA was introduced to the inner compartment.

Voltammograms obtained for hexaminecobalt(III)/hexaminecobalt(II) and chromic/chromous couples are shown, respectively, Fig. 3 (upper) and Fig. 3 (lower) with opposite asymmetry to those observed in the case of AscA (not shown). A higher current (absolute value) is flowing through the BLM

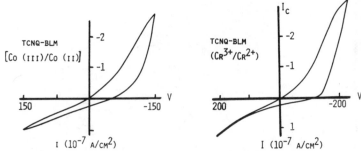

Fig. 3. Voltammograms of TCNQ-containing BLM separating two different redox couples in 0.1 M KCl and 0.1 M Na acetate at pH 5.6, one of which was equal molar (3.6 x 10^{-2} M) $Fe(CN)_6^{3-}/Fe(CN)_6^{4-}$. Upper: 680 1 of a saturated $Co(NH_3)_6^{3+}$ solution added to 10^6 ml bathing solution. C_m = 0.37 F/cm^2. Scan rate = 2.5 mV/s. Lower: 2.2 x 10^{-3} M $CrCl_3$. C_m = 0.45 F/cm^2. Scan rate = 3.7 mV/s.

under negatively applied voltages. Voltammograms obtained here exhibit very interesting shapes; they display a loop on the negative side of the scan (i.e., the BLM becomes a stronger reducing entity).

In both cases shown in Fig. 3 the polarity of generated voltage V_{oc} was negative $(Cr^{3+}/Cr^{2+} V_{oc}$ = -1.5 mV, $Co(NH_3)_6^{3+}/Co(NH_3)_6^{2+}$ = -2.5 mV). Additionally, we performed experiments when the outer compartment contained nitrate/nitrite (2.5 10^{-2} M or 6.3 10^{-2} M) instead of ferri-/ferrocyanide while the inner compartment contained AscA or

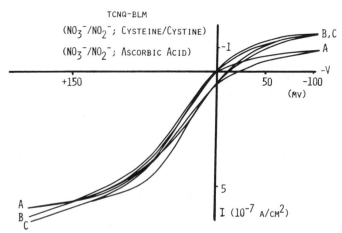

Fig. 4. Voltammograms of TCNQ-containing BLM separating two different redox couples in 0.1 M KCl and 0.1 M Na acetate at pH 5.6. One side of the bathing solution always contained equal molar solution of NO_3^-/NO_2^-. Scan rate: 2.5 mV/s. Curve A: 2.5 x 10^{-2} M NO_3^-/NO_2^- // TCNQ-BLM // 3.1 x 10^{-3} M ascorbic acid. Curve B: 6.3 x 10^{-2} M NO_3^-/NO_2^- // TCNQ-BLM // 3.8 x 10^{-3} M cystine/cysteine. Curve C: same as Curve B except the cystine/cysteine concentration was 8.7 x 10^{-4} M.

cystine/cysteine. Voltammograms obtained for these systems (Fig. 4) are similar to those using ascorbic acid.

In the second series of experiments, modified BLMs were examined as a function of either TCNQ or TTF concentrations in the membrane. Upon the addition of TCNQ or TTF to the BLM forming solution, the membrane resistance remained unaffected regardless TCNQ (or TTF) contents. This situation changed drastically when ascorbic acid (electron donor) was present on one side, and $K_3Fe(CN)_6$ (electron acceptor) was present on the other side of the membrane. Typical voltammograms, obtained for various contents of TCNQ or TTF in the BLM, but constant concentrations of donor and acceptor in aqueous solutions, are shown in Fig. 5 and Fig. 6, respectively. Observed large asymmetry of voltammograms was apparently due to asymmetry of the system in which electrons pass through the BLM only from the donor side to the acceptor side (see Fig. 1).

DISCUSSION

In order to apply the powerful CV technique to the BLM system, a conceptual effort has to be made, that is to consider one side of the membrane as the working electrode, while the other side is providing the connection to the external circuit as shown in Fig. 1. The usual picture of a BLM separating two aqueous solutions consists of a liquid hydrocarbon phase sandwiched between two layers of hydrophilic groups of lipids. The electrical properties of an unmodified BLM (i.e., a BLM formed from common phospholipids or oxidized cholesterol dissolved in n-octane in 0.1 M KCl) possess typical values of membrane resistance (R_m) greater than 10^8 ohm cm^2, membrane capacitance (C_m) about 5000 pF, membrane potential (E_m) about 0, breakdown voltage (V_b) 200 $\pm$ 50 mV, and current/voltage (I/V) curves obeying Ohm's Law. Owing to its ultrathinness, it is worth noting that an electric field gradient of 100,000 volts per cm is easily developed across the BLM. The structure of the BLM is considered to be a bimolecular liquid crystal in two dimensions having a fluid hydrocarbon core of about 50 Å thick. This liquid crystalline structure of BLM is essentially an excellent insulator, whose electrical properties, however, can be drastically altered by incorporating a variety of compounds such as the antibiotic valinomycin for K^+ specificity and chloroplst pigments and their derivatives (eg. meso-tetraphenylporphyrins, TPP). Incorporation of chloroplast pigments into BLM has made the membrane photoelectric in that light-induced electron-transfer and redox reaction have been demonstrated (Tien, 1968; Antolini et al., 1982).

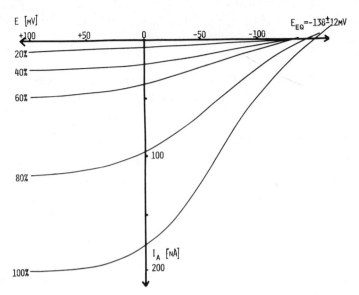

Fig. 5.   Voltammograms of BLMs containing various concentrations of TCNQ.
          2.5 mM ferric cyanide in 0.1 M KCl (pH 5.5) on one side; 2.5 mM
          ascorbic acid in 0.1 M KCl (pH 5.5) on the other side. E_{eq} =
          equilibrium potential (I = 0).

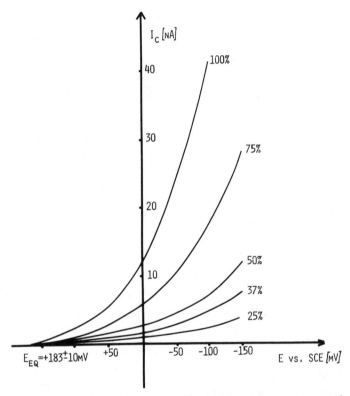

Fig. 6.   Voltammograms of BLM's containing various concentrations of TTF.
          0.5 mM ferric cyanide in 0.1 M KCl (pH 5.5) on one side; 0.5 mM
          ascorbic acid in 0.1 M KCl (pH 5.5) on the other side. E_{eq} =
          equilibrium potential (I = 0).

158

In contrast, evidence for electronic processes in BLMs in the absence of light has been difficult to demonstrate until recently. Two factors are believed to be mainly responsible for the lack of success: (1) modified BLM capable of electronic processes has not been found, and (2) a suitable experimental method with established theoretical basis has not been tried. By incorporating molecular conductors such as TCNQ and TTF into BLM, the membrane became electron-conducting in the presence of redox compounds in the bathing solutions. The selection of TCNQ (or TTF) for incorporation into the BLM is motivated by the fact that TCNQ and TTF belongs to a class of organic metals whose most unusual properties have great technological potential in the construction of molecular electronic devices and as components of solar cells (Tien, 1984).

As mentioned above, an unmodified BLM behaves essentially as an excellent insulator (specific resistivity $>$ 10^{14} ohms) and does not function as a working electrode. However, upon incorporation of TCNQ (or TTF) into the BLM and in the presence of ascorbic acid in the donor side and equal molar (0.001 to 0.01) $K_3Fe(CN)_6/K_4Fe(CN)_6$ in the acceptor side, the observed electrical properties are such (R_m decreasing from 10^8 to 10^6 ohms-cm^2, C_m increasing from 0.4 to 0.5 μFcm^{-2}, E_m developing to about 180 mV with the donor side positive, and asymmetrical I/V curves) as to leave little doubt that transmembrane redox reactions are taking place at the BLM/solution interfaces with electrons moving from the donor side of the BLM to the acceptor side. The I-V characteristic for this system is diode-like until external voltage around +200 mV is reached. The high conductivity of TCNQ-BLM with ferri/ferrocyanide in one aqueous phase and AscA in the other may be due solely to the fact that redox potentials of TCNQ in membrane phase and redox couples on both sides of BLM are well matched.

From Fig. 1 it may be seen that, depending on the redox property of species added, the direction of electron movement is governed by the relative redox power of the compounds present. Voltammograms obtained for these redox couples as chromic/chromous and hexaminecobalt(III)/ hexaminecobalt(II) chloride display a loop on the negative side of the scan (Fig. 3). One explanation of the loop may be based on the following assumptions: on the basis of our results and those obtained in a recent study (Colombetti and Lenci, 1984), it is possible to consider the influence of positively charge ion species on the electrostatic potential at the BLM surface. Further, each BLM solution interface can be considered as a Gouy-Chapman double layer (Tien, 1985). In presence of negatively charged phospholipid (in our case phosphatidylserine) in the BLM, a screening effect of the positively charged redox species ($Co(NH_3)_6^{+3}$, Cr^{+3}) would be expected. This assumption may explain the asymmetrical capacitance-voltage relationships in our case as well as in the aforementioned study. It should be emphasized that in our study (a) the change in BLM capacitance in these systems was more pronounced under external negative voltages applied to the membrane than under positive voltages, which caused the loops formation, and (b) it is reasonable to suggest that the added redox species may penetrate into the polar regions of the membrane phase perturbing the membrane structure, which is accompanied by electrostatic potential change (McLaughlin and Eisenberg, 1975; Ohki, 1976).

In order to explain the results shown in Figs. 5 and 6 in terms of voltammetry, the following conceptual effort has to be made, that is to consider the BLM interior as an "organic solvent" containing TCNQ or TTF as a solute, two membrane surfaces as electrodes and bathing solutions as connectors of these electrodes to the external circuit. This concept is based on disproportion of charged species concentrations, mobilities and dielectric constant differences between membrane interior and bathing solutions, respectively, which resembles the situation observed in classical voltammetry of electrolyte solution/metal interface. Furthermore, due to the electrochemical behavior of the compound dissolved in the BLM, it has to be assumed that one of the BLM/aqueous solution can be considered as a "working electrode", determining the voltammograms, whereas the other one can be considered as a "reference electrode". Keeping in mind the asymmetry of bathing solutions, e.g. good donor (ascorbic acid) present on one side of the BLM and good acceptor (ferric cyanide) on the other side, it seems evident that if a compound having donor properties is introduced into the BLM, the transfer of electrons would be much easier from this compound to the acceptor side of the membrane than from the donor side to the compound present in the BLM. This means that the interface BLM/ascorbic acid solution should be considered as the "working electrode" and the BLM/ferric cyanide solution interface should be considered as the "reference electrode". If a compound having acceptor properties is present in the BLM, the situation is just the opposite. In our experimental system the situations for TCNQ containing BLM and TTF containing BLM are depicted in Fig. 7 (a and b). Thus, in the case of TCNQ containing BLM the recorded voltammograms represent the anodic current-voltage relation resulting from oxidation of

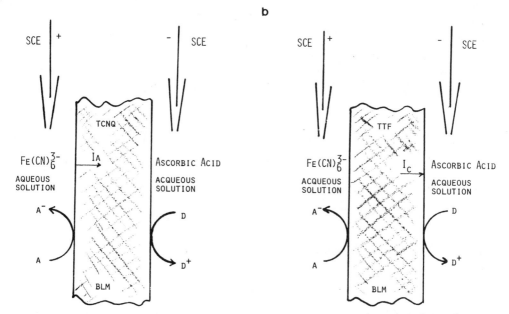

Fig. 7. Mechanism of electron-transfer and redox reactions in (a) TCNQ-BLM
and (b) TTF-BLM. WE = working electrode, AE = auxiliary
electrode, I_A = anodic current, I_c = cathodic current, SCE =
saturated calomel electrode.

TCNQ (see Fig. 5). The cathodic current branch may be neglected because,
due to asymmetry of the system, reduction of TCNQ cannot produce a
noticeable current. In the case of TTF, voltammograms represent cathodic
current-voltage relations and the anodic current branch is negligible for
the same reason as above.

As already mentioned, from the obtained voltammograms several parameters characterizing the electron transfer processes, for TCNQ and TTF,
can be deduced. According to the generally recognized theory of electrode
processes, the following equation for current I can be written (Bard and
Faulkner, 1980; Koryta, 1982):

$$I = I_0 \left\{ \frac{C_0(0,t)}{C_0^*} \exp\left[\frac{-\alpha nF}{RT}(E-E_{eq})\right] - \frac{C_R(0,t)}{C_R^*} \exp\left[\frac{(1-\alpha)nF}{RT}(E-E_{eq})\right] \right\} \quad (1)$$

Here: $C_0(0,t)$ and C_0^* are the oxidized forms of TCNQ (or TTF) at the BLM/
aqueous solution interface and in the BLM interior, respectively; $C_R(0,t)$

161

and C_R^* are the same reactant reduced forms in these regions. E is the applied potential and E_{eq} is equilibrium potential, where no net current I occurs, although balanced faradaic activity still exists and can be expressed in terms of exchange current I_o. $(E-E_{eq})$ is called overpotential and denoted η in further considerations. α is the charge transfer coefficient for cathodic process, describing asymmetry of energy barriers for electron transfer through the interface; $(1 - \alpha)$ is the charge transfer coefficient for anodic process. n is the number of electrons of redox process. R,T,F have their usual meaning. The first term of this equation describes the cathodic component I_c at any applied potential, and the second gives the anodic contribution I_a.

Under the experimental conditions of our system, for TCNQ-containing BLM the first term can be neglected (see Fig. 5) and the total current is essentially the same as I_a. Thus:

$$I = I_a = -I_o \; \frac{C_R(0,t)}{C_R^*} \; \exp \left[(1 - \alpha) \; \frac{nF}{RT} \eta \right] \tag{2}$$

In going toward a positive direction from E_{eq}, the magnitude of current rises rapidly because the exponential factor dominates behavior, but for sufficiently large overpotentials, the current levels off. In this level region, the current is limited by mass transfer of TCNQ to the interface rather than by the kinetics of electron transfer through the interface. For the case of TTF-containing BLM the anodic current is negligible (Fig. 6). Thus, the total current is equal I_c:

$$I = I_c = I_o \; \frac{C_o(0,t)}{C_o^*} \; \exp \left[-\alpha \frac{nF}{RT} \eta \right] \tag{3}$$

Our results presented in Fig. 6 show that in the applied potential range cathodic current does not reach saturation. This implies that the exponential factor dominates the behavior of the system. In other words, the heterogeneous kinetics of electron transfer through the ascorbic acid solution/BLM interface is the rate limiting step.

From voltammograms that are shown in Figs. 5 and 6 the parameters characterizing different properties of TCNQ and TTF as electron carrier in the BLM were calculated. The results are presented in Table 1.

TABLE 1 - PROPERTIES OF TCNQ-AND TTF BLM

Compound	E_{eq} mV)	I_o (nA)	α
TCNQ	-138 ± 12	3.6 ± 0.5	0.18 ± 0.03
TTF	183 ± 10	1.3 ± 0.2	0.32 ± 0.03

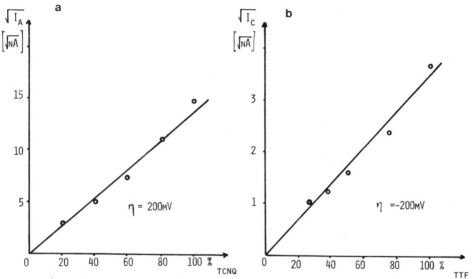

Fig. 8. The plot of square root of the current at 200 mV overpotential vs. modifier concentration; (a) for TCNQ, (b) for TTF. Circles – experimental points; continuous line – least square fit.

Further implications of voltammograms, for both TCNQ and TTF BLMs, are presented in Fig. 8 (a and b). This figure shows that the square root of the anodic (TCNQ) or cathodic (TTF) current, at constant overpotential (200 mV), linearly depends on the concentration of the compound in the forming solution. Because the magnitude of each current depends linearly on reactant concentration at the interface (see Eqs. 2 and 3 and expression for I_o (Bard and Faulkner, 1980), and on the other hand our results imply their quadratic dependence on TCNQ and TTF forming solution concentration, thus:

$$c_o(0,t) \; \alpha \; c_{TCNQ}^2; \qquad c_R(0,t) \; \alpha \; c_{TTF}^2, \qquad\qquad (4)$$

Where: c_{TCNQ} = concentration of TCNQ in BLM-forming solution, c_{TTF} = concentration of TTF in BLM-forming solution. This, in turn, suggests that the molecules of both compounds are randomly distributed in the BLM interior, with higher concentration at the interface. This seems reasonable in view of the specific structure of BLM.

In conclusion, the main findings of our experiments have been that redox reactions can occur in electron-conducting BLMs as demonstrated by CV. The emphasis of future experiments will be centered on the study of certain key components of electron transport chains involved in the oxidative and photo-phosphorylation (Bolton and Hall, 1979; Metzler, 1977; Mitchell, 1979). In particular, we plan to incorporate proteins and co-enzymes such as cytochromes and unbiquinone into BLMs and to investigate their redox properties (e.g., measurements of $E^{o'}$ values) by cyclic voltammetry and/or by photoelectrospectrometry. In the case of cytochromes, for instance, they are the most important redox enzymes occurring in all aerobic organisms. These cytochromes play a key role in plant and animal respiration (Tabushi et al., 1984). In mitochondria, cytochrome c carries electrons to cytochrome oxidase (Chan et al., 1982). $E^{o'}$ values in the literature range from -15 to 19 mV vs. SCE at pH 7 (Kreishman et al., 1980). These values were obtained by the classical potentiometric technique without "membrane" participation. The potentiometric technique is frequently complicated by the low electron exchange currents at Pt electrodes. As discussed by Heineman et al. (1972; 1980), such low exchange currents prohibit good coupling of the electrode to the potential of the bioredox couple. To overcome this difficulty, a so-called electron mediator is added to the bathing solution, which indirectly couples the redox enzyme to the electrode. However, the potential instability often appears even in the presence of the electron mediator. Alternatively, another technique involving evaluation of the position of equilibrium between mixtures of the redox species and other enzymes of known $E^{o'}$ is used. Obviously $E^{o'}$ values determined by this method are subject to uncertainties or variations of the referenced $E^{o'}$ value.

Since accurate $E^{o'}$ values are important in determining the sequence of electron transfers among redox enzymes in electron transport chains, their measurements in "membrane-bound" states are self-evident. Up to now the measurement of $E^{o'}$ values of electron transport chain components in association with lipid bilayers has not been possible due simply to the fact that no such technique existed. With the availability of electron-conducting BLMs coupled with the cyclic voltammetry technique described (Tien, 1985), an entirely new approach is open for unraveling the complicated electron-transfer and redox reactions in the processes of oxidative and photo-phosphorylation.

The significant advantage of the new technique of cyclic voltammetry of BLMs is that, besides its simplicity and the good precision of measurement, the involvement of the lipid bilayer and the capability for future development. Since the electron-transfer chain components are known to be closely associated with the lipid bilayer, the values thus determined hitherto by the usual Pt electrode may be quite different from their actual values in the membrane. Conceivably, therefore, the technique described here offers a new approach to the determiantion of E^{o}' of biomembrane-bound molecules such as the cytochromes and other redox enzymes using modified BLM as the working electrode. Further, this new type of electronically conducting BLMs coupled with the CV technique may provide unique opportunities to probe the mechanisms of biological redox reactions.

ACKNOWLEDGEMENTS: This work was supported by grants from the NIH (GM-14971), Provost Office, Dean's Offices of Colleges of Natural Science and Human Medicine of MSU. We thank Ms. Sharon Shaft for her expert typing.

References

Antolini, R., Gliozzi, A., and Gorio, A., eds., 1982, Transport in Biomembranes: Model Systems and Reconstitution, Raven Press, New York.
Bader, H., Dorn, K., Hupfer, B., and Ringsdorf, H., 1985, Adv. Polym. Sci., 64:1.
Barber, J., ed., 1979, Photosynthesis in Relation to Model Systems, Elsevier, New York, pp. 116-170.
Bard, A. J. and Faulkner, L. R., 1980, Electrochemical Methods, Fundamentals and Applications, Wiley, New York, Chap 3.
Bockris, J., O'M., 1979, in: Bioelectrochemistry, H. Keyzer and F. Gutmann, eds., Plenum Press, New York, pp. 5-17.
Blank, M., ed., 1970, Surface Chemistry of Biological Systems, Plenum Press, New York.
Blank, M., ed., 1980, Bioelectrochemistry: Ions, Surfaces, Membranes, Adv. Chem., Series 188, Washington, D.C.

Blumenthal, R. and Klausner, R. D., 1982, in: Membrane Reconstitution, G. Poste and G. L. Nicholson, eds., Elsevier Biomedical Press, Amsterdam, pp. 43-82.

Bolton, J. R. and Hall, D. O., 1979, Ann. Rev. Energy, 4:353.

Bryce, M. R. and Murphy, L. C., 1984, Nature, 309:119.

Chan, S. I., Brudvig, G. W., Martin, C. T., and Steven, T. H., 1982, in: Electron Transport and Oxygen Utilization, C. Ho, ed., Elsevier, Amsterdam.

Gerischer, H., 1979. Top Appl. Phys., 31:115.

Gliozzi, A. and Rolandi, R., 1984, in: Membranes and Sensory Transduction, G. Colombetti and F. Lenci, eds., Plenum Press, New York, pp. 1-68.

Hauska, G., and Orlich, G., 1980, J. Memb. Sci., 6:7.

Heineman, W. R., Kuwana, T., and Hartzell, C. R., 1972, Biochem. Biophys. Res. Commun., 49:1.

Hong, F. T., 1980, in: Bioelectrochemistry, M. Blank, ed., Adv. Chem. Ser., 188:211.

Koryta, J., 1982, Ions, Membranes and Electrodes, Wiley, New York.

Kreishman, G. P., Su, C-H., Anderson, W., Halsall, H. B., and Heineman, W. R., 1980, in: Bioelectrochemistry, M. Blank, ed., ASC, Washington, D.C., pp. 169-185.

Martonosi, A., ed., 1982, Membranes and Transport, Plenum Press, New York.

McLaughlin, S. and Eisenberg, M., 1975, Ann. Rev. Biophys. Bioeng., 11:231.

Metzler, D. E., 1977, Biochemistry: The Chemical Reactions of Living Cells, Academic Press, New York.

Milazzo, G., ed., 1983, Topics in Bioelectrochemistry and Bioenergetics, Vol. 5, Wiley, New York.

Miller, I. R., 1983, Bioelectrochm. Bioenerg., 11:231.

Mitchell, P., 1979, Science, 206:1148.

Mueller, P., Rudin, D. O., Tien, H. T., and Wescott, W. C., 1963, J. Phys. Chem., 67:534.

Ohki, S., 1976, Prog. Surf. Memb. Sci., 10: 117.

Perlstein, J. H., 1977, Angew Chem. Int. Ed. Engl., 16:519.

Robertson, J. D., 1981, J. Cell Biol., 91:189s.

Scott, A. C., 1977, Neurophysics, Wiley-Interscience, New York, Chap 3.

Tabushi, I., Nishiya, T., Shimomura, M., Kunitake, T., Inokuchi, H., and Yagi, T., 1984, J. Am. Chem. Soc., 106:219.

Tien, H. T., 1968, J. Phys. Chem., 72:4512.

Tien, H. T., 1974, Bilayer Lipid Membranes (BLM): Theory and Practice, Dekker, New York.

Tien, H. T., 1984, J. Phys. Chem., 88:3172.

Tien, H. T., 1985, Bioelectrochem. Bioenerget., 13 in press.

Tien, H. T., 1985, Planar Bilayer Lipid Membranes in Prog. in Surface Science, 19,#3 (S. G. Davison, ed.) Pergamon Press, N. Y.

ELECTRON TRANSFER AT BIOLOGICAL INTERFACES

Agnes Rejou-Michel, M. Ahsan Habib and John O'M. Bockris

Department of Chemistry
Texas A&M University
College Station, Texas   77843 (USA)

ABSTRACT

The electron transfer rates were electrochemically measured across the manifested bio-membrane-solution interface. Three or five layers of lipid membranes (dipalmitoyl phosphatidylcholine with or without gramicidin) were formed on the metal substrate (gold, platinum or tin dioxide) and were brought into contact with aqueous solution containing redox ions (Q/H_2Q). Tafel line characteristics with biolipid-gramicidin were radically different from those of the biolipid (which resembled those on the substrate).

It was concluded that electron exchange occurred between the protein containing membrane surface and the redox system in solution.

INTRODUCTION

The basis of the classical view of electrical potential difference in a biological system has traditionally been explained in terms of Nernst-Planckian concepts, i.e., in terms of selective ionic permeability across the membrane (Goldman, 1943· Hodgkin, 1952). However, these concepts have met with many inconsistencies as experimental data has become available (Shaw, 1956; Jahn, 1962). Moreover, this traditional view was influenced by the concepts of the time that electronic conduction could only take place in metals and semiconductors.

Homogeneous electron transfer processes in biology were suggested by Lund (1928) and a decade later by Stiehler and Flexner (1939). At the same period, Szent Gyorgyi (1941) stressed the possibility that solid proteins were electronic conductors and by inference that electron transfer took place in the solid phase in biological processes. In 1949, Katz theorized that the primary process in photosynthesis could be a migration of an electron inside chlorophyll. At the same time, King and Medley (1949) showed experimentally that protons are the conducting entities in very highly hydrated proteins.

In 1961, Williams portrayed biological reactions as involving charge transfer between solid and liquid phases. Correspondingly, Jahn (1962) pointed out that membrane potentials may be primarily oxidation-reduction

167

potentials, resulting from the presence of oxidation-reduction enzymes on the two sides of the membrane. Shortly afterwards, Cope (1963) put forward a model in which enzymes can act like electrodes. Bockris (1969) summarized the evidence for the general hypothesis that interfacial electron transfer* at biological surfaces controls the rate of metabolism.

Eley (1962), Rosenberg (1962, 1970, 1971) and more recently, Gascoyne, Pethig and Szent-Gyorgyi (1981) have shown that when proteins are wet, their conductivity is increased by several orders of magnitude (cf. King and Medley 1949). But the conductivity changes when the protein is deuterated (Gascoyne, et al., 1981). Thus, electronic conductivity in the sense originally suggested by Szent-Gyorgyi (1941) is to be replaced by a more complex mechanism which involves protons as well as electrons.

Between 1964 and recent times, several indirect attempts have been made to manifest a solid-liquid electron-transfer in biological mechanisms. Thus, Bethe (1914), Digby (1965), and Tien (1977) all reported the deposition of thin films of metals on biolipid membranes from solution. Pohl and Sauer (1978) observed the reduction of Nile Blue A on the outside of an isolated tick salivary gland, when cathodic current was passed across the gland. Habib and Bockris (1984) found that the current across a bilayer lipid membrane varied exponentially with the potential applied across the membrane, as expected if the membrane potentials were controlled by redox processes with solid-liquid electron transfer.

In this paper, experiments are reported which aim to show interfacial electron transfer between biolipid membranes containing proteins and solutions containing redox ions.

EXPERIMENTAL METHODS

The major difficulty in devising convincing experiments with the above objective in fiew is the making of a satisfactory electrical contact between the membranes and the outside potential source. A lack of equipotential inside the membrane would arise from conventional arrangements, e.g., attachment of one part of the membrane to an outside electronic circuit.

However, if a few monolayers of a biolipic membrane covers the smooth surface of a metallic conductor, then it may become possible to use the combination metal-membrane directly for electrochemical experiments, where the membrane alone is in contact with the solution. The aim is to make an effectively pinhole-free membrane of phospholipid bilayers which contain proteins. These lipids are deposited upon the metal by the use of a Langmuir-Blodgett trough. By repeated dipping of the metallic substrate in the subphase, one or more layers may be deposited: in our work 3 or 5 layers were made. One side of the biolipid membrane is in contact with the metal, the other side being exposed to solutions containing electron donor-acceptor couples. The electron transfer reaction rates were then studied at the membrane-solution interface.

---

*The famous "chemiosmotic" model due to Mitchell (1966) represents a different view from those described here: it involves proton gradients across the membrane and these are seen as involving Nernstian equilibrium potentials.

Langmuir and Blodgett showed that a variety of organic compounds which have a low degree of solubility in water, but possess polar functional groups, can be spread at the water-air interface to form a simple layer of oriented molecules (1935). There are several variables which play a part in the fabrication of Langmuir-Blodgett multilayers and in the deposition process, i.e., the nature of the liquid subphase, the solid substrate, the temperature, the surface pressure and the rate of dipping. For our experiments, the aqueous subphase was a solution of calcium carbonate (5.10^{-3} mole l^{-1}). For the purpose of adding ultra-pure water to the trough, a teflon tube pierced with small holes was fixed on the compression barrier of the trough. This tube was connected to a nitrogen cylinder and directed a gentle stream onto the subphase (Angerstein-Kozlowska, et al., 1979). When the barrier was then moved, impurities were expelled, being removed with the tip of a capillary, connected to a suction line, which was moved along the barrier.

During the transfer of a monolayer onto a solid substrate, the monolayer must be kept under a constant surface pressure. Our Langmuir trough was equipped with a Wilhelmy balance. In this method, the force due to the surface tension acting on a sheet of filter paper partially immersed, is determined. The balance was used for monitoring the surface pressure during the deposition of monolayers on solid substrates; it was part of a servo-mechanism maintaining constant surface pressure.

Four different substrates were used, i.e., glass coated by a thin layer of gold (thickness of gold 5000 Å) and by tin oxide, respectively (Fig. 1a); gold foil and platinum foil (thickness 0.25mm) were also used, enclosed in a PVC support (Fig. 1b).

The substrates were held with stainless-steel clips to a vertically movable rod which was lowered or raised at a speed of 1-5 cm/min. In our experiments, the substrate was immersed and the water surface cleaned as described above before the monolayer was formed. After the monolayer was formed, the substrate was raised. This procedure was repeated several times depending on the number of monolayers needed.

The substrate was cleaned by immersing it in concentrated H_2SO_4 overnight and successive washings with tri-distilled water. In order to verify a sufficiently clean status of the electrode, cyclic voltammograms were recorded for gold and platinum electrodes in 1N H_2SO_4 and the results were compared with those obtained in the literature (Angerstein-Kozlowska, et al., 1979).

The phospholipid used was DL-α-dipalmitoyl phophatidylcholine (Sigma No. P-6769) dissolved in a mixture of chloroform-ethanol (4:1 in volume); the concentration was 1 mg/ml. The protein was gramicidin NF (Sigma No. G-5002) dissolved in ultrapure ethanol; the concentration was 1 mg/ml. For those membranes containing proteins, both solutions were mixed in equal portions by weight. As long as the protein is not soluble in the subphase, then it becomes possible to spread a mixed solution of phospholipids and protein at the water-air interface of the trough (Colacicco, 1969).

The four substrates were covered either by 3 or by 5 monolayers of pure phospholipids and mixtures of phospholipids and proteins.

The electron transfer reaction rates were measured in an electrochemical cell using a calomel electrode as a reference electrode, a platinum electrode as a counter electrode and one of the four substrates (coated or

uncoated as the working electrode (Fig. 2). These 3 electrodes were connected to a galvanostat (ECO Model 549-potentiostat). The supporting
electrolyte used was 1 mole l^{-1} KCl and the redox couple chosen was quinone
(Q) and hydroquinone (QH$_2$) from Sigma. The concentrations of both of the
elements of this couple were 10^{-3} mole l^{-1}. This solution (KCl + Q + QH$_2$)
was buffered with 1 m mole^{-1} of a sodium salt of piperazine N-N' ethane
sulfonide (Na$_{1.5}$ PIPES). The pH was 6.82.

The experiments were carried out at a series of constant currents and
the potential of the working electrode at each of these currents was
measured with respect to the reference electrode. The potentials recorded
here are on the Normal Hydrogen Scale. The range of current densities
applied was from 5.10^{-7} to 10^{-5} amp cm^{-2}. The upper limits were chosen to
avoid evolution of oxygen on the anodic side and hydrogen on the cathodic
side. Stirring was carried out conventionally.

RESULTS

Examination of the Effect of Pinholes and Diffusion

An important point of these experiments was to examine whether
pinholes in the membrane or diffusion through it would allow electron

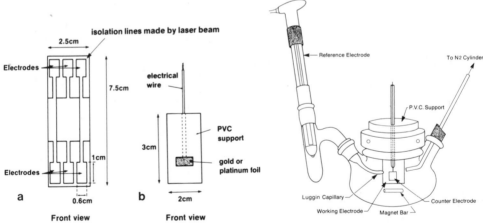

Fig. 1.
  (a) Working electrode. This consisted of
  a glass base onto which had been evaporated a gold layer of 5000 Å thick. After
  the formation of the layer, a laser beam
  was used to isolate 3 sections. Each
  section of continuous metal layer could
  be used as tow electrodes by inverting
  the plate. Electrical connections were
  made to the outside circuit by silver
  solder onto copper wire and this solution by the PVC holder of the cell top
  (see Fig. 2).
  (b) Working electrode. A gold or platinum foil was enclosed in a PVC support.

Fig. 2. Electrochemical cell. In
  this drawing the working
  electrode is a glass
  slide covered by gold
  (see Fig. 1a).

transfer at the metallic underlying substrate as an alternative to electron-transfer at the membrane-solution interface.

In the "pinhole model", the membrane is an impervious layer on the electrode surface which contains many small pinholes filled with solution which allows the redox ions to contact the metal. In the "membrane diffusion model," the membrane can be considered as a uniform phase through which ions diffuse and contact the metal (Leddy and Bard, 1983). The third alternative, that desired, was the transfer of electrons to the membrane and then a further (rate-determining) transfer from the membrane to redox ions in solution (see Fig. 3). Several relevant experiments to decide between these alternatives were performed.

Use of Fourier Transform Infrared (FTIR) Spectroscopy Method

Two experiments were performed. In one experiment a drop of a solution of dipalmitoyl phosphatidylcholine was deposited on the surface of a small polished copper block. The solution was evaporated and the spectrum recorded. In the second experiment a monolayer of DPPC was deposited on the surface of the same block by a Langmuir-Blodgett method and the spectrum was recorded (Golden and Saperstein, 1983). The spectra (Fig. 4) were indistinguishable from each other and compared reasonably well with a spectrum of stearic acid (Sadtler, 1962).

Current-potential Plot (Cyclic Voltammogram)

A current potential plot (cyclic voltammogram) was obtained with an EG&G/P.A.R.C. Model 175 Universal Programmer and a Model 363-Potentiostat/Galvanostat. The voltammogram of a gold foil electrode was recorded at 100 mV/sec between 0.240V and 1.450V (NHE scale); the counter electrode was a platinum foil and a reference electrode of calomel was used. These electrodes were dipped in an aqueous solution of 1N H_2SO_4. The normal features of the current potential relation in the anodic and cathodic directions showing gold oxide formation and the corresponding reductions are seen in Fig. 5a. The voltammogram was recorded for the gold electrode covered by 5 layers of DPPC by the Langmuir-Blodgett method (Fig. 5b) between 0.240V out 1.450V. The upper limit was chosen to avoid the evolution of oxygen. No anodic or cathodic currents corresponding to

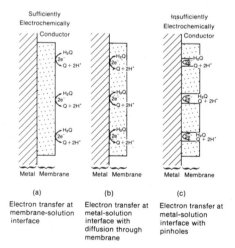

Fig. 3. Schematic diagram showing the three possible alternatives for the location of electron transfer.

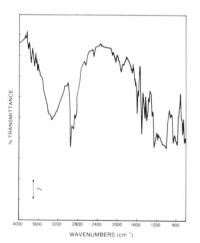

Fig. 4. IR spectra of dipalmitoyl phosphatidyl-choline.

the free gold surface phenomena were observed. At this time the membrane had been in contact with the solution for 10 to 15 minutes.

Dissolution of Silver Substrate

Another experiment was carried out to ascertain the integrity of the membrane. A silver substrate was covered by 3 layers of DPPC and dipped in 1N H_2SO_4. Then, an anodic current of 1µA cm^{-2} was passed for 10 hrs between this silver electrode, using a platinum counter electrode. The concentration of silver in the solution was determined by an Inductively Coupled plasma technique. For a non-covered silver substrate the concentration was 10 µmol l^{-1}. For a silver substrate covered by 3 layers the concentration was 8.9 µmol l^{-1}.

Underpotential Deposition of Copper on Gold

The underpotential deposition describes the formation of a sub-monolayer of metal at potentials positive to the reversible Nernst potential, for the standard state of the substrate. A quantitative description of underpotential deposition can be made in terms of anodic stripping peaks (Kolb, 1978).

For these experiments, the working electrode was a gold foil which could be covered by up to 5 layers of DPPC or DPPC-gramicidin mixture. The solution was an aqueous solution of 1N H_2SO_4 to which cupric sulfate (concentration 10^{-3} mol l^{-1}) was added.

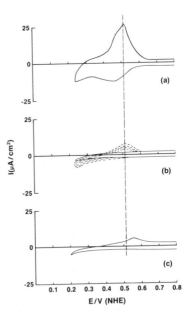

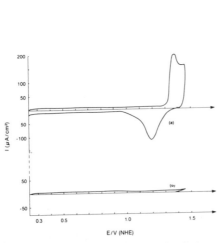

Fig. 5.   Current-potential plot.
a. for a bare gold elec-
trode; b. for a gold
electrode covered by 5
layers of DPPC.

Fig. 6.   Underpotential deposition of Cu^{2+} on: a.
bare gold electrode;
b. gold electrode cov-
ered by 5 layers of
DPPC; and c. gold elec-
trode covered by 5
layers of DPPC +
gramicidin.

Figure 6a shows the cathodic deposition and anodic stripping curves of copper on a bare gold electrode. The anodic stripping peak is at 0.510V. This curve represents 10 sweeps recorded at 50 mV/s.

Figure 6b shows the same phenomena but with a gold electrode covered by 5 layers of pure DPPC. The first sweep was made a few seconds after the addition of Cu^{2+} ions to the solution and showed neither cathodic deposition nor anodic stripping. The second sweep was made about half a minute after the first and showed a small cathodic current followed by a corresponding small anodic peak at 0.510V, the same as that on bare gold (Fig. 6a). These peaks increased during cycling for 2-3 minutes until the fifth sweep, after which they became constant. The observations were continued for around 10 minutes

Figure 6c shows the voltammograms for an electrode covered by 5 layers of DPPC-gramicidin mixture. From the first sweep to the tenth, the cathodic deposition and the anodic stripping do not change, in contrast to the behavior of the DPPC covered electrode (Fig. 6b). However, the potential of the peak of the anodic stripping on the gramicidin-DPPC covered electrode is shifted from 0.510V to 0.550V.

Current-Potential Relationship

The variations of the logarithm of current (log i) with applied electrode potential (V) for bare electrodes in solution containing quinone/hydroquinone redox couples are shown in Fig. 7. For the anodic direction, when V is greater than 0.330V, the log i-V plots (Tafel plots) are linear. The values of kinetic parameters are shown in Table 1. They agree substantially with those recorded earlier both for the cathodic and anodic directions (Vetter, 1967).

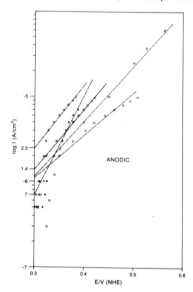

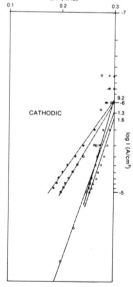

Fig. 7a. Logarithm of the anodic current density-potential curves of the Q/QH_2 redox couple on (Δ) glass coated by gold; (ɴ) glass coated by tin dioxide; (X) gold foil; (.) platinum foil; (o) Vetter's results (1967).

Fig. 7b. Logarithm of the cathodic current density-potential curves of the Q/QH_2 redox couple on (Δ) glass coated by gold; (ɴ) glass coated by tin dioxide; (X) gold foil; (.) platinum foil; (o) Vetter's results (1967).

Table 1. Kinetic parameters for the redox reaction involving Q/QH_2 on bare electrodes.

	glass coated with gold	gold foil	tin dioxide	platinum foil
slope for anodic direction (V)	0.140	0.132	0.200	0.096
slope for cathodic direction (V)	0.047	0.049	0.098	0.127
$i_{0,a}$ (A cm^{-2})	$1.4*10^{-6}$	$2.5*10^{-6}$	10^{-6}	$7.0*10^{-7}$
$i_{0,c}$ (A cm^{-2})	$9.2*10^{-7}$	$1.6*10^{-6}$	$9.6*10^{-7}$	10^{-6}

Figure 8 shows the results for a gold electrode covered by 3 or 5 layers of DPPC. On 3 layers the results are quite similar to those on gold. With 5 layers a diffusion limiting current is indicated.

Figure 9 shows the results for a gold electrode (gold foil or glass covered by gold) covered by 3 or 5 layers of phospholipid containing 50% by weight of gramicidin. The exchange current density and Tafel slope remain essentially the same with the 3 or the 5 layers. The $i_{0,a}$ is about three times less than the bare substrate and the slope about 1.5 times

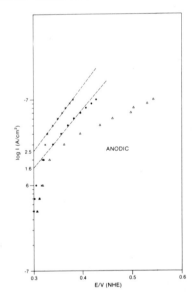

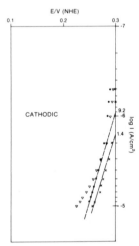

Fig. 8a. Logarithm of the anodic current density-potential curves of the Q/QH_2 redox couple on (X) gold electrode; (.) gold electrode covered by 3 layers of DPPC; (Δ) gold electrode covered by 5 layers of DPPC.

Fig. 8b. Logarithm of the cathodic current density-potential curves of the Q/QH_2 redox couple on (X) gold electrode; (.) gold electrode covered by 3 layers of DPPC; (Δ) gold electrode covered by 5 layers of DPPC

higher. No limiting current is observed. The results differ significantly
from those corresponding to DPPC layers not containing gramicidin.

Figures 10a and 10b show the results for tin dioxide covered by 3 or
5 layers of the phospholipid-protein mixture. The $i_{o,a}$ approximates to
that observed with layers of DPPC and gramicidin on gold. The Tafel slope
is 1.3 times greater than that with a gold substrate.

Figures 11a and 11b show the results for platinum covered by 3 or 5
layers of phospholipid-protein mixture. Here the $i_{o,a}$ is about twice the
value for the same situation with the gold substrate. The slope is about
the same as for gold.

The kinetic parameters for the substrates covered by 3 layers are
shown in Table 2 and by 5 layers in Table 3.

Table 2. Kinetic parameters for the redox reaction Q/QH_2 on electrodes
covered by 3 layers of protein-phospholipid mixture.

	glass coated with gold	tin dioxide	platinum foil
slope for anodic direction (V)	0.190	0.260	0.194
slope for cathodic direction (V)		0.104	0.147
$i_{o,a}$ (A cm^{-2})	$5.4*10^{-7}$	$7.8*10^{-7}$	$1.2*10^{-6}$
$i_{o,c}$ (A cm^{-2})		$3.8*10^{-7}$	10^{-6}

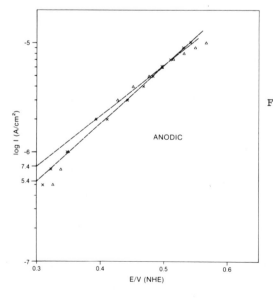

Fig. 9. Logarithm of the anodic
current density-potential
curves of the Q/QH_2 redox
couple on gold electrode
covered by: (X) 3 layers
of DPPC + gramicidin; (Δ)
5 layers of DPPC + gramicidin.

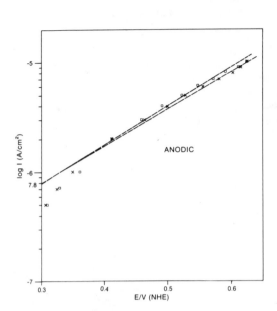

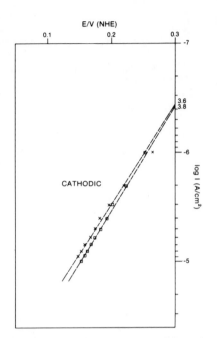

Fig. 10a. Logarithm of the anodic current density-potential curves of the Q/QH_2 redox couple on tin dioxide covered by: (□) 3 layers of DPPC + gramicidin; (X) 5 layers of DPPC + gramicidin.

Fig. 10b. Logarithm of the cathodic current density-potential curves of the Q/QH_2 redox couple on tin dioxide covered by: (□) 3 layers of DPPC + gramicidin; (X) 5 layers of DPPC + gramicidin.

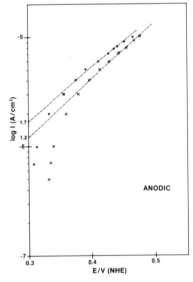

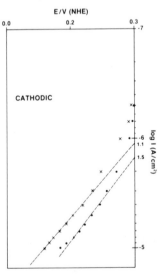

Fig. 11a. Logarithm of the anodic current density-potential curves of the Q/QH_2 redox couple on platinum electrode covered by: (●) 3 layers of DPPC + gramicidin; (X) 5 layers of DPPC + gramicidin.

Fig. 11b. Logarithm of the cathodic current density-potential curves of the Q/QH_2 redox couple on platinum electrode covered by: (●) 3 layers of DPPC + gramicidin; (X) 5 layers of DPPC + gramicidin.

176

Table 3. Kinetic parameters for the redox reaction Q/QH_2 on electrodes covered by 5 layers of protein-phospholipid mixture.

	glass coated with gold	gold foil	tin dioxide	platinum foil
slope for anodic direction (V)		0.213	0.260	0.220
slope for cathodic direction (V)			0.110	0.135
$i_{o,a}$ (A cm^{-2})		$7.4*10^{-7}$	$7.8*10^{-7}$	$1.7*10^{-6}$
$i_{o,c}$ (A cm^{-2})			$3.6*10^{-7}$	$1.5*10^{-6}$

DISCUSSION

The first, and most important, matter in discussion of the material presented above concerns the desiderata of the location of the electron transfer. Intended as an experiment to demonstrate electron transfer at the interface of the protein-containing membrane with the solution, there is throughout the danger that the electron transfer at a possible under-lying metal solution interface is faster than that at the membrane-solution interface, either through pinholes, or through diffusion of the electrolyte through the membrane.

Among the initial approaches to these problems were attempts to examine the membranes with various forms of microscopy. Thus Fig. 12 shows a picture taken of a monolayer of stearic acid deposited on a copper block used in place of gold for slide-matching reasons by the Langmuir-Blodgett method; before the deposition, the monolayer was spread on the surface of a Mn Cl_2 bath (concentration 10^{-3} mole l^{-1}) (Pomerantz, 1980). However, the magnification with the available SEM was only 20,000. Attempts to see the membrane in greater magnification involve transmission electron microscopy which cannot be used with the membrane-metal system, for it is too thick. Correspondingly, carbon compounds do not X-ray as well as compounds containing atoms of higher atomic weight, e.g., platinum.

However, it is possible to prove by the use of FTIR data that a membrane of DPPC was in fact deposited on the substrate metal (Fig. 4).

Such evidence has little bearing upon the question of pinholes and diffusion.

In order to throw light upon the question of the existence of pinholes and diffusion, the experiments described above were carried out (current-potential plot; dissolution of silver substrate; under potential deposi-tion of copper on gold).

Thus, when 5 layers of DPPC (containing no gramicidin) are deposited upon the gold electrode, the potential sweep pattern obtained on this electrode (Fig. 5b) differed from that obtained on a normal gold electrode (Fig. 5a). In the former case, the anodic current is diminished by about 5 times and no peaks were obtained. In agreement with the expectation, this membrane showed no signs of electronic conductivity; the small current

observed may have been due to pinhole effects although no cathodic peak
was observed.

Underpotential deposition experiments (see Kolb, 1978) were carried
out with the following aim in view: the cathodic phase of the underpotential deposition is followed by a corresponding anodic dissolution peak.
The character of this peak is influenced by the characteristics of the
underpotential deposition, which is itself substrate dependent. Thus, one
test of whether an electrode reaction is taking place on the membrane, or
the underlying gold, is to compare the underpotential dissolution potential
peak obtained upon a pure gold surface, e.g., with the deposition of
copper, with the corresponding peak obtained in the presence of the membrane.

Complex phenomena were observed in this experiment. In the case of
the membranes of DPPC containing no protein the phenomena were time dependent (Fig. 6b). Thus, when the copper was first introduced into the solution,
almost no current was observed during sweeps in both anodic and cathodic
directions. However, in times of the order of tens of seconds, peaks
increasingly appeared with time and these corresponded well in respect to
electrode potential (± 3 millivolts) with those which were obtained upon a
gold substrate without the membrane. The implication of these experiments
is clear. For these pure DPPC membranes, diffusion of solution had taken
place through the membrane and made contact with the underlying metal substrate. The fact that the copper peaks are less high than those which are
obtained on pure gold is consistent with this view.

When gramicidin was introduced into the biolipid membrane, it is
reasonable to expect (Capaldi, 1977) that enhanced electronic conductivity
makes it possible for electrons to pass from the metal-membrane contact
to the membrane-solution interface. In accordance with this view are the
results shown in Fig. 6c. Here, when the potential sweep is carried out,
the current observed no longer varies with time of contact of the membrane
with the solution. Moreover, the anodic peak is shifted by about 40 mV
in the anodic direction.

It seems reasonable to interpret the experiments of Fig. 6c as
indicating that electron transfer is occurring at the membrane-solution
interface, and that copper is being deposited thereon and dissolved

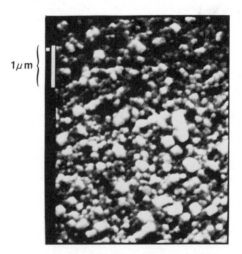

1μm

Fig. 12. Scanning electron micrograph of a layer of
stearic acid on a metal-
lic support (x 20000).

therefrom. This is consistent with the fact that no variation of the current with time is seen (corresponding to increasing diffusion through the membrane) and with the fact that the current peak for copper dissolution is shifted corresponding to underpotential deposition of copper on a different substrate from gold.

The potential at which the anodic stripping peak is observed will depend on the rate constant at some standard state for the dissolution. A more anodic peak (compared to membrane-free gold) means a more difficult dissolution, i.e., stronger bonding of the deposited copper adions to the gramicidin surface than to the gold. Were the surface metal species to be copper atoms this should be unlikely: indeed gold alloys form easily with copper in deposition (Conway and Bockris, 1961). However, the nature of the initial surface species before the formation of crystals on the metal surface is ionic, i.e., the adatoms are in fact adions $Cu^{\delta+}$. In this case, it seems acceptable to suggest that a binding of $Cu^{\delta+}$ to the protein substrate occurs and that it may be greater by 0.04 (δ); F ($\cong$ 0.45 kcal for δ = 0.5) than that to gold.

An alternative explanation in terms of pinholes seems unlikely. Thus, the current density observed at the same potential is about 1/5 that for dissolution from the membrane-free gold, i.e., the total pinhole area would have to be at least 20% of the total area (an improbably large contribution). Diffusion through the membrane has been shown to have time effects in the order of minutes (see Fig. 6b) and these are not observed here (Fig. 6c).

It is reasonable to conclude therefore that the introduction of gramicidin into the biolipid membrane did, indeed, increase the conductivity and allow charge transfer at the interface, and that the biolipid membrane has a greater stability and resistance to diffusion of the electrolyte than the membrane without protein. Thus, Papahadjopoulos et al. (1975) found that addition of proteins stabilized lipid membranes. In respect to the experiment with silver dissolution through a membrane containing no protein, this clearly represents a situation in which the silver ions diffuse from the silver substrate through the membrane, i.e., the latter lacks integrity and electronic conductivity. It is noteworthy that such a membrane did not contain gramicidin and that the contact of silver with the biolipid membrane was observed not to be a good one. Thus, upon raising the silver substrate through the biolipid monolayer at the water-air interface of the Langmuir trough, only abnormally small changes in surface tension were seen.

Coming now to the measurement of Tafel lines, the early experiments were to re-establish the phenomenology of the reduction and oxidation of quinone and hydroquinone at an inert electrode, as originally done in the classical study of Vetter (1967). The results are shown in Fig. 7 where the present measurements are compared with those done by the original investigator - it is seen that the comparison is fairly good.

In Figs. 8 and 9 are shown the results for the corresponding oxidation of hydroquinone on membranes without (Fig. 8) and with (Fig. 9) gramicidin on gold electrodes. It is seen that, corresponding to the underpotential deposition, different behavior was observed in situations in which only the biolipid membrane was present and in which the biolipid membrane containing protein was deposited on the substrate.

In the former case (no gramicidin) the results are quite similar for a 3-layer deposit to those obtained on gold, and suggest that diffusion of the electrolyte took place through the biolipid membrane. This is further confirmed by the fact that when 5 layers were deposited, the Tafel line showed characteristics similar to those obtained with diffusion-controlled situations, i.e., 5 layers of the membrane acted to retard the access of ions to the gold electrode which was still the dominant partner in the electron exchange.

However, when the reaction was examined on the gramicidin-containing biolipid membrane, the results changed. The most important result is that there was no change in the slopes for 3 or 5 layers. Had diffusion or pinhole effects been in operation, one would clearly have obtained a different result for these two, as shown in the case of the membrane without gramicidin (Fig. 8). In addition, the characteristic of the line changed so that the Tafel slope now became 0.20V, compared with the original slope on gold of 0.140V. The i_o values changed from 7.10^{-7} Amp cm^{-2} to $1.4 \ 10^{-6}$ Amp cm^{-2} (see Tables 2 and 3).

It appears, therefore, that for the gramicidin-biolipid membrane, electron transfer in the oxidation of hydroquinone occurs at the membrane-liquid interface.

Corresponding experiments were then carried out on a tin dioxide base. The results are shown in Fig. 10. In favor of concluding that these experiments represent electron transfer at the membrane-solution interface are the identical results obtained on both 3 and 5 layers of the protein-containing biolipid membrane. The anodic Tafel slope (0.260V) from these experiments is also characteristic of the membrane-solution interface and substantially greater than that on the tin dioxide substrate (0.20V).

The cathodic data are less indicative, being about the same on the tin dioxide as on the membrane.

Lastly, on platinum (Fig. 11), the anodic slopes are clearly different for experiments of the oxidation of hydroquinone on the membrane compared with those on bare platinum substrate, being essentially 0.2V on the membrane, and 0.1V on the platinum. The i_o values on the membrane are about twice those on the platinum substrate alone. Correspondingly, the 3- and 5-layer results are practically equal, in consistence with the conclusion that here, too, electron transfer occurs at the membrane-solution interface.

In considering these matters, it has been assumed that the mixing of protein with the biolipid membrane increases the electronic conductivity of the ensemble to a degree such that electrons can pass relatively easily through a layer of about 100 Å in thickness. It is interesting to attempt to estimate the specific conductivity of the ensemble produced. Thus, no measurements of the electronic conductivity of wet gramicidin appear in the literature, but there are many measurements of the conductance of proteins, the most comprehensive collection being that of Gutmann, Keyzer and Lyons (1983). Here, values for the specific conductivity of proteins vary between 10^{-8} and 10^{-3} mho.cm^{-1}.

In considering the potential difference which would arise by the passage of current from the metal to the membrane, with the lowest value of the above κ range, it is relevant to refer to a deduction which has been given by Bockris (1954) for the potential difference in the passage of a current through a layer of thickness L and conductivity κ. It is $\eta_{ohmic} = iL/\kappa$, where i is the current density. Thus, as the maximum i value used here was 10^{-5} Amp.cm^{-2}, the ohmic potential difference would be about 1 mV, i.e., negligible.

Another assumption made in the above considerations concerns the contact between the metal and the membrane. It has been taken to be ohmic.

Evidence for such an assumption is not available from experimental studies. However, the fact that the results obtained did not differ greatly during the substantial change between, say, a gold substrate and a tin oxide substrate, give rise to some support for the assumption of an ohmic contact.

There remains the question of the interpretation of the different Tafel slopes obtained with electron transfer at the membrane solution interface, compared with those at the substrate-solution interface. In general, there was an increase in the slopes obtained, and this was between 50 and 100%.

A marked property of the Tafel lines on the DPPC-gramicidin surfaces is that although the i_o values remain fairly close to these characteristics of the underlying metal, i.e., around 10^{-6} amp cm^{-2}, the values of the Tafel slopes are radically different: they increase roughly double, in the transition from redox reactions on the metal substrate to those carried out where the DPPC-gramicidin membrane may provide the source or sink for electrons. This change in the transfer coefficient, α, provides strong support for the contention that a radical change of reacting surface occurs with introduction of gramicidin.

Further, the numerical values of the Tafel slopes observed on the noble metal substrates ($\cong 0.1$ volts/decade) are reminiscent of the results observed when a metal becomes covered by a relatively thick film of oxide and electron transfer occurs at the oxide-solution interface. The changes of the transfer coefficient here, have been described by Meyer (1960) to the functioning of both the film and the double layer at the oxide-solution interface in control of the current. Meyer's theory (1960) results in a value of the Tafel slope of 4 RT/F. In decadic units, this is about 0.240 volts/decade at 25°C in agreement with the present results on the membrane.

An alternative explanation would involve taking into account the fact that quinone-like structures, which are in use here, bond to the protein active sites (Morrison, et al., 1982), whereas, with metallic substrates, this does not occur. Quinone was thought to exist as a mobile electron carrier, shuttling the electron between electron-transfer complexes in the phospholipid membrane. Recent discovery and isolation (Yu, Yu and King, 1977) of a protein that binds quinone shows that protein and quinone can form complexes, but their structure and function are not yet fully understood. It is true that benzene is strongly adsorbed on platinum (Heiland, et al., 1966) but it is known that such adsorption ceases if the molecule loses the alternative ethylene linkages in the ring characteristic of aromatic structure (Blomgren, et al., 1961). Gileadi and Stoner (1971) have shown that when water dipoles are displaced from a surface during adsorption of a reactant prior to the electron transfer step, the Tafel slope is then given by: 2 RT/F (1-0.14 n), where n is the number of water molecules displaced by one reactant molecule from the surface at which electron transfer occurs. The above equation gives n = 3.6 for Tafel slope of b = 4 RT/F. Flat adsorption of the structure would displace around 9 water molecules (Heiland, et al., 1966).

However, the amino group of the amino acids of a protein can react with the carboxyl group of an aldehyde or ketone to form stable tertiary amine (Rejou-Michel, 1983). So, quinone should preferentially bind end on to the gramicidin molecule which contains at least 2 molecules of tryptophane and a secondary amine at the periphery of the phospholipid layer (Stryer, 1975). Moreover, one can assume that if a molecule of quinone is bound end on, the radius of the occupying surface is 2.35 Å (Tables of Interatomic Distances, 1958, London). Taking into account that the radius of a molecule of water is 1.38 Å, one molecule of quinone should replace 3 molecules of water. This is in a better accordance with the experimental results than a flat adsorption.

These suggestions are speculative but serve to increase the consistency of the case model here for protein-solution charge transfer.

The present results suggest that one may observe some electrochemical activity directly at biological-solution interfaces, particularly those containing enzymes. A number of matters in interpretive mechanistic biology may be influenced by the results.

Thus, for example, it would be possible to see metabolic action occurring at the mitochondria-solution interface in terms of a fuel cell-like model (Bockris, et al., 1984), a number of cathodic sites being bunched together on one section of the membrane, while anodic sites are at another. Such a result would be consistent with the very high energy conversion efficiencies obtained in biological materials (Mitsui and Kumazawa, 1977), which are not consistent with another energy.

CONCLUSIONS

When a DPPC membrane is deposited onto the surface of a metal by means of a Langmuir-Blodgett technique, and the ensemble introduced into solution, the system behaves in consistency with the view that the ions transfer electrons to the <u>metal</u> substrate; i.e., the conductance of the membrane is <u>insufficient</u> for electron transfer, and that the constitution of the membrane is such that diffusion from the solution to the substrate may easily occur through it.

Conversely, when gramicidin is introduced into a DPPC membrane, the behavior changes in consistency with the view that protein introduces sufficient electronic conductivity for the dominant reaction to occur at the membrane-solution interface, with insufficient diffusion to disturb this transfer.

ACKNOWLEDGMENTS

The authors are grateful to the National Foundation for Cancer Research for financial support of this work.

The authors are also grateful for discussion and help during the work, particularly to Mr. Flamarion Diniz in respect to the experimental technique, to Dr. Felix Gutmann in respect to concepts of electronic conductivity in protein, to Dr. Charles Martin in respect to diffusion through coatings on electrodes, to Mr. Scon Prodea of Flinders University, Adelaide, Australia and to Messrs. Joyner R. Faulk and Greg Morris of the Electronics shop of Texas A&M University, in respect to the operation of the Langmuir-Blodgett trough. Thanks are due to the authorities of the Flinders University for the gift of the trough.

REFERENCES

Angerstein-Kozlowska, H., Conway, B. E., Barnett, B. and Mozota, J. (1979), J. Electroanal. Chem. 100, 417-446.
Bethe, A. and Toropoff, T. (1914). Z. Phys. Chem. 88, 686-742.
Blodgett, K. B. (1935). J. A. C. S. 57, 1007-1022.
Blomgren, E., Bockris, J. O'M. and Jesch, C. (1961). J. Phys. Chem. 65, 2000-2010.

Bockris, J. O'M. (1954). in "Modern Aspects of Electrochemistry," 1, pp. 180-276, Butterworths, London.

Bockris, J. O'M. (1969). Nature, $\underline{224}$, 775-777.

Capaldi, R. A. (1977). In "Membrane Proteins in Energy Transduction," Vol. 1.

Colacicco, G. (1969). J. Colloid. Sci. $\underline{29}$, 345-364.

Conway, B. E. and Bockris, J. O'M. (1961). Electrochimica Acta $\underline{3}$, 340-366.

Cope, F. W. (1963). Bull. Math. Biophys. $\underline{25}$, 165-176.

Digby, P. S. B. (1965). Proc. Roy. Soc. (London) $\underline{8161}$, 504-525.

Eley, D. D. (1962). In "Horizons in Biochemistry" (Kasha M and Pullman B. Eds.), pp. 341-380, Acad. Press, New York.

Gascoyne, P. R. C., Pethig, R. and Szent-Gyorgyi, A. (1981). Proc. Nat. Acad. Sci. USA $\underline{78}$, 261-265.

Gileadi, E. and Stoner, G. E. (1971). J. Electrochem. Soc. $\underline{118}$, 1316-1319.

Golden, W. G. and Saperstein, D. D. (1983). J. Electron. Spectroscopy $\underline{30}$, 43-50.

Goldman, D. E. (1943). J. Gen. Physiol. $\underline{27}$, 37-60.

Gutmann, F., Keyzer, H. and Lyons, L. E. (1983). In "Organic Semiconductors," Part B. R. E. Krieger Publishing Company, Malabar, Florida.

Habib, M. A. and Bockris, J. O'M. (1984). Bioelectricity $\underline{3}$, 247-280.

Heiland, W. Gileadi, E., and Bockris, J. O'M. (1966). J. Chem. Phys. $\underline{70}$, 1207-1216.

Hodgkin, A. and Huxley, A. L. (1952). J. Physiol. $\underline{117}$, 500-544.

Jahn, T. L. (1962). J. Theoret. Biol. $\underline{2}$, 129-138.

Katz, E. (1949). In "Photosynthesis in Plants," p. 291 Iowa State College, Ames.

King, G. and Medley, J. A. (1949). J. Colloid. Sci. $\underline{4}$, 1-7.

Kolb, D. M. (1978). In "Advances in Electrochemistry and Electrochemical Engineering," 11 pp. 125-271, J. Wiley Chichester.

Leddy, J. and Bard, A. J. (1983). J. Electroanal. Chem. $\underline{153}$, 223-242.

Lund, E. J. (1928). J. Exp. Zool. $\underline{51}$, 265-290.

Meyer, R. E. (1960). J. Electrochem. Soc. $\underline{107}$, 847-853.

Michel, A. Rejou. (1983). PhD dissertation, University of Paris, 6 Paris.

Mitchell, P. (1966). Biol. Rev. Cambridge Philos. Soc. $\underline{41}$, 445-502.

Mitsui, A. and Kumazawa, S. (1977). In "Biological Solar Energy Conversion, pp. 23-51, Academic Press, New York.

Morrison, L. E., Schelhorn, J. E., Cotton, T. M., Bering, C. L. and Loach, P. A. (1982). In "Function of quinones in Energy Conserving Systems" pp. 35-58. Academic Press, New York.

Pant, H. C. and Rosenberg, B. (1971). Chem. Phys. Lipids $\underline{6}$, 39-45.

Papahadjopoulos, D., Moscarello, M., Eylar, E. H. and Isac, T. (1975). B.B.A. $\underline{401}$, 317-335.

Pohl, H. and Sauer, J. (1978). J. Biol. Phys. $\underline{6}$, 118-123.

Pomerantz, M. (1980). In "Phase Transitions in Surface Films," pp. 317-346, Plenum Press, New York.

Rosenberg, B. (1962). J. Chem. Phys. $\underline{36}$, 816-823.

Rosenberg, B. and Pant, H. C. (1970). Chem. Phys. Lipids $\underline{4}$, 203-207.

Sadtler. (1962). In "Standard IR Spectra," O. Spectrum No. 50.

Shaw, F. H., Simon, S. E. and Johnstone, B. M. (1956). J. Gen. Physiol. $\underline{40}$, 1-17.

Stiehler, R. D. and Flexner, L. B. (1939). J. Chem. Biol. $\underline{126}$, 603-617.

Stryer, L. (1975). "Biochemistry" (Freeman and Co. Eds.). San Francisco.

Szent-Gyorgyi, A. (1941). Nature $\underline{148}$, 157-159.

Tables of Interatomic Distances and Configuration in Molecules and Ions, 1958, London.

Tien, H. T., Karvaly, B. and Shieh, P. K. (1977). J. Colloid. Interface. Sci. $\underline{62}$, 185-188.

Vetter, K. J. (1967). In "Electrochemical Kinetics: Theoretical and Experimental Aspects, pp. 483-487, Acad. Press, New York.

Williams, R. J. P. (1961). J. Theoret. Biol. $\underline{1}$, 1-17.

Yu, C. A., Yu, L. and King, T. E. (1977). Biochem. Biophys. Res. Commun. $\underline{78}$, 259-265.

ALTERATIONS IN ELECTRICAL DOUBLE LAYER STRUCTURE DUE

TO ELECTROMAGNETIC COUPLING TO MEMBRANE BOUND ENZYMES

James D. Bond and N. Convers Wyeth

Science Applications International Corporation
1710 Goodridge Drive
McLean, Virginia 22102

INTRODUCTION

We shall discuss the role that the electrical double layer formed at the interface of a charged biological membrane and the membrane's extracellular ionic environment might play in gaining an understanding of how time-varying electromagnetic fields of very low intensity interact with the membrane per se. The analysis we shall present originated in an effort to explain what have generally become known as nonthermal interactions; that is coupling between an external field and the biological system of interest that cannot be attributed to the thermalization of the energy carried by the field. Alternatively such responses cannot be elicited by simple heating processes, for example via some other experimental means of heating.

We employ the double layer concept in several ways in order to understand field-membrane coupling. As we shall subsequently discuss, the interaction between the field and the membrane can manifest itself through alterations in both the potential profile and the ionic profile that extend from the surface of the membrane to the bulk electrolyte. In other words field-membrane coupling alters the structure of the electrical double layer.

Also, by systematically starting out with different double layer configurations, for example by varying the pH and/or ionic strength of the extracellular fluid, changes in the initial charge state of the membrane can be brought about, and the effects of such changes on the interaction between a field of given intensity and frequency can be studied. That is, a kind of electrostatic modulation of the response to the external perturbation can be accomplished in this way.

Although this symposium is primarily concerned with the role of double layers in biology, we should like to make a few brief comments about the nature of the electromagnetic coupling to a macromolecular system such as a biological membrane. For the case of membranes the interaction is, in all probability, of a cooperative or collective nature. These ideas have been discussed extensively by Frohlich (1980) and Bond and Huth (1985). There exists, however, no direct experimental evidence that the coupling of weak external electromagnetic fields to biological membranes is cooperative. There do exist sound theoretical bases to suspect that such is the case, as discussed in the aforementioned references. We shall in the next section briefly indicate how a cooperative response could be triggered by the application of an external field.

FIELD-MEMBRANE COUPLING

For reasons previously mentioned we shall only outline the manner in which an external electromagnetic field can couple (interact) with a biological membrane to initiate (trigger) a cooperative (autocatalytic) response. Bond and Huth (1985), based on an earlier lattice statistics analysis by Hill (1967), have discussed how an external field could trigger a cooperative response. The membrane is imagined to be a two-dimensional lattice where each lattice site can reside in one of two states, A or B. Each lattice site is characterized by its own internal partition function, j_i, where i = A or B.

The membrane, in addition to any external time varying fields, is in the presence of a static field, E, usually very large ($\simeq 1 \times 10^7$ V/M), due to the asymmetric distribution of ions on either side. In addition ligands can bind to these lattice sites. Each bound ligand is characterized by a partition function, q_i. Each site is also characterized by a polarizability α_i. If interactions among lattice sites are allowed, then the system can exhibit cooperative behavior. This is accomplished by introducing an interaction free energy, w_{ij}, between neighboring sites. Analytically this is handled most simply by using the Bragg-Williams approximation.

The net result for the condition of a phase transition (a highly co-operative event) to occur is

$$\frac{j_A \exp(-cw_{AA}/2kT)}{j_B \exp(-cw_{BB}/2kT)} \; \frac{1 + q_A \lambda}{1 + q_B \lambda} \; \exp(\alpha_A - \alpha_B) \; E^2/2kT \; (\exp(w_A - w_B)/kT) = y, \quad (1)$$

where c is the number of nearest neighbor sites, λ is the absolute activity of the ligand species, for a given site, k is Boltzmann's constant, T is the absolute temperature, and y identically equal to unity for a phase transition (Hill (1967)). The w_A and w_B in Equation 1 correspond to the interaction energy between a site and the perturbing field, E'. We do not attempt to write an explicit analytic form for the w_i. Suffice it to say that both are functions of E', i.e. $w_i = w_i(E')$. The point to be made is that λ, E, and E' can all serve to trigger a phase transition. A more detailed discussion is provided in Bond and Huth (1985) and Hill (1967).

It should be pointed out that Equation 1 was derived using arguments from equilibrium statistical mechanics. Therefore, as it stands, the interaction energy of the system with any external field, as characterized by the $w_i(E')$, is necessarily going to be thermalized no matter what functional form $w_i(E')$ might take. Therefore Equation 1, while it does show that a very weak field, or perhaps more appropriately the correct combination of λ, E, and E', can serve to initiate a cooperative response, strictly speaking cannot serve to explain nonthermal responses.

We can at this point only speculate about how the activity of a membrane bound enzyme could be altered by an external field. There could exist direct cooperative coupling to a protein embedded within the lipid matrix such that a conformational change is initiated resulting in altered activity. Also direct cooperative coupling to the lipid such that its physicochemical state is altered could affect enzyme activity since integral membrane protein function is sensitive to the specific lipid that envelopes it.

DOUBLE LAYER ANALYSIS

DeSimone (1977), using classical perturbation theory, has demonstrated in a general way that alterations in the structure of the electrical

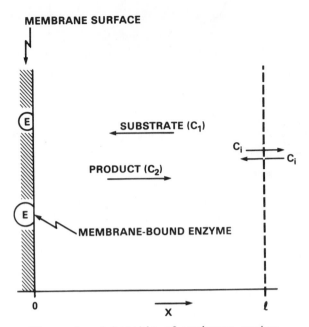

Figure 1.   Schematic of membrane system.

double layer can occur when there exist active enzymes embedded in a charged matrix such as a biological membrane.   Pennline et al. (1977) subsequently verified the DeSimone calculation using an alternative technique by which they showed the governing solution for the electrostatic potential profile extending from the charged membrane surface into the surrounding extracellular electrolyte is a solution to Poisson's equation related to a function known as the second Painleve transcendent.   Our basic premise in this analysis is that an external electromagnetic perturbation interacts with the cell membrane such that the intrinsic activity of certain membrane bound enzymes is altered.   This alteration in enzymic activity then leads to a change in the concentration profiles of those ionic species that comprise the so-called electrical double layer.   In fact, as shall subsequently be demonstrated, an explanation for reported alterations in Ca^{+2} efflux from the cell surface could be related to these changes in double layer structure.

Our system of interest then is the plasma membrane plus its external electrolytic environment: Figure 1. The essential physical and chemical features of such a system can be retained by employing the same basic assumptions as DeSimone (1977) with some slight modifications. Those assumptions are briefly reviewed for the sake of completeness and continuity of thought. We consider the membrane to be a uniformly negatively charged planar surface of infinite extent with five distinct ionic species forming the ionic environment. The assumption of a negatively charged surface is different from the positively charged surface assumed by DeSimone (1977) and Pennline et al. (1977); however, a negatively charged surface is more consistent with cell surfaces encountered in real biological membranes. Normal to the planar membrane in the x-direction is a diffusion zone of thickness ℓ. For $x > \ell$, electroneutrality prevails, and the electrolyte consists of a bulk solution containing five different ion species. Species 1 represents the substrate and has charge $|z|$. Species 2 is the product released when species 1 reacts with a membrane bound enzyme, and it also has charge $|z|$. Species 3 with charge $-z$ is the counterion common to Species 1 and 2. Species 4 and 5 are considered nonreacting ions with charges $-|z|$ and $+|z|$ respectively.

It is assumed here also that the electrodiffusion of all ions in the diffusion zone is governed by the Nernst-Planck electrodiffusion equations. That is,

$$J_i = - D_i \left(\frac{dc_i}{dx} + \frac{z_i e}{kT} c_i \frac{d\psi}{dx} \right) , \tag{2}$$

where the subscript i corresponds to any one of the five ionic species under consideration and

$$z_i = |z| , \quad i = 1, 2, 5$$
$$z_i = -|z| , \quad i = 3, 4. \tag{3}$$

The notation is that normally used where $D_i \equiv$ diffusion coefficient of species i, $c_i \equiv$ concentration of species i, $k \equiv$ Boltzmann's constant, $T \equiv$ absolute temperature, $e \equiv$ proton charge, and $\psi \equiv$ electrostatic potential extending from the charged membrane surface out into the bulk solution.

Equation 2 can be solved in a straightforward manner for all i by using the appropriate integrating factor. The solutions for a positively charged membrane surface are given in DeSimone (1977) and Pennline et al. (1977) and for negatively charged surface in the Appendix to this chapter. For simplicity the reduced potential, ϕ, is defined such that

$$\phi(x) = \frac{|z|e\psi(x)}{kT} . \tag{4}$$

With this definition Poisson's equation takes the form

$$\frac{d^2\phi}{dx^2} = \frac{-4\pi e|z|}{\varepsilon kT} \rho(x), \tag{5}$$

where

$$\rho(x) = \sum_{i=1}^{5} z_i e c_i(x) \tag{6}$$

for $0 < x < \ell$, and

$$\rho(x) = \sum_{i=1}^{5} z_i e c_i(x) = 0 \tag{7}$$

for $x \geq \ell$. Using the following reduced variables introduced by Pennline et al. (1977), we define

$$y = \kappa x, \qquad \Phi(y) = \phi(y/\kappa) , \qquad L = \kappa\ell, \tag{8}$$

$$\mu = \frac{J_R}{2\kappa c} \left(\frac{1}{D_2} - \frac{1}{D_1}\right), \text{ with } J_R = J_2 = -J_1, \tag{9}$$

and

$$\kappa^2 = \frac{8\pi|z|^2 e^2 c}{\varepsilon kT} \tag{10}$$

The parameter c is defined as

$$c = c_3(\ell) + c_4(\ell) = c_1(\ell) + c_2(\ell) + c_5(\ell), \tag{11}$$

which follows from Equation 7. Substituting the solutions for the concentration profiles, $c_i(x)$, given in the Appendix, into Equation 6, then substituting Equation 6 into Poisson's Equation, and making use of Equations 8-10, yields the following integrodifferential equation analogous to that arrived at by Pennline et al. (1977).

190

$$\frac{d^2\phi}{dy^2} = \sinh\phi(y) - \mu\, e^{-\phi(y)} \int_{y}^{L} e^{\phi(\xi)}\, d\xi. \qquad (12)$$

Note especially that the second term in Equation 12, i.e., the nonequilibrium contribution to the potential, ϕ, is preceded in our case by a negative sign. The signs of the exponential arguments here are reversed from those for a positively charged membrane surface. Equation 12 is the basic equation that must be solved in order to obtain the potential profile, i.e., $\phi = \phi(y)$. This equation can be written in a more tractable form for solution according to the method outlined by Pennline et al. (1977). For our numerical analysis, however, we chose to employ a shooting method to obtain the potential profile and not a finite difference technique.

Numerical solutions for the potential $\phi(y)$ in Equation 12 are shown in Figure 2 for a surface potential $\phi(o) = -4$, a diffusion zone corresponding to $L = 10$, and the family of μ values indicated. The boundary condition $\phi(o) = -4$ corresponds to a real surface potential of approximately -130 mV, a reasonable value for a typical biological membrane. This value was obtained for $T = 310$ K and $|z| = 1$ from Equations 4 and 8. If the reaction rate, J_R, is assumed constant, the μ-parameterized curves in Figure 2 can be interpreted as showing the different potential profiles produced by varying the relative magnitudes of D_1 and D_2. In the case of $\mu = 0$ $(D_1 = D_2)$, the substrate and product ionic species are essentially indistinguishable as far as electrodiffusion effects, and the resulting potential curve corresponds to the Gouy-Chapman solution. The curves for $\mu > 0$ are representative of D_1 (substrate) $> D_2$ (product) and vice versa for $\mu < 0$.

Of interest here are the ionic concentration profiles for various physiologically reasonable values of μ. Once the potential profile has been determined for a given set of parameters, the ionic concentration profiles can be obtained from Equations A.1-A.3 (or in reduced variable form, Equations A.4-A.8. As shown in the equations, the non-reacting ion concentrations are given by an appropriately signed exponential function of the potential (Boltzmann factor) normalized to the bulk concentration $c_i(\ell)$. In this way a set of potential curves, such as those in Figure 2, have a directly related set of non-reacting ionic concentration profiles (anions and cations) which can possess extrema depending on the behavior of ϕ.

The equations governing the reacting species (substrate and product) are more complicated and involve a direct dependence on u and hence on J_R. To examine the consequences of changes in membrane bound enzyme activity on the reacting ion concentration profiles, we consider μ as a function of J_R, i.e. keep D_1, D_2, etc. constant, and rewrite Equation 9 as

$$\mu = g \, J_R \tag{13}$$

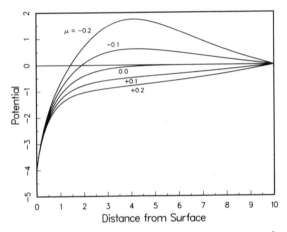

Figure 2.   Potential versus distance (reduced variables) for different values of the parameter μ with L=10 and $\Phi(0)=-4$.

where g is a constant. We further assume that the enzyme mediated reaction can be characterized by Michaelis-Menten type kinetics in the linear region, i.e. that the reaction rate is directly proportional to the substrate concentration at the membrane surface:

$$J_R = b \, c_1(0) \tag{14}$$

where b is the proportionality constant. Now Equation A.4 can be used to find $c_1(y=0)$. Using $c'_1(y)$ as the substrate concentration normalized to its value in bulk (y=L) and with some rearrangement, Equation A.4 becomes

$$c'_1(0) = e^{-\Phi(0)} \left[1 - hJ_R \int_0^L e^{\Phi(\eta)} d\eta \right], \qquad (15)$$

where h is a constant containing g from Equation 13 and all the other parametric constants multiplying the integral in Equation A.4. The integral in Equation 15 has a fixed value for a given potential profile; this value can be labeled S_p. The relative membrane enzyme activity level is represented by the parameter b in Equation 14. To see how the substrate concentration profile varies as the relative activity b is changed, we use the same values of L and $\Phi(0)$ used in Figure 2 and proceed as follows. We choose a baseline value of μ equal to -0.1. This determines $\Phi(y)$ and hence S_p, which is calculated numerically to be 1.263. Since J_R must be relatively small to be in the linear region (Equation 14), we arbitrarily set $hJ_R S_p = 0.01$. Using this and $\Phi(0) = -4$, we find that $c'_1(0) = 5.405$, and, since $hJ_R = hbc'_1(0)$ (where b is now used with normalized concentrations), we know that $hb = 1.465 \times 10^{-3}$. We now choose h such that b = 1.0 for this baseline case. Writing Equation A.4 as

$$c'_1(y) = e^{-\Phi(y)} \left[1 - hJ_R \int_y^L e^{\Phi(\eta)} d\eta \right], \qquad (16)$$

this expression was used with the specified value of $hbc'_1(0)$ and $\Phi(y)$ for $\mu = -0.1$ to calculate $c'_1(y)$ over the layer width ($0 \leq y \leq L$); the results are shown in Figure 3. Then, using numerical trial and error iterations, the potential functions resulting from changing b by a factor of 2 and 4 were found, and the $c'_1(y)$ functions for each case are shown in Figure 3. These curves demonstrate how changing the enzyme activity affects the substrate concentration profile in the double layer region. We should mention that the range of μ values chosen are compatible with physically reasonable values of the parameters, e.g. the D_i, on which depends.

Recall that one of our goals is to ascertain the effect of an external electromagnetic perturbation on the electrolytic cellular environment assuming the perturbation manifests itself by causing a change in enzymic activity, J_R. We imagine the system residing prior to the perturbation in some steady-state configuration, characterized by an apparent equilibrium ($\mu = 0$) or a true nonequilibrium ($\mu \neq 0$). The effect of the perturbation, through ΔJ_R, is to initiate a transition to some new steady-state configuration now characterized by a value of μ different from the initial value. For example we could imagine the external field serving as a stimulus to activate or "turn-on" a previous-

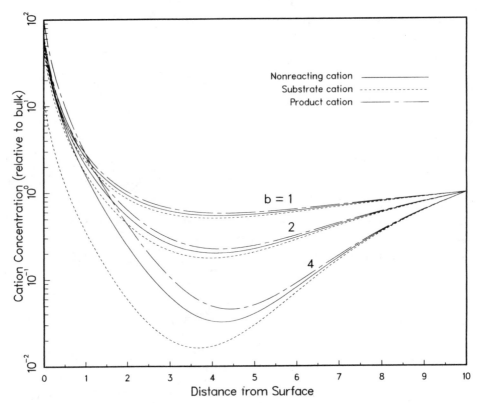

Figure 3. Normalized cation concentration profiles for different values of relative enzyme activity b with L=10, $\Phi(0) = -4$, and a baseline value of μ =-0.1 for b=1.

ly inactive enzyme, i.e. $J_R = 0$, such that $J_R \neq 0$ in the final state. Assuming $D_1 \neq D_2$ such a response would manifest itself as an alteration in the double-layer structure as illustrated in Figure 3. In other words the double-layer structure in the initial state is characterized by a given value of μ and in the final state by a different value of μ.

RELEVANCE TO FIELD INDUCED Ca^{+2} EFFLUX CHANGES

It has been previously suggested that the source of divalent calcium ions responsible for the observed alterations in calcium efflux caused by weak external fields might be ions that are not specifically bound, i.e. contact adsorbed, but ions that are adsorbed by merely being sequestered within the electrical double layer formed at the membrane-electrolyte interface (Bond and Jordan (1984)). By the arguments presented above, an electromagnetically induced ΔJ_R, and thus a $\Delta \mu$, could be sufficient to "release" calcium ions from the state characterized by $b = 1$ and achieve a new steady state, e.g. $b = 4$, as illustrated in Figure 3. A redistribution of calcium ions in the initial double-layer after the formation of a "new" double-layer as a result of ΔJ_R might be sufficient to account for the type of bioassays reported.

The extant data for the calcium efflux experiments does not provide information that allows for the determination of the amount of calcium ions released, and so quantitative comparisons between our model and those experiments cannot be made. Needless to say, it would be advantageous to have such experimental data, especially on systems where a discrimination can be made between the two possible types of adsorption processes, i.e. specific (or contact) adsorption and the kind of nonspecific adsorption we discussed above. The nonequilibrium contribution, μ, to the potential profile, Φ, is a function of the parameter, κ, which is dependent on the ionic strength through c (Equations 9 and 10). Thus the magnitude of μ can be systematically increased or decreased experimentally, by respectively decreasing or increasing the ambient ionic strength. This provides us with a means of varying the initial state in which the system resides. In fact if our hypothesis concerning electromagnetic alterations in J_R is correct, a possible explanation for the extreme difficulty encountered in replicating many of the reported electromagnetic effects at the cellular level might be the effects of slight variations in the bathing medium, i.e., a different κ. It is well known that variations in ionic strength influence the state of membrane surface potential via electrostatic screening of membrane surface charge. Such effects, too, obviously influence the dissociation equilibria of acidic groups at the membrane surface since the degree of dissociation of such groups is a function of membrane surface potential. In the context of membrane-bound enzymic activity it is also known that both local hydrogen ion concentration and membrane surface charge density can influence enzyme activity in the plasma membrane, endoplasmic reticulum membrane, and the mitochondrial inner and outer membrane. In this analysis we have not considered the effects of local pH changes due to the release of hydrogen ions as the product. That could have important consequences as well. Alternatively, systematically manipulating the extracellular pH would permit a determination of the response of the system as a function of different $\Phi(0)$. Considering the structural heterogeneity as well as both structural and functional complexity of many of the biological systems in which effects of very low intensity electromagnetic fields have been reported, but where such effects have either been impossible or very difficult to replicate, a possible explanation for the difficulties in analyzing those systems where the coupling between field and system is enzyme mediated might be in the nonestablishment of the "correct" initial state.

ANALYSIS OF DOUBLE LAYER CAPACITANCE

By integrating over the cation concentration curves shown in Figure 3, we can calculate the total amount of each species stored in the double layer. The anion concentration (not shown) is just the reciprocal of the nonreacting cation curve (relative to the bulk value) and can also be integrated. Using these totals we can calculate the net charge stored in the double layer by taking the (monovalent) cation total and subtracting its associated anion total because neutrality requires that their bulk concentrations be equal. These results are shown in Table 1; the values were found with the same parameter values used in deriving Figure 3, e.g. the baseline value of μ is -0.1 for $b = 1$. For all three types of cation under these conditions, the net charge (cation minus associated anion) in the double layer quickly becomes negative as the enzyme activity is increased. This can be understood physically by referring to the potential curves of Figure 2. As μ becomes negative (substrate diffusion coefficient less than that of the product), the potential slope approaching the membrane surface becomes steeper to help drive the more slowly responding substrate cations toward the reaction site and keep their flow equal to the outward flowing product cation stream. This results in a region of positive potential in the double layer (not present for $\mu \geq 0$) and the consequent depletion of cations in this region as shown in Figure 3. Increasing the enzyme activity b causes μ to be more negative (by increasing J_R) and further depletes the double layer of cations (and increases the anion concentration). Thus the shift to net negative charge seen in Table 1. Although the three cation types are treated independently in Table 1, the results do not imply that a change in enzyme activity can produce such large changes in nonreacting cation concentration under any circumstances. The expression for μ contains c, the total bulk ionic concentration, in the denominator. Thus if the non-reacting species pair were the dominant concentration in the bulk with $c \gg J_R/\kappa D_{1,2}$, the baseline value of $|\mu| = 0.1$ would be inappropriate, and changes in b would not affect the potential profile or the net charge in the double layer very much.

The differential capacitance of the double layer, defined here as the incremental change in net charge in the double layer divided by the incremental change in surface potential (with bulk potential equal zero), can also be calculated. The results for the case considered are listed in Table 1. The differential capacitance is given in units of normalized charge (relative to bulk) per volt, and as such is not directly relatable to experimentally measureable values. The contribution of each cation type to the differential capacitance is shown versus increasing enzyme activity. As mentioned above, the results shown are dependent upon the particular parameter set used, and the relative importance of each species' capacitance contribution will depend on the relative size of the species' bulk concentration. However we can conclude that the differential capacitance of the layer at a fixed voltage will increase significantly with increasing enzyme activity; this follows from the increases in capacitance shown in Table 1 for all three cation contributions. Differential capacitance of a double layer is in general also a function of voltage or net potential change across the layer. To provide a comparison to the change in capacitance with relative activity, we also show the differential capacitance of the layer when the voltage is changed by 100 mV, i.e. to -4.1 volts. This was calculated to provide a measure of the relative size of the changes in capacitance expected. The numbers in Table 1 show that, for the layer parameters assumed, the capacitance change from a doubling of enzyme activity (14-29%) is much greater than expected from voltage drifts if the voltage is held fixed to an accuracy of 10 mV or less.

CONCLUSIONS

Experiments in which structurally and functionally well character-ized systems are subjected to weak electromagnetic fields would allow for more realistic attempts at modeling and developing a framework for theories that would explain such phenomena at the molecular level. Experimental systems such as monolayers of lipid and lipid plus incorporated functional enzymes could serve as models to test some of the ideas concerning changes in double-layer structure secondary to field-enzyme coupling. Such effects, if indeed they exist and are sufficiently large, would manifest themselves thermodynamically through a change in the spreading pressure of the monolayer. Artificial lipid bilayers and artificial lipid bilayers incorporating functional enzymes might prove to be another experimental system worthy of consideration. It is not clear, however, whether the kinds of alterations in double-layer structure that we are suggesting and which might be biologically and physiologically important, are large enough to be detected, for example via influencing the electrophoretic mobility of such systems.

TABLE 1   VARIATION OF NET CHARGE AND DIFFERENTIAL CAPACITANCE OF DOUBLE LAYER VS ENZYME ACTIVITY (b)

Surface Potential (volts)	Relative Activity	Net Charge (normalized to bulk)	Differential Capacitance (relative)
Nonreacting Cation:			
−4.00	1	+0.622	0.716
	2	−0.895	0.875
	4	−8.487	1.836
−4.10	1	+0.700	0.849
	4	−8.304	1.825
Substrate Cation:			
−4.00	1	+0.468	0.651
	2	−1.153	0.745
	4	−9.387	1.382
−4.10	1	+0.539	0.772
	4	−9.250	1.344
Product Cation:			
−4.00	1	+0.775	0.780
	2	−0.637	1.004
	4	−7.588	2.289
−4.10	1	+0.860	0.926
	4	−7.358	2.307

REFERENCES

1. Bond, J.D., and Huth, G.C., 1985, Electrostatic Modulation of Electromagnetically Induced Nonthermal Responses in Biological Membranes, in: "Modern Bioelectrochemistry," F. Gutman and H. Keyser, eds., Plenum Press, New York.

2. Bond, J.D., and Jordan, C.A., 1984, Electrostatic Influences on Electromagnetically Induced Calcium Ion Displacement from the Cell Surface, Bioelectrochem. and Bioenergt., 12:177.

3. DeSimone, J., 1977, Perturbations in the Structure of the Double Layer at an Enzymic Surface, J. Theor. Biol., 68:225.

4. Frohlich, H., 1980, The Biological Effects of Microwaves and Related Questions, Adv. in Electronics and Electron Physics, 53:85.

5. Hill, T.L., 1967, Electric Fields and the Cooperativity of Biological Membranes, Proc. Natl. Acad. Sci., 58:111.

6. Pennline, J.A., Rosenbaum, J.S., DeSimone, J.A., and Mikulecky, D.C., 1977, A Nonlinear Boundary Value Problem Arising in the Structure of the Double Layer at an Enzymatic Surface, Math. Biosci., 37:1.

Appendix

The solutions to Equation 1 are easily obtained by using an integrating factor since the equation is of first order and is linear in the c_i and all derivatives of c_i.

$$c_i(x) = c_i(\ell)e^{-\phi(x)} + \frac{J_i}{D_i}e^{-\phi(x)} \int_x^\ell e^{\phi(x)}dx, \quad i = 1, 2; \qquad (A.1)$$

$$c_i(x) = c_i(\ell)e^{\phi(x)}, \qquad i = 3, 4; \qquad (A.2)$$

$$c_i(x) = c_i(\ell)e^{-\phi(x)}, \quad i = 5. \qquad (A.3)$$

In terms of the reduced variables introduced in Equations 8 and 9, Equations A.1-A.3 take the form

$$c_1(y) = c_1(L)e^{-\phi(y)} \qquad (A.4)$$

$$- 2c\mu\left(\frac{D_2}{D_1 - D_2}\right)e^{-\phi(y)} \int_y^L e^{\phi(\eta)}d\eta,$$

$$c_2(y) = c_2(L)e^{-\phi(y)} \qquad (A.5)$$

$$+ 2c\mu\left(\frac{D_2}{D_1 - D_2}\right)e^{-\phi(y)} \int_y^L e^{\phi(\eta)}d\eta,$$

$$c_3(y) = c_3(L)e^{\phi(y)}, \qquad (A.6)$$

$$c_4(y) = c_4(L)e^{\phi(y)}, \qquad (A.7)$$

$$c_5(y) = c_5(L)e^{-\phi(y)}. \qquad (A.8)$$

200

TRANSIENT IMPEDANCE MEASUREMENTS ON BIOLOGICAL MEMBRANES:

APPLICATION TO RED BLOOD CELLS AND MELANOMA CELLS

R. Schmukler, J.J. Kaufman, P.C. Maccaro, J.T. Ryaby and
A.A. Pilla
Bioelectrochemistry Laboratory, Department of
Orthopaedics, Mount Sinai School of Medicine, One Gustave
L. Levy Place, New York, N.Y. 10029

Introduction

A technique has previously been described (1-5) by which the
impedance of isolated cells can be evaluated over a wide frequency range
(DC -20 MHz). The methodology centers about the use of a polycarbonate
filter containing well defined and uniform cylindrical pores. A
pseudo-epithelium is achieved by the entrapment of cells from a
suspension in these pores using hydrostatic pressure. Data analysis by
discrete Laplace transformation of the input and output waveforms allows
electrochemical kinetic models to be tested via aperiodic equivalent
circuit representation. In this study the effect of the state of the
cell surface on the impedance parameters is examined. This is achieved
by the use of the enzyme neuraminidase on the human red blood cell
(HRBC) to reduce the surface charge and by melanocyte stimulating
hormone (MSH) to alter the state of differentiation of melanoma cells.

Experimental Methods

The experimental electrochemical chamber consists of a vertical
cylinder of polycarbonate with a 1 cm inside diameter, and 3 cm length.
The upper and lower half of the chamber each contains: 1 axially placed
porous carbon electrode for current input, 1 radially placed voltage
measuring electrode of porous tantalum, 1 water jacket, and 1 teflon
flat ring for sealing between the two halves. Four electrodes, two
carbon and two tantalum, have large surface areas which eliminate
electrode polarization effects at the voltage measuring (tantalum)
electrodes. The carbon electrodes, for the current input, form the
upper and lower ends of the cylinder and are 3 cm apart. The vertical
separation, between the upper and lower voltage measuring electrodes is
2 mm. Both tantalum electrodes are recessed into the polycarbonate so
that they lie completely outside the center of the chamber. This is
done in order that perturbations to the current distribution in the
chamber are miminized (6). When the current electrodes are separated by
a distance 2X the diameter of the cylinder, the current flux
distribution at the point of measurement in the center of the cylinder
is uniform and linear (6-9).

Between the upper and lower halves of the chamber, a polycarbonate
Nuclepore (TM) filter is clamped between the two teflon rings. When the

cells are embedded into the filter a psuedo-epthelium is created. The filter is a sheet of approximately 10 μm thick polycarbonate, transversed by true cylindrical pores. The resistance of the polycarbonate film is very high (10^{12} Ω), so that the only pathway for current conduction between the upper and lower halves of the chamber is confined to the electrolyte filled pores of the filter. Edge sealing of the filter to the teflon is checked by using a polycarbonate sheet with no holes. When this is measured the resistance at the edge seal was found to be greater than 10^8 Ω. Both halves are water jacketed so that temperature control of the cells and the electrolyte inside the chamber is possible. The access ports provide a means of changing or adding to the electrolyte around the psuedo-epithelium.

Care is taken to insure that the filter is fully wetted and that no air bubbles exist in the system. First a measurement is taken without the filter. This determines Re, the IR ohmic drop between the measurement electrodes due to the electrolyte in the chamber. The chamber is carefully disassembled and then reassembled with a filter between the two halves and another measurement taken. This determines RL^o which is the electrolyte filled filter resistance. Both of these measurements also provide a check on the linearity of the system. There are no frequency dependent terms at the frequencies of interest in an electrolyte solution impedance, just a simple resistance, so that the voltage and current waveforms should be identical except in amplitude. This was found to be the case down to about 40 nanoseconds.

The final step in preparation of the pseudo-epithelium and subsequent measurement of the cell impedance is to occupy the pores of the filter with the cells. A suspension of washed cells is placed in the upper half of the chamber and a hydrostatic pressure head attached from below via the lowest access port. The cells are layered above the filter and allowed to move under the influence of gravity (sedimentation) and the pressure differential (convective volume flow) until they occupy the pores of the filter. The pressure head forces the cells into the pores and since the current pathways across the filter are restricted to the pores, this forces the majority of the current through the cells. There is thus an increase of several fold in the impedance measurement sensitivity compared to suspension techniques, due to the large increase in leak resistance around the cells.

The determination of the relative contribution of multiple cell layers to the measured electrical impedance of the psuedo-epithelium is extremely important because of current pathway considerations (10). The multiple cell layers are formed as a result of possible sedimentation of cells on top of the filter, and are in series with respect to current pathways with the pore-embedded cells. This can result in the inclusion of a multiple cell layer component to the measured impedance. If, however, the conductivity of the leak pathways around the cells in the multiple cell layers above the pseudo-epithelium is suffiently high, the relative contribution to the measured impedance should be small. This was determined by allowing cells to sediment without the presence of the hydrostatic pressure head, and the presence of these multiple layers of cells above the filter resulted in an identical impedance to that seen without any cells present. During an experiment the multiple cell layers above the filter are subjected to an isotropic pressure due to the depth of fluid above the filter. Since the pressure head is applied from below, the isotropic pressure is small and this condition is analogous to the absence of a pressure head. Thus, the cell layers above the filter during an impedance measurement do not contribute to the measured impedance and can be neglected (1).

The instrumentation used in the experiment consists of a high speed (2ns risetime) Hewlett-Packard 8007B pulse generator, appropriate circuitry, the electrochemical chamber, 2 high speed (30MHz) Comlinear differential amplifiers and a Gould Biomation 4500 Digital Oscilloscope. The digital oscilloscope has a dual time base capability with a maximum sampling rate of 100MHz (10ns/pt). The output of the pulse generator is terminated with 50Ω directly into a 0.1% reference 100Ω resistor (for input current measurement). The other end of the 100Ω resistor is connected to one carbon electrode of the chamber and the other carbon electrode is grounded. The amplifiers are connected directly to the tantalum electrodes (A.C. coupled .06Hz) for the voltage output measurement. The digital oscilloscope is interfaced to a Hewlett Packard 1000E minicomputer for data aquisition and processing.

Human red blood cells for the neuraminidase experiments are obtained by venipuncture and collected in a heparinized blood tube. The cells were washed twice with Tris/Albumen Saline buffer (pH 7.4), then resuspended in the same buffer to the original volume. Ten percent V/V Neuramindase was added (1 unit/ml, Calbiochem-Behring) and the cells were incubated at 37°C for 1 hour (4). The cells were then washed twice in the buffer and resuspended in the same buffer to 20% of the original volume.

Melanoma cells (Cloudman S-91 Clone M3) are obtained from the American Type Culture Collection. The cells are maintained in Ham's F-10 Nutrient medium containing 15% Horse Serum, and 2.5% Fetal Bovine Serum (HyClone Laboratory) under antibiotic-free conditions. Cells are harvested from tissue culture flasks by the addition of 0.5mM EDTA calcium, magnesium free Hanks Balanced Salt Solution (HBSS). Cells are centrifuged at 400g, and resuspended at a density of 3×10^6 cells/ml of tissue culture medium (as above). The above procedure was used for the preparation of both controls and experimental cells. Experimental cells were induced to differentiate by the addition of 10^{-8}M MSH and incubated for 48 hours. It has been previously reported that addition of MSH to Melanoma cells results in the stimulation of tyrosinase activity, the rate limiting enzyme in the cellular biosynthesis of melanin (11). Melanin production is an indication of a differentiated melanocyte. Tyrosinase activity is routinely measured to ascertain the degree of differentiation. At 48 hours the cells showed a threefold increase in tyrosinase activity and melanin synthesis.

Results and Discussion

The technique used to analyze which of the parameter of the generalized model is present, is performed by first converting the data to a frequency function via a real axis digital Laplace Transform (12,13). The transforms of the measured input voltage and current signals are obtained by numerically evaluating the Laplace Transform integral for values of s on the real axis (s=σ). A trapezoidal approximation to the integral is employed, viz,

$$V(\sigma_i) = \Sigma 1/2 \ \{V(t_{j+1}) + V(t_j)\} \ e^{-\sigma_i t_j} \ (t_{j+1} - t_j) \qquad (1)$$

In general, impedance parameters are de-embedded by plotting the impedance as a function of σ, 1/σ and σ 1/2. As each parameter is estimated, it is subtracted from the impedance function Z(σ) in order to look for the next parameter. In order to construct the appropriate model to be used in the analysis, it is necessary to utilize all available pertinent information. A model and the approach for data analysis will be illustrated below for the RBC system used here.

From the physical construction of the filter (see Figure 1) with the embedded RBC, there exists an ohmic drop Re which is always present in series with the model due to the resistivity of the solution between the measuring electrodes and the filter surface. In the model there is, in parallel to the cells in the pores, a leak pathway RL, due to the unfilled pores, if any, and the leaks between the RBC and the pore wall. In series with both the leak pathway around the cells and the cells themselves is the resistance Rf due to the unfilled pore length, if the cells do not penetrate the filter pore completely. An important point to remember is that Rf can never be larger than RL^o, the filter resistance without cells and must be some fraction of it. The cell model consists of a resistance due to the cytoplasmic resistance Ri in series with the membrane impedance Zc (see Figure 2 below):

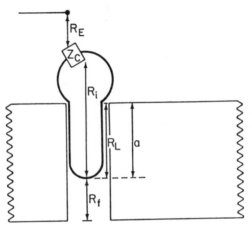

Fig. 1 Schematic representation of a living cell embedded in the cylindrical pore of the insulating filter. See text for details.

As stated in Experimental Methods, Re as well as RL^o, which is the filter resistance without cells, can be determined from a measurement so that Zw/o cells = Re + RL^o. The resistance of a single electrolye filled pore, Rp, is given by; Rp = $\rho eL/A$, where ρe is the resistivity of the electrolyte in ohm-cm, L is the path length in cm(10-13 µm), and A is the cross sectional area in cm^2. RL^o is simply Rp/N, where N is the total number of pores in the filter filled with electrolyte.

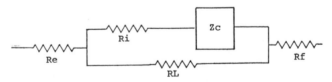

Fig. 2 Equivalent electrical circuit representing the impedance of a living cell embedded in the pore of an insulating filter.

Inspection of Figure 2 shows that the limiting behavior at both high and low frequency limits can be used to evaluate some of the components of the model. The equivalent circuit high frequency limit is shown in figure 3A and that at the low frequency limit is shown in figure 3B.

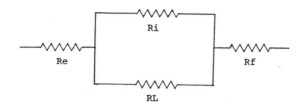

Fig. 3A   Equivalent electrical circuit representing the high frequency limiting behavior of the living cell/pore complex.

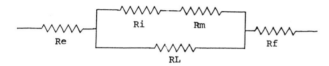

Fig. 3B   Equivalent electrical circuit representing the low frequency limiting behavior of the living cell/pore complex.

To relate the leak pathway resistance to the cytoplasmic resistance and the unfilled portion of the filter pore, it can be shown (1) that $RL = (aRL^o)/(1-x)$, $Ri = aBRL^o/x$ and $Rf = (1-a)RL^o$ where a = fraction of total pore length, L, into which the cell penetrates, or the ratio of the path length the current traverses inside the cell to the path length of a pore; x = fraction of cross sectional area for current conduction of the filter obstructed by the cells; $B = \rho c / \rho e$ is the ratio of relative resistivities of the cytoplasm to the extracellular fluid electrolyte (figure 1).

By using a sensitive (to .01 ml) buret as the end of the hydrostatic pressure head the entrapment of the cells by the filter can be measured from the volume flow at steady state pressure before and after the cells are added to the impedance measurement chamber. If the filter has pores left open, i.e., unfilled, this will change by a small increment the value of x, but will not add any error to the data analysis using the model. This is because all the available conduction pathways through the filter are predetermined by the measuremnt of RL^o.

Zc contains at least one capacitance due to the membrane dielectric structure (see figure 2) and one or more due to processes such as specific adsorption of various ions. In addition, there is in parallel a membrane resistance Rm (or D.C. pathway) for ionic phase transfer across the membrane, so that, assuming at least one specific adsorption time constant, the membrane impedance model can be shown as in figure 4.

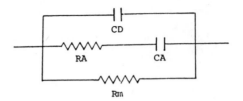

Fig. 4   Electrical equivalent circuit representing
          the impedance model of a cell membrane.

If the value of Rm is much greater than RL, the low frequency
circuit limit becomes Re + RL_2+ Rf (see fig. 3B). The value for Rm for
the HRBC is approximately 10^{12} Ω based on the smallest available surface
area of 5.4 μm^2 and a ρm of from 10^{-5} Ω-cm^2 (14). If we use the most
frequent value in the literature, of ρm of about 10^{3} Ω-cm^2, the lower
limit for the measurement resistance for one cell is still 1.9 x 10^{10} Ω.
The resistance of a single fluid filled 1.85μm pore is Rp= (ρeL)/A=
2.9 x 10^{8} Ω where ρe= 70Ω-cm, L=11μm, and A=2.7μm^2. The RL value is
normally 10X the RL^{o} value, so that the resistance for a single pore is
approximately 3 x 10^{7} Ω. Thus, it can be seen from figure 3B that the
pathway containing Rm will have a resistance of at least 3, but closer
to 5, orders of magnitude greater than the RL pathway. This means that
less than 1% of the current at low frequencies will go through Ri and Rm
(see fig. 3B), which is the limit of experimental accuracy of the
measurement. The model for a RBC filled filter then becomes that shown
in figure 5.

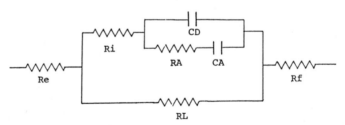

Fig. 5   Final equivalent electrical circuit representation
          of the cell/pore complex assuming a simple specific
          adsorption impedance model.

From the low frequency data, the factor x can be determined if Rf is
known. Rf can be determined by measuring the cells in the filter with
either the scanning electron microscope (SEM) or the transmission
electron microscope (TEM) after the impedance measurement to determine
the relative penetration of the cells into pores (15-17). Once x has
been determined, it is possible to calculate the factor B due to the
relative resistivity of cytoplasm with respect to electrolyte solution.
This provides a check on the assumptions made, since Ri has been
determined in several other studies (18-20) for HRBC.

    It can be seen from the above that, using the high and low
frequency limiting impedance behavior, along with a physical measurement

206

of the depth of penetration of the cell into the pore, it is possible to evaluate Re, RL and Rf (see fig. 5). The remaining impedance parameters may be evaluated by using the graphical de-embedding technique as mentioned above. To illustrate this, the total impedance, $Z(\sigma)$, for the model shown in figure 5 may be written as:

$$1/[Z(\sigma)-(Re + Rf)] = 1\{/Ri + 1/[CD\sigma + 1/Zm\ (\sigma)]\} + 1/RL \qquad (2)$$

where:

$$1/Zm(\sigma) = 1/[RA + 1/CA\sigma] \qquad (3)$$

if the cell system exhibits a single specific adsorption pathway and :

$$1/Zm(\sigma) = 1/[RA + 1/CA\sigma + 1/RB + 1/CB\sigma] \qquad (4)$$

if two specific adsorption pathways are present.

Inspection of equations 3 and 4 shows that the equivalent circuit topology for specific adsorption is a series configuration for the equivalent adsorption resistance (RA, RB) and capacitance (CA, CB). This results directly from a consideration of the kinetics of ion binding which has been treated in detail elsewhere (21,22). The analysis shows that RA and RB are proportional to the respective rate constants for adsorption, whereas, CA and CB are proportional to the initial surface concentrations of the adsorbing species (i.e. to that portion of the surface charge due to specific adsorption).

The graphical de-embedding procedure is as follows: if a sufficiently high frequency range is examined such that $CD\sigma \gg 1/Zm(\sigma)$ then equation 2 becomes

$$1/[1/Z(\sigma) - Re-Rf) - 1/RL] = Ri + 1/CD\sigma \qquad (5)$$

which shows that a plot of the left hand side (lhs) vs $1/\sigma$ is a straight line (over 1 decade of frequency at least) having an intercept Ri and slope 1/CD. The remaining parameters are contained in $Zm(\sigma)$ which can now be isolated by using the known values for Ri and CD in equation 2 and rewriting it such that the right hand side (rhs) is: $RA + 1/CA\sigma$. A second plot is then made vs $1/\sigma$ and RA and CA are evaluated as above. If a third pathway exists then the known values of Ri, CD, RA and CA are again employed in equation 2 which can then be rewritten so that the rhs is: $RB + 1/CB\sigma$. A final plot is then made vs $1/\sigma$ and RB and CB isolated.

Of relevance to the above mentioned impedance identification technique is the problem of determining system order, i.e., number of relaxation pathways present. For this purpose, we use a method based on the F-statistic. Specifically, define a number f as

$$f=(MSE1 - MSE2)/MSE2\ [(N-p2)/(p2-p1)] \sim F(N-p2,p2-p1) \qquad (6)$$

where MSE1 and MSE2 are the mean square errors (i.e., the residuals - either in the time or frequency domains) between experimental data and models with p1 and p2 parameters, respectively, and N is the number of data points. As indicated in equation 6, f is distributed as an F-statistic with N-p2 and p2-p1 degrees of freedom, and may be used to determine the statistical significance of using models of successively higher order.

Using the above approach, the results for HRBC both untreated and neuraminidase treated are shown in Table 1. The graphical de-embedding technique along with F statistics showed the presence of two time constants (Ri, CD and RA, CA) for the untreated control HRBC. However, using an identical procedure for the neuraminidase treated HRBC, a third specific adsorption-like relaxation (RB,CB) was isolated. Examination

Table 1

IMPEDANCE VALUES FOR NEURAMINIDASE PRETREATED
AND CONTROL HUMAN RED BLOOD CELLS

	CD (μF/cm^2)	RA (Ω-cm^2)	CA (μF/cm^2)	RB(Ω-cm^2)	CA(μF)cm^2)
Treated	0.65 ± .03	1.65 ± .13	0.70 ± .04	249 ± 19	0.22 ± .01
Control	0.95 ± .03	1.78 ± .09	0.62 ± .05	-	-

of Table 1 shows that the dielectric capacitance CD is substantially changed after neuraminidase treatment. This enzyme cleaves the sialic acid residues on the glycocalyx of the external membrane surface. The neuraminidase used in this study was characterized (23) and shown to have a low proteolytic activity and therefore a high specificity for sialic acid. The enzyme treatment produces an associated decrease in surface charge (24-27). This effectively lowers the charge storage capacity of the entire membrane and should therefore result in a decrease in CD as shown. The next isolatable time constant (RA,CA) appears unaffected by neuraminidase treatment (Table 1). If RA and CA represent the kinetics of ion binding as suspected (4), then it is most probable that the binding sites are located on the lipid bilayer surfaces which are not affected by the removal of sialic acid residues that reside some distance from the lipid bilayer surface. The appearance of a third pathway (RB,CB) for the neuraminidase treated HRBC is of interest since it has been observed in previous studies in this laboratory, however its resolution has not been adequate. It is possible that the HRBC is "seated" more tightly in the filter pore after cleavage of the sialic acid residues, or that the change in the HRBC surface charge increases the leak resistance, RL, by changing the structure of the thin aqueous film between the cell and the pore wall. Evidence for this is the fact that RL (see fig. 5) is substantially higher for the treated vs untreated HRBC (>50%).

Table 2

IMPEDANCE VALUES FOR CONTROL AND MSH TREATED MELANOMA CELLS

	CD (μF/cm^2)	RA (Ω-cm^2)	CA (μF/cm^2)	RB (Ω-cm^2)	CB (μF/cm^2)
Control	1.1 ± .06	0.22 ± .01	13 ± 2	3.07 ± .25	13.5 ± 2.3
Treated	0.55 ± .04	0.29 ± .02	12.5 ± .8	3.28 ± .3	15 ± 2.8

The results for the melanoma cells are shown in Table 2. Here, it can be seen that the graphical de-embedding technique has isolated three time constants. The MSH treated cells again exhibit a decrease in CD, with all other membrane impedance parameters remaining essentially unchanged vs. control cells. The effect of MSH is to further differentiate the melanoma cells. This is normally associated with a decrease in the membrane surface charge (28). In this case, however, it is not known which membrane entities are involved. Of interest is the observation that, if the remaining time constants are associated with ion binding, these also are unaffected by the change in membrane components associated with differentiation.

Acknowledgements

The authors wish to express their sincerest appreciation to Professor Kung-Ming Jan of the Department of Physiology of Columbia University, College of Physicians and Surgeons for his kind assistance in the neuraminidase experiments. We would also like to thank Gary Johnson of Columbia University, College of Physicians and Surgeons for his assistance in the construction of the cell impedance chamber.

Part of this work was submitted by P. Maccaro in partial fulfillment of the B.S. degree at Polytechnic Institute of N.Y. under the direction of Professor Shirley Motzkin.

REFERENCES

1. Schmukler, R., E.Sc.D. Thesis Columbia University, N.Y. 1981.
2. Schmukler, R. and Pilla, A.A., J. Electrochem. Soc., 129: pp. 526-528, 1982.
3. Schmukler, R., Pilla, A.A., Cerf, G., and Lee, M., Proceeding 10th Northeast Bioengineering Conference, ed. E.W. Hansen, pp. 213-216, 1982.
4. Pilla, A.A., Schmukler, R.E., Kaufman, J.J. and Rein, G. in Interactions Between Electromagnetic Fields and Cells, Chiabrera, A., Nicolini,C and Schwan, H., eds. Plenum Press, N.Y. pp. 423-435, 1985.
5. Pilla, A.A., Kaufman, J.J. and Schmukler, R.E., Proc. 11th N.E. Bioengineering Conf., March, 1985.
6. Schwan, H.P., Physical Techniques in Biological Research, vol. 6, ed. W.L. Nastuk, Academic, N.Y. 1963.
7. Rall, W., Biophysical Journal, 9: pp. 1509-1541, 1969.
8. Eisenberg, R.S. and Johnson, E.A., Prog. Biophys. Mol. Biol., 20: pp. 1-65, 1970.

11. Hadley, M.C., Anderson, B., Heward, C.B., Science, 213:, pp. 1025-1027, 1981.
12. Pilla, A.A., in Information Chemistry, eds., Fujiwara, S. and Mark, H.B. Jr., pp. 181-193, University of Tokyo Press, Tokyo, 1975.
13. Pilla, A.A., in Calculations, Stimulation and Intrumentation, ed. Mattson, J., Mark, H., Jr. and McDonald, H., Jr. Ch.6, Marcel Dekker, N.Y., 1972.
14. Lassen, V.V., in Membrane Transport in Red Cells, eds., Ellory, J. and Lew.V., Academic Press, N.Y., 1977, pp. 137-174.
15. Brailsford, J.D., et al., Blood Cells, 3: pp.25-38, 1977.
16. Bull, B.S., et al., Blood Cells, 3: pp..39-54, 1977.
17. Brailsford, J.D., Personal Communications, .
18. Cole, K.S., Membranes, Ion and Impulses, pt 1, Univ. of California Press, Berkley, L.A., 1968.
19. Pauly, H., Nature, 183: pp. 333-334, 1959.
20. Pauly, H. and Schwan, H.P., Biophysical Journal, 6: pp. 621-639, 1966.
21. Pilla, A.A., J. Bioelectrochem. and Bioenergetic, 1: pp. 227-243, 1974.
22. Pilla, A.A., Ann. N.Y. Acad. Sci., 238:pp. 149-170, 1974.
23. Jan, K.M., Personal Communication; see Chapter this volume.
24. Svennerholm, L., Acta Soc. Med. Ups, 61: pp. 75, 1956.
25. Langer, G.A., Frank, S.S., Nudd, L.M., and Seraydarian, K., Science, 193, pp. 1013-1015, 1976.
26. Jan, K.M., Chien, S., J. Gen. Physiol., 61, pp.638-654, 1973.
27. Butterfield, D.A., Farmer, B.T.II, Feix, J.B., Ann. N.Y.A.S., 414: pp. 169-179, 1983.
28. Danon, D., Marikovsky, Y., and Fischler, H., Ann. N.Y.A.S., 416: pp. 149-158, 1983.

MEASUREMENT OF FREQUENCY-DEPENDENT IMPEDANCE

ACROSS NATURAL CELL MEMBRANES

Stephen M. Ross

Medical Research Council Group in Periodontal Physiology
Faculty of Dentistry, University of Toronto
Toronto, Ontario, Canada M5S 1A8

INTRODUCTION

A property of cell membranes which is intimately bound up with electrical double layers is that of the frequency-dependent membrane impedance. The
aim of this paper is to outline the basics of the Fourier transform method of
measuring impedance across a spectrum of frequencies simultaneously, and to
describe practical experimental techniques which the author has used in obtaining these measurements across the intact membranes of two very different
kinds of cell.

It has been common to model the electrical properties of the cell membrane with equivalent circuits of a resistor and capacitor connected in parallel and to express the membrane impedance in terms of the "conductance" and
"capacitance" of these hypothetical circuit elements. This makes intuitive
sense, as illustrated in Fig. 1, if we consider the membrane as a charge-
storing plane penetrated by elements, ion-channel and ion-carrier molecules,
which are purely conductive. If we consider the structure of a membrane at
the grossest level to be a thin layer of lipid approximately 10 nm thick,
then using the standard relationship for the capacitance, C, of a thin sheet
area:

$$\frac{C}{a} = \frac{\epsilon}{d} \tag{1}$$

where a = area, ϵ = dielectric constant $\simeq 10^{-10}$ F/m for lipids and d = thickness; we obtain a specific capacitance of 10 mF/m^2. This is the value which
was typically measured in early experiments in which a square pulse of current was applied to a cell membrane and the average membrane capacitance was
calculated from the resulting exponential rise in transmembrane potential
difference (PD); see, for example, Williams et al. (1964).

When an engineer designs a capacitor for use in a circuit he designs it
to approximate as closely as possible an "ideal" circuit element which has
the same capacitance value at all frequencies. Modern man-made capacitors
approximate these properties very well, but biological membranes are far from
ideal. Some workers have measured the complex membrane impedance of membranes at discrete frequencies using injected sinusoidal signals and measuring
the amplitude and phase shift of the response (Cole, 1968; Coster and Smith,
1974, 1977; Coster et al., 1980) or by using variations of the Fourier transform technique to be described in this article (Poussart et al., 1977; Wills
and Clausen, 1982), and it turns out that in living material the measured

membrane capacitance varies greatly with frequency. Coster et al. (1980) measured the frequency-dependent capacitance of artificial lipid bilayers in the absence of intrinsic ion conducting proteins. They found that the capacitance of these bilayers was high at low frequencies and declined toward higher frequencies, while conductance increased toward higher frequencies. The capacitance could be adequately described by means of Maxwell-Wagner dispersion, in which each structural sub-layer of the lipid bilayer contributes a frequency-dependent component with a characteristic time constant.

When ion transport proteins are added to a bilayer the situation becomes more complex. Voltage-dependent channel proteins can result in time-varying resistances, which can produce pseudoinductive or pseudocapacitative effects in which current lags or leads an applied voltage (Mauro, 1961; DeFelice 1981; Ferrier et al., 1985). The term pseudoinductance is used in this paper because it is not possible for biological membranes to significantly exhibit true inductance, in which current lags voltage because of a self-induced electromotive force; current lags voltage in the case of channels which tend to open in response to an applied voltage signal. Similarly the term pseudocapacitance is used in this case because an effect in which current leads voltage is produced by channels which tend to close in response to an applied voltage signal, and not by the storage of charge. Measurement of membrane impedance provides information which can be of practical use to electrophysiologists; for example, the membrane impedance may be used to convert from voltage noise spectra to current noise spectra and vice versa in noise analysis (see Ross, 1982), but as we shall see, the measurements are interesting in their own right, and may provide some information about processes ocurring in the membrane, or at the membrane surface.

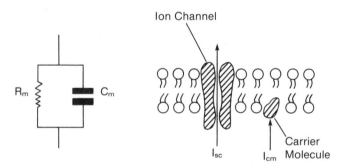

Fig. 1. An illustration of the equivalent circuit often used to represent a biological membrane. This representation makes sense when we consider the physical arrangement of a typical biological membrane. The lipid bilayer acts as a charge-storing plane which is penetrated by purely conductive elements, ion channel proteins and carrier molecules. Lipid bilayers without ion channels or carriers have a typical resistivity of $10^4 \, \Omega\text{-m}^2$ (Tien, 1974). Most of the ionic current leaks through the channels (I_{sc}) and carriers (I_{cm}), and the plasmalemma of a typical living cell would have a membrane resistivity of $0.1 \, \Omega\text{-m}^2$, although wide variations are possible from cell type to cell type depending on the kinetics and permeability of transporting molecules.

THEORY OF IMPEDANCE MEASUREMENT

It is possible to measure the frequency-dependent impedance of a bilayer membrane accross a spectrum of frequencies simultaneously using the techniques of Fourier analysis (Poussart et al. 1977; Ross, 1982). The discrete, finite Fourier transform, $F_x(f)$, at frequency, f, is calculated from a set of n sampled data values, x_k, k= 0, . . ., n-1, as follows:

$$F_x(f) = \Delta t \sum_{k=0}^{n-1} x_k \exp(-j2\pi fk) = \Delta t \sum_{k=0}^{n-1} x_k[\cos(2\pi fk) - j\sin(2\pi fk)] \qquad (2)$$

where Δt = period between samples and $j = \sqrt{-1}$.

The function of the Fourier transform is to break a signal down into the contributions made by each frequency, and the signal can be reconstructed from these Fourier components by performing the inverse Fourier transform operation (see Bendat and Piersol, 1971; or Otnes and Enochson, 1978). The Fourier transform used is neccessarily finite, and is discrete because the analysis is performed digitally on a computer with data sampled by means of an analog-to-digital (A/D) converter (it is possible to perform continuous Fourier transformation in analog fashion using, for example, optical systems, see Lipson, 1972; or Duffieux, 1983). When using equally-spaced, discrete sampled data points the frequency values at which $F_x(f)$ is evaluated are determined by the sampling frequency, $f_s = 1/\Delta t$, and the number of sampled data points. It is apparent that at least two sampled data points per cycle are required to determine the amplitude and phase of a sinusoidal Fourier component, and so the maximum frequency value at which F can be evaluated is $f_n = f_s/2$, where f_n is called the "Nyquist frequency" in honour of H. Nyquist, who made many fundamental contributions to the theory of noise and communications. $F_x(f)$ is evaluated for n/2+1 points, at f = 0, $2f_n/n$, $4f_n/n$, $6f_n/n$, . . ., f_n, in the case of an even number of sampled data points.

The expression in eq. (2) is quite cumbersome to compute, since the number of arithmetic operations required to calculate the complete transform is proportional to n^2. It turns out, however, that many of the arithmetic operations are redundant and that an improved algorithm, the Fast Fourier Transform (FFT), can perform the transformation much more rapidly. If r_k, k = 1, . . ., m is the set of prime factors of n, then the number of operations required by the FFT, p, is:

$$p = n \sum_{k=1}^{m} r_k \qquad (3)$$

Eq. (3) provides the most effective saving of arithmetic steps if n is a power of 2 (i.e. $n = 2^m$, in which case p = 2mn), and so nowadays most digital Fourier analysis is performed on data sets of 256, 512, 1024 points, and so on. A very concise explanation of how the FFT works can be found in Theilheimer (1969).

The power spectral density (PSD), $S_{xx}(f)$, is calculated at each frequency from the magnitude of the complex Fourier transform:

$$S_{xx}(f) = \frac{2}{n\Delta t}\left|F_x(f)\right|^2 \qquad (4)$$

this quantity will be useful later. Since the object of Fourier analysis is to simultaneously measure a spectrum of values of a frequency-dependent quantity, a signal must be sampled which has components at all frequencies of interest. To measure the impedance, Z(f), over a range of frequencies, we must feed in a voltage or current signal having measurable power at each frequency of interest, measure the current or voltage response, and then determine Z(f) from the Fourier transforms of the voltage, $F_v(f)$, and

213

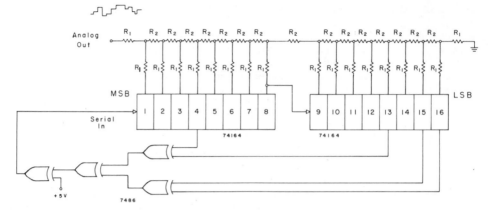

Fig. 2. Schematic of the circuit used for generation of pseudorandom white
noise. Two 8-bit 74164 TTL shift registers are connected in series
to create a 16-bit binary number which is shifted to the right at a
rate from 1 Hz to 100 kHz under the control of a clock circuit which
uses the 555 chip. Bits 4, 13, 15 and 16 are fed back through the
gating logic to provide the most significant bit input, which will
result in a cycling through of every permutation of bits except the
all 1's combination, which would latch up were it to occur. The
circuit is easily initialized to start at a standard state by using
the clear pins of the shift registers to set all 0's. The binary
number contained in the shift registers is digital-to-analog converted by means of the resistor ladder illustrated at the top of the
figure.

current, $F_i(f)$, by the relation:

$$Z(f) = F_v(f)/F_i(f) \tag{5}$$

An ideal signal for most situations is the "white noise" signal, which
is a random signal having equal power spectral density at all frequencies.
White noise may be generated by amplifying the voltage noise produced by
thermal motion of charge carriers across a large resistance, or another
popular technique is to amplify the "shot noise" produced by the passage of
electrons in a vacuum-tube diode; these signals are truly random. It is also
possible to generate white noise signals digitally by shifting binary bits to
the right in a shift register and using feedback of the values from certain
bits to determine the input to the register, as illustrated in Fig. 2 (see
also Ross, 1982). If the right feedback bits are chosen the binary number
contained in the shift register will pass through every possible permutation
of bits. The sequence of numbers generated must then be digital-to-analog
converted and amplified for use as a stimulating signal. This signal is
called pseudorandom because although it has a white PSD and a uniform probability density when values are sampled from it at a rate less than the
shifting frequency, it is really deterministic, and will repeat the same
pattern every 2^m-1 shifts, where m is the number of bits in the shift register.

The relationship of eq. (5) is only useful practically when a deterministic signal is used. One can take advantage of the pseudorandom nature of
digitally generated signals, however, if one wishes to use eq. (5) to calculate impedance. If one wishes to sample 2^m data points one simply generates
a white noise signal from a shift register having m bits and matches the
sampling of current and voltage signals to the shift rate; this was the

214

approach taken by Poussart et al. (1977). If one finds the impedance spectrum a bit noisy one can introduce averaging by resetting the shift register to the same initial value and accumulating successive identical runs in the sampling memory. There is the problem of needing to sample 2^m points from a signal which is 2^m-1 points long, but there are various ways to handle it. Perhaps the easiest way is to make the last point of each data record zero, since there is nothing wrong with adding zeroes to a data record to make the number of points up to a power of two before Fourier transforming (see Bendat and Piersol, 1971; or Otnes and Enochson, 1978), this results in values which are interpolations at the calculated frequencies.

If one wishes to use white noise signals which are truly random, or to sample impedance spectra over several different frequency ranges from the same pseudorandom signal, then a different method of computing the impedance which introduces spectral averaging must be used. To perform this calculation we must first compute the cross spectrum, $S_{iv}(f)$:

$$S_{iv}(f) = \frac{2}{n\Delta t}F_i^*(f)F_v(f) \tag{6}$$

where $F_i^*(f)$ is the complex conjugate of $F_i(f)$.

Now to improve the consistency of the impedance estimate we may average several estimates of $S_{iv}(f)$ and of $S_{ii}(f)$ to obtain the new averaged spectra, $\overline{S}_{iv}(f)$ and $\overline{S}_{ii}(f)$, and calculate the impedance by the relation:

$$Z(f) = \overline{S}_{iv}(f)/\overline{S}_{ii}(f) \tag{7}$$

Another advantage of this method is that the coherence function, $H(f)$, can be calculated:

$$H(f) = \left|\overline{S}_{iv}(f)\right|^2/\overline{S}_{ii}(f)\overline{S}_{vv}(f) \tag{8}$$

In the case of data which are known to be dependent on one another, and where there is low extraneous noise in the output signal, this function, which must have a value between 0 and 1, provides a measure of the linearity of the impedance relation between i and v, with a value of 1 indicating linear impedance, and values less than 1 indicating nonlinearity. It is a kind of frequency-domain equivalent of the correlation coefficient, and in the case where one is not sure that there is no extraneous noise included in the output signal, a value of less than one may indicate a the presence of noise not resulting from the input signal (see Otnes and Enochson, 1978).

Impedance can be plotted in various ways. Magnitude, $\left|Z(f)\right|^2$, and phase shift, $\phi(f) = \tan^{-1}[Re(Z(f))/Im(Z(f))]$, can be plotted against frequency, or alternatively the conductance, $G(f)$, and capacitance, $C(f)$, of the equivalent circuit shown in Fig. 1 can be calculated from the real and imaginary parts of the impedance according to the equation:

$$1/Z(f) = G(f) + j2\pi fC(f) \tag{9}$$

and plotted against frequency. Another type of plot which has been popular is the so-called Cole-Cole plot in which a locus of points is constructed by plotting the imaginary part of the measured impedance value against the real part at each frequency of interest (Cole, 1968), see Fischbarg and Lim (1973) for an example of an application of this type of plot.

Some cautions must be observed when the finite, discrete Fourier transform is used for frequency analysis. One problem, leakage, is a smearing of the power at a given frequency over a band of adjacent frequencies. Leakage problems may be reduced by the application of a "window function" to the data, an excellent discussion of the merits of various types of window functions may be found in Otnes and Enochson (1978). Another problem which may arise is that of aliasing. Frequency components above the Nyquist frequency will still be sampled during A/D conversion and will contribute components to the spectrum at frequencies below f_n. Each frequency, f, which is estimated in the power spectrum, $S_{xx}(f)$ will include power from all frequencies, f_a, given by the following relation:

$$f_a = 2kf_n \pm f, \qquad k = 1, \ldots, \infty \tag{10}$$

(see Bendat and Piersol, 1971). To solve the problem of aliasing we must use a low pass frequency filter to remove all frequencies above f_n before

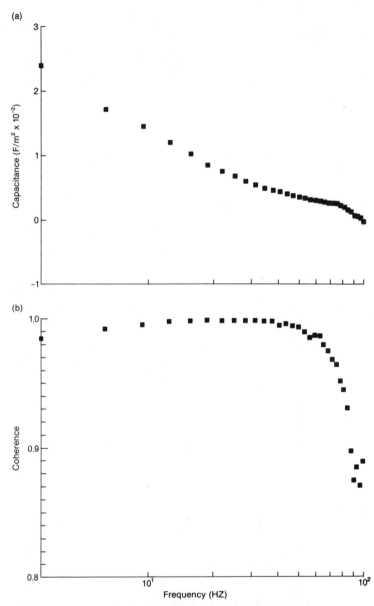

Fig. 3. An illustration of the effect of aliasing on the impedance estimate. The negative capacitance values at the upper end of the spectrum in (a) are spurious, due to aliasing of frequency components just beyond the Nyquist frequency which were passed by the antialiasing filter. The loss of coherence in (b) near the Nyquist frequency indicates the presence of an aliasing problem as well.

discrete sampling is performed by A/D conversion. It turns out that aliasing can be a severe problem in Fourier impedance analysis, even when low-pass frequency filtering is employed with the cutoff frequency set equal to f_n before sampling data. The low-pass cutoff frequency, f_c, is commonly defined as the half-power point in electronic frequency filters, and there is a narrow band beyond $f_c = f_n$ where a significant amount of signal is passed by the filter. This can cause severe problems in the estimate as illustrated in Fig. 3. If it is feasible, the best solution is to simply discard the points which are affected by aliasing. In the author's analyses a spectrum twice as long as required is calculated, then the upper half of the impedance spectrum is discarded so that the upper frequency value plotted is $f_s/4$.

EXPERIMENTAL TECHNIQUES

The experimental apparatus for measuring current and voltage signals across the intact cell membrane for impedance analysis must either allow simultaneous measurement of transmembrane current and PD, or provide for current clamping or voltage clamping to follow a white noise command signal. When clamping is performed only the voltage or current response needs to be recorded from the cell simultaneously with the command signal. In either case, one must ensure that the transmembrane currents and PD's are uniform over the whole cell surface.

It is neccessary to consider the geometry and approximate resistance

Fig. 4. Morphology of the green alga <u>Chara</u> <u>corallina</u>. The stalk is composed of long, cylindrical, single cells called internodes. The lower internode in this photograph is approximately 60 mm long. Whorls of smaller cells called lateral internodes radiate from the stalk at each node, or joint between stalk internodes.

of the cell membrane when designing experiments. For example, inserting a glass micropipette to inject current at a point along a long, cylindrical cell such as a giant axon or giant algal cell of the type to be described shortly would result in non-uniform spatial current distribution, with most of the current leaking out of the cell in a short band centred about the pipette. Injection of current by micropipette would provide uniform current density across the membrane of a spherical cell, but microelectrodes for impalement of small cells must be very fine-tipped, and consequently have high resistances. Accurate voltage clamping through a single high resistance pipette is difficult even at the low frequencies required for membrane impedance analysis, and a second impalement for measuring transmembrane PD would be required. A large cell might survive such double impalement, but it is difficult for small cells to remain intact after such an operation.

Chara corallina is a multicellular freshwater green alga which has the morphology illustrated in Fig. 4. The stalk of the plant is composed of giant cylindrical cells which can reach 100 mm long and 2 mm in diameter in

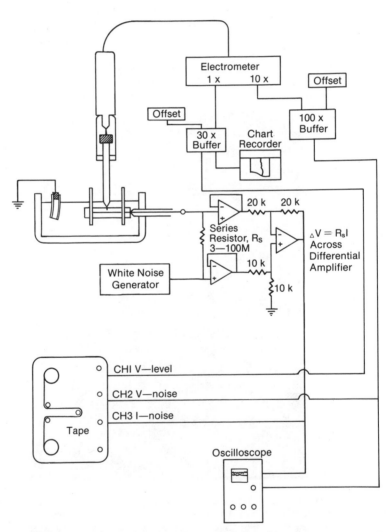

Fig. 5. Schematic diagram of apparatus used for impedance measurements in internodes of Chara corallina.

plants growing in the wild. These plants can easily be cultured in the laboratory as described in Ross et al. (1985). Young stalk internodes approximately 10 mm in length were dissected away from adjacent internodes and whorls and placed in a pair of stocks in a chamber which was illuminated by means of a halogen lamp at an intensity of 30 W/m^2 (Fig. 5). These cells have a large, central vacuole, and an axial electrode made from 75% Pt/ 25% Ir wire was manipulated down the length of the cell through a previously inserted glass micropipette. The axial electrode did not cause any major disturbance to the cell, since the central vacuole is filled with a dilute, uniform sap, and it ensured that injected current was evenly distributed over the cell surface. The cells had a total membrane resistance of approximately 100 kΩ, and so an approximate current clamp could be achieved by injecting the current through a large-value series resistor (> 3 MΩ). Measurement of the voltage drop across this resistor provided a measure of the current signal being fed into the cell, which could be recorded simultaneously with the voltage response on an instrumentation tape recorder for later analysis. Transmembrane PD was measured by means of a micropipette inserted vertically into the vacuole of the cell, so that this experimental arrangement measured the impedance of the tonoplast in series with the plasmalemma of these cells. The layer of cytoplasm around the periphery of the cell is very thin, and the cytoplasm produces a material which occludes the tip of the pipette, therefore cytoplasmic insertions, which would be required for measurement of plasmalemma impedance alone, are difficult to perform and are short-lived.

The author's current research program involves studies of the electrophysiology of bone-derived cell clones grown in long-term culture. These cells typically have a diameter of 20 - 50 μm with an irregular, flattened morphology (Fig. 6), and an entirely different experimental approach must be taken in order to obtain accurate estimates of membrane impedance. The particular cell used to date in experiments by the author, a clone of rat osteosarcoma (ROS) cells which have been used as a model for the osteoblast (Grigoriadis et al., 1985), have been successfully impaled (Dixon et al., 1984), but double impalements are not reliable, and so the author turned to the "whole cell recording" patch clamp technique (Marty and Neher, 1983) which is illustrated in Fig. 7. A large-tipped, low impedance (1 MΩ) micropipette is pressed against the membrane of the cell (Fig. 7 a), and suction is applied (Fig. 7 b). Upon application of suction a tight seal of extremely high resistance (approximately 2.5 GΩ) forms between the membrane and the

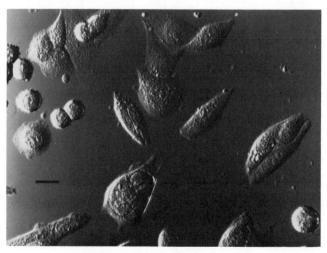

Fig. 6. Nomarski differential interference contrast photomicrograph of cloned rat osteosarcoma cells (ROS 17/2). The scale bar indicates 20 μm.

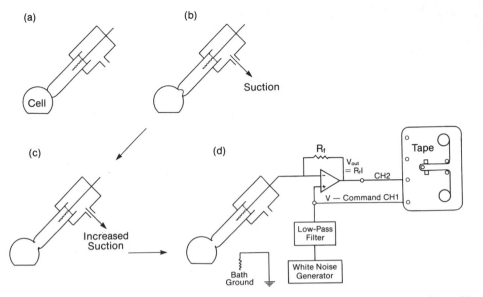

Fig. 7. Schematic diagram of the procedure used to voltage clamp small cells by means of the whole cell recording method.

glass of the pipette (Hamill et al., 1981). Suction is increased until rupture of the membrane "patch" under the pipette occurs (Fig. 7 c). This rupture is signaled by a drop in resistance at the pipette tip to 20 - 30 MΩ the typical whole cell membrane resistance of these cells. There is now a low-resistance connection to the cell interior through which the cell may be voltage clamped by means of the circuit diagrammed schematically in Fig. 7 d. Any deviations of transmembrane PD, which is detected at the inverting input of the operational amplifier, from the command voltage will result in the feeding back of a compensating current to restore the internal PD of the cell to the command voltage. This diagram represents only the basic princip-le of the voltage-clamp circuit used for patch clamping; the electronic sys-tem actually used by the author is the model 8900 patch clamp manufactured by Dagan Instruments Corp. of Minneapolis, which has a frequency-compensation circuit to expand the bandwidth of the instrument. For details of the prac-tical considerations involved in the design of electronics for patch clamping, see Sigworth (1983). The patch electrodes used typically have a tip diameter of 1 μm and so the contents of such small cells rapidly equilibrate with the pipette filling solution by diffusion. Pipettes were therefore filled with the solution described by Marty and Neher (1983) which uses EGTA to buffer calcium to a low level.

Once the simultaneous current and voltage signals have been captured on tape they can be sampled on a minicomputer system. Low-pass frequency fil-ters are connected between each channel of the tape recorder and the A/D con-verter to avoid the aliasing problem described above. An advantage of tape recording is that the same portion of an experiment can be analyzed over sev-eral different frequency ranges. Details of the computer program used to perform the analysis described in the previous section of this paper are provided in Ross (1982).

RESULTS

Fig. 8 presents the frequency-dependent impedance of the plasmalemma an tonoplast in series in <u>Chara corallina</u>. Of particular interest was the large negative capacitance, or pseudoinductance, which occurred at low frequencies; note the difference in vertical scale between Fig. 8 d and Fig. 8 e. The frequency at which the capacitance becomes negative is simply the frequency at which the phase shift crosses the zero axis in Fig 8 b.

Preliminary results from the measurements of impedance in the ROS cell

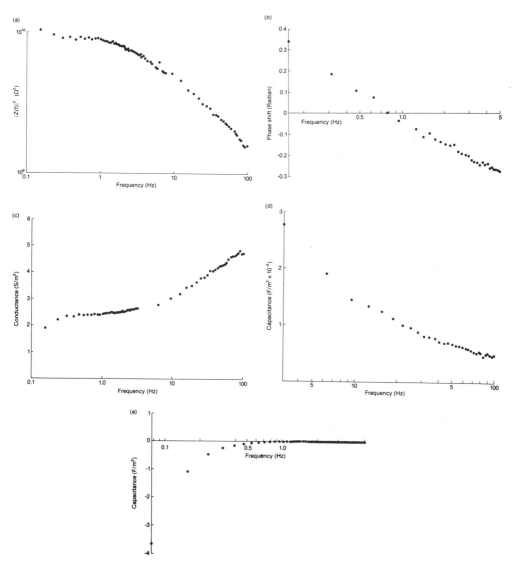

Fig. 8. (a) Impedance magnitude measured in an internode of <u>Chara corallina</u> in the light (intensity 30 W/m²) in artificial pond water buffered to pH 7 with 5 mM MOPS. The artificial pond water medium used was that described by Hope and Walker (1975). (b) Phase shift, (c) conductance, (d) capacitance over the frequency range 3.125 - 100 Hz and (e) capacitance over the low frequency range 0.0781 - 5 Hz, notice the difference in vertical scale between (d) and (e).

type using the whole cell recording technique are illustrated in Fig. 9. The results are not corrected for the pipette impedance, and are not divided by cell area to obtain area-specific measurements. Conductance was fairly constant from 31.25 Hz to 125 Hz, then began the rise typical of biological membranes at higher frequencies. Membrane capacitance showed the typical decline toward higher frequencies.

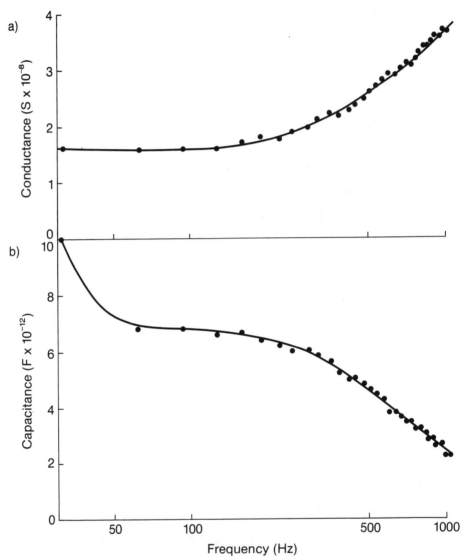

Fig. 9 (a) Conductance and (b) capacitance of an ROS 17/2 cell sampled by means of the whole cell recording technique illustrated in Fig. 7. Cells were in a 60 mm tissue culture dish in α-minimal essential medium buffered to pH 7.4 with 25 mM HEPES and supplemented with 15% foetal bovine serum. The cells were maintained at 37° C during the course of the experiment by means of a heated microscope stage.

DISCUSSION

The Fourier transform method of estimating frequency-dependent membrane impedance offers advantages of speed and flexibililty. The more traditional analysis of sinusoidal signals is slower, and is limited to the particular set of signal frequencies applied during the experiment. With Fourier analysis, however, a spectrum of membrane impedance values can be estimated simultaneously and any number of different frequency ranges can be analyzed from the same signals, limited only by the bandwidth of the signals and the equipment used for recording them.

Membrane conductance seems to follow a fairly typical pattern in both cell types studied by the author, changing relatively little at low frequencies, then commencing to rise at higher frequencies. The point at which the rise began was different in the two cell types, commencing at 5 Hz in Chara and at 125 Hz in the ROS cell, but otherwise the shape of the conductance versus frequency curves is similar.

The capacitance at higher frequencies of both the Chara internode and the ROS cell likewise seems to follow a similar pattern, having a high value at lower frequencies, declining toward lower frequencies. It seems that a pattern of capacitance dispersion in which membrane capacitance declines from a high at frequencies below 100 Hz to lower values at high frequency is a general feature of all biological membranes which results from the structural sublayers of the lipid bilayer itself (Coster et al., 1980). The measurements of impedance in C. corallina were performed on the plasmalemma and tonoplast in series, but the impedance of the vacuolar membrane is very small relative to that of the plasmalemma, and impedance results from the plasmalemma and tonoplast in series are not much different from those from the plasmalemma alone (Coster and Smith, 1977). The measurements of impedance in the ROS cells using the whole cell recording technique have not been corrected for the impedance of the pipette. The most significant source of resistance in the pipette-headstage combination is the pipette tip, which, as was mentioned previously results in approximately a 1 MΩ resistance. Capacitance results from charge storage at the surface of the glass of the pipette and from charge storage in the various surfaces of the headstage and pipette holder. The headstage has a driven shield to minimize any capacitance problems in the headstage and pipette holder, and so we can see that the cell-pipette combination can be considered to consist of the pipette tip resistance in series with the cell membrane resistance, and the pipette capacitance in parallel with the cell capacitance. It is intended to modify the computer analysis program so that when the impedance of the pipette is measured before attachment to the cell under study, compensation can be made to yield the corrected cell impedance. An improvement which the author plans to use as a matter of course for future runs is coating of the pipette with Sylgard® resin, which is hydrophobic and increases the effective thickness of the glass, resulting in greatly reduced pipette capacitance (Corey and Stevens, 1983). Estimation of membrane area in ROS cells can also present problems. The cells often have blebs or folds on their surfaces, and usually have an irregular shape which does not allow for calculation of area by means of a geometric formula, as is the case for cylindrical Chara internodes. It seems that the best approach to estimating the area of these cells will involve measuring surface area of the dish covered by the spread cells from photographs such as the one in Fig. 6 by means of a digitizing tablet connected to the computer, and relating the surface area to average proportional area of cells as determined from electron micrographs.

The large negative capacitance illustrated in Fig. 8 e was also found in the experiments performed by Coster and Smith (1977) on small lateral internodes of Chara corallina. They found, however, that many cells had a positive capacitance of equally large magnitude at low frequencies, and a few cells were even observed to change from one state to the other during the course of an experiment. In the author's experiments all cells studied (9 cells) displayed the large pseudoinductance at low frequencies without exception. Pseudoinductance has been found in numerous excitable cell types (Cole, 1968; DeFelice, 1981), but it occurs at higher frequencies. In the case of the squid giant axon this pseudoinductance can be attributed to the time-varying resistances of the voltage-dependent Na^+ and K^+ channel systems (see Mauro et al., 1970). It seems, however, that voltage-dependent channel kinetics cannot explain the extremely low-frequency pseudoinductance observed in internodes of Chara corallina (Ferrier et al., 1985).

There are physical transport effects which can produce a pseudoinductive effect in biological membranes. Cole pioneered in this area of theory with his studies of electrodiffusion models (see Cole, 1968). Electroosmosis can produce a pseudoinductance, but the effect is relatively too small to explain the pseudoinductance observed in these cells. There is another situation in which pseudoinductance can arise. This occurs in the case of transport of a single ion species which has a much higher diffusion coefficient than the average for the majority of ions free to diffuse in the surrounding medium. When a steady state ion current crosses a membrane, current is transported away from the membrane in the unstirred layer which surrounds the cell by means of concentration gradient-driven diffusion (Ferrier and Lucas, 1979; Ferrier, 1981, 1983). When the transported ion has a substantially greater diffusion coefficient than that typical of other ions in the extracellular fluid, as is the case for protons and hydroxyl ions, the diffusion-driven component of a sinusoidal current will lag the total current. This means that the component of current driven by the electrical potential gradient will lead the total current, so that the applied voltage will lead the total current, producing a pseudoinductive effect.

The reader will notice upon glancing back at Fig. 4 that there are whitish bands at regularly spaced intervals along the internodes in the photograph. These result from deposition of calcium carbonate on areas of the cell surface where a proton influx results in a local rise in pH. A hydroxyl efflux might be occurring, but at the moment it is not possible to distinguish between the two experimentally, and so for convenience a proton flux will be referred to. The clear bands on the cell surface correspond to areas of localized lowering of pH, so that there are circulating proton currents between the "acid" and the "alkaline" bands along the cell (Walker and Smith, 1977). There is good evidence that a proton or hydroxyl transport system is the predominant factor in determining membrane conductance in this type of giant algal cell (Spanswick, 1972; Saito and Senda, 1974; Lucas and Ferrier, 1980). It turns out that the extracellular negative capacitance effect can quantitatively explain the low frequency pseudoinductance observed in Chara internodes if one assumes that most of the current driven across the cell membrane during impedance measurements consists of a flux of proton or hydroxyl ions (Ferrier et al., 1985).

REFERENCES

Bendat, J. S., and Piersol, A. G., 1971, "Random Data: Analysis and Measurement Procedures," Wiley-Interscience, New York.
Cole, K. S., 1968, "Membranes, Ions and Impulses," University of California Press, Berkeley.
Corey, D. P. and Stevens, C. F., 1983, Science and technology of patch-recording electrodes, in: "Single-Channel Recording," B. Sakmann and E. Neher, eds., Plenum, New York.
Coster, H. G. L., Laver, D. R., and Smith, J. R., 1980, On a molecular basis of anaesthesia, in: "Bioelectrochemistry," H. Keyzer and F. Gutman, eds., Plenum Press, New York.
Coster, H. G. L., and Smith, J. R., 1974, The effect of pH on the low frequency capacitance of the membranes in Chara corallina, in: "Membrane Transport in Plants," U. Zimmerman and J. Dainty, eds., Springer-Verlag Heidelberg.

Coster, H. G. L., and Smith, J. R., 1977, Low-frequency impedance of Chara corallina: simultaneous measurements of the separate plasmalemma and tonoplast capacitance and conductance, Aust. J. Plant Physiol., 4:667.

DeFelice, L. J., 1981, "Introduction to Membrane Noise," Plenum Press, New York.

Dixon, S. J., Aubin, J. E., and Dainty, J., 1984, Electrophysiology of a clonal osteoblast-like cell line: evidence for the existence of a Ca^{2+}-Activated K^+ conductance, J. Membrane Biol. 80:49.

Duffieux, P. M., 1983, "The Fourier Transform and its Applications to Optics," John Wiley & Sons, Inc., New York.

Ferrier, J. M., 1981, Time-dependent extracellular ion transport, J. Theor. Biol. 92:363.

Ferrier, J. M., 1983, A mechanism for the regulation of ligament width based on the resonance frequency of ion concentration waves, J. Theor. Biol. 102:477.

Ferrier, J. M., Dainty, J., and Ross, S. M., 1985, Theory of negative capacitance in membrane impedance measurements, J. Membrane Biol., in press.

Ferrier, J. M. and Lucas, W. J., 1979, Plasmalemma transport of OH^- in Chara corallina. II. Further analysis of the transport system associated with OH^- efflux, J. Exp. Bot. 30:705.

Fischbarg, J. and Lim, J. J., 1973, Determination of the impedance locus of rabbit corneal endothelium, Biophys. J. 13:595.

Grigoriadis, A. E., Petkovich, P. M., Ber, R., Aubin, J. E. and Heersche, J. N. M., 1985, Subclone heterogeneity in a clonally-derived osteoblast-like cell line, Bone 6:193.

Hamill, O. P., Marty, A., Neher, E., Sakmann, B., and Sigworth, F. J., 1981, Improved patch-clamp techniques for high-resolution current recording from cells and cell-free membrane patches, Pflügers Arch. 391:85.

Hope, A. B. and Walker, N. A., 1975, "The Physiology of Giant Algal Cells," Cambridge University Press, London.

Lipson, H., 1972, "Optical Transforms," Academic Press, London.

Lucas, W. J. and Ferrier, J. M., 1980, Plasmalemma transport of OH^- in Chara corallina. III. Further studies on transport substrate and directionality, Plant Physiol. 66:46.

Marty, A., and Neher, E., 1983, Tight-seal whole cell recording, in: "Single-Channel Recording," B. Sakmann and E. Neher, eds., Plenum, New York.

Mauro, A., 1961, Anomalous impedance, a phenomenological property of time-variant resistance, Biophys. J. 1:353.

Mauro, A., Conti, F., Dodge, F. and Schor, R., 1970, Subthreshold behavior and phenomenological impedance in the squid axon, J. Gen. Physiol. 55: 497.

Otnes, R. K. and Enochson, L., 1978, "Applied Time Series Analysis. Volume I: Basic Techniques," Wiley-Interscience, New York.

Poussart, D., Moore, L. E., and Fishman, H. M., 1977, Ion movements and kinetics in squid axon I. complex admittance, Ann. N.Y. Acad Sci. 303: 355.

Ross, S. M., 1982, NOISE: an interactive program for time series analysis of physiological data, Comp. Prog. Biomedicine 15:217.

Ross, S. M., Ferrier, J. M. and Dainty, J., 1985, Frequency-depepndent impedance in Chara corallina estimated by Fourier analysis, J. Membrane Biol., in press.

Saito, K. and Senda, M., 1974, The electrogenic pump revealed by the external pH effect on the membrane potential of Nitella. Influence of external ions and electric current on the pH effect, Plant Cell Physiol. 15:1007.

Sigworth, F. J., 1983, Electronic design of the patch clamp, in: "Single-Channel Recording," B. Sakmann and E. Neher, eds., Plenum, New York.

Spanswick, R. M., 1972, Evidence for an electrogenic pump in Nitella translucens. I. The effects of pH, K^+, Na^+, light and temperature on the membrane potential and resistance, Biochim. Biophys. Acta 288:73.

Theilheimer, F., 1969, A matrix version of the fast Fourier transform, <u>IEEE</u> <u>Trans</u>. <u>Aud</u>. <u>Electroacoust</u>. AU-17(2):158.

Tien, H. T., 1974, "Bilayer Lipid Membranes (BLM): Theory and Practice," Marcel Dekker Inc., New York.

Walker, N. A. and Smith, F. A., 1977, Circulating electric currents between acid and alkaline zones associated with HCO_3^- assimilation in <u>Chara</u>, <u>J</u>. <u>Exp</u>. <u>Bot</u>. 28:1190.

Williams, E. J., Johnston, R. J. and Dainty, J., 1964, The electrical resistance and capacitance of the membranes of <u>Nitella</u> <u>translucens</u>, <u>J</u>. <u>Exp</u>. <u>Bot</u>. 15:1.

Wills, N. and Clausen, C., 1982, Impedance properties of the rabbit descending colon, <u>Biophys</u>. <u>J</u>. 37(2, pt.2):279a.

ELECTROROTATION OF SINGLE CELLS -

A NEW METHOD FOR ASSESSMENT OF MEMBRANE PROPERTIES

Roland Glaser and Günter Fuhr

Department of Biology,
Humboldt-Universität zu Berlin, GDR

Abstract

Cells placed in a rotating high-frequency field spin slowly at two characteristic frequencies f_{c1} and f_{c2} against and according to the rotatation of the field, respectively. This method, designated here as ELECTROROTATION, allows to measure membrane capacity, membrane conductivity and internal conductivity of single cells. Basic equations determining this process are given and correlated with experimental data. The dependence of electrorotation on external, membraneous, and internal conditions during the experiment is demonstrated. For practical use of this method it is sufficient to measure the first characteristic frequency (f_{c1}), and additionally the frequency where the spin of the cells converts (f_0). Measurements on plant protoplasts, erthrocytes and platelets indicate a lower relative membrane capacity as expected.

1. Introduction

Investigations on the behaviour of cells in high-frequency alternating electric fields frequently indicate cells starting to spin under defined conditions. Holzapfel et al. (1982) explained this phenomenon as resulting from direct influence of the applied alternating field on the cells and, additionally, from interaction of dipoles generated in adjacent cells. Based on this theory, Arnold and Zimmermann (1982a) indicated that even single cells rotated if they were placed in a rotating high-frequency field. Such a field can be generated in a chamber consisting of three or four electrodes driven by sinusoidal voltage with progressive 120° resp. 90° phase shift. In a series of more recent papers that method has been considered a useful tool to measure cellular and membraneous parameters, such as membrane conductivity, capacity, and conductivity of the intracellular medium (Arnold and Zimmermann 1982b, Arnold et al. 1985, Fuhr 1985, Fuhr et al. 1985, Fuhr et al. 1984, Glaser et al. 1983, Hagedorn and Fuhr 1984, Hub et al. 1982). The term "e l e c t r o r o t a t i o n" will be used in this paper to characterise that method.

Notwithstanding the fact electrorotation was first applied to biological objects by Arnold and Zimmermann, the first hint on rotating dielectric bodies in alternating fields had been given by Heinrich Hertz, as early as 1881. Lertes (1921a, 1921b) investigated electrorotation of spherical bodies in air, and he calculated this phenomenon. It was on the basis of these formulae that we derived an equation which describes the situation of a cell of particular internal conductivity (G_i) and membrane conductance (G_m) surrounded by a solution with conductivity (G_e). The dielectric constants ($\varepsilon_i, \varepsilon_e, \varepsilon_m$) were additionally taken into account (Fuhr et al. 1985, Fuhr 1985$^=$).

The method is briefly introduced in this paper and some results are presented as measured from animal and plant cells. We shall discuss, in this context, the advantages of the method in relation to measurements of passive electric properties of cells in suspension.

2. Theoretical considerations

In general, the rotational behaviour of biological particles can be interpreted as rotation of dielectric bodies in rotating electric fields. That phenomenon was first calculated and experimentally verified by Heydweiller (1897), Lampa (1906), and Lertes (1921). Particle rotation was found to be attributable to polarisation phenomena resulting from charge separations on dielectric boundary layers. In Fig. 1A, the principle of rotation is demonstrated on a homogeneous sphere in an external medium (liquid or gaseous).

Particle polarisation can be described by the resulting electrical dipole (m). At low angular frequencies, the dipole (m) follows the field vector (E) without delay. Charge separation and charge dispersion, therefore, are much faster than the movement of the field vector. In other words, $\omega \tau \ll 1$ (where τ is the time constant of the polarisation process and ω the angular frequency of the rotating field.). With ω increased, dipole orientation is slower and follows the field vector by a definite angle (φ) (Fig. 1B). In such case the relaxation time of the polarisation process close to the dielectric barriers cannot be neglected ($\omega \tau \approx 1$). A definite torque (N) will occur in this frequency range and lead to rotation of the particle. A maximum of angular velocity can be observed if $\omega \tau = 1$. Further increase in angular frequency is followed by drop in polarisation, and at $\omega \tau \gg 1$ no torque acts at all.

Rotational behaviour is much more complicated when it comes to biological particles (e.g., single cells, protoplasts, isolated vacuoles). The model of a homogeneous sphere is not applicable to such cases and must be replaced by a model of a dielectric sphere surrounded by a shell (Fig. 2). Such a sphere is the simplest model of a cell, and the electric properties of the membrane system are considered (the shell representing the membrane and the interior of the sphere resp. the cytoplasm). The principle of torque derivation for this model has been calculated in detail in previous papers (Fuhr 1985, Fuhr et al. 1985, Hagedorn and Fuhr 1984). A general derivation of the torque (N) of dielectric spheres and cylinders under introduction of Maxwell's stress tensor was recently developed by Sauer and Schlögel (1985), but the results for single-shell spheres are nearly identical with our equations.

It has been assumed that the dielectrics (membrane, interior etc.) are homogeneous, with the dielectric values and conductivities of all materials (including the external medium) being constant. Nonlinearities have been ruled out. For calculation, cells are considered spheres with smooth surface. The torque (N) leading to the rotation of a single-shell sphere is given in Eqn. 1.

At two characteristic angular frequencies (ω_{c1}, ω_{c2}), the function shows extreme values of torque. Both relaxation processes are

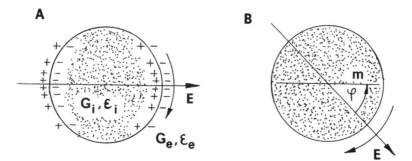

Fig. 1: (A) Polarized dielectric sphere in an external medium (G_i, G_e - conductivities, ε_i, ε_e - dielectric constants)
(B) Vector diagram for $\omega\tau \simeq 1$ (see text for more information).

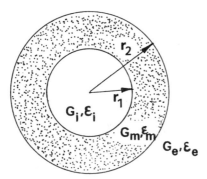

Fig. 2: Model of a dielectric sphere surrounded by a shell
G_i; ε_i - electric constants of the internal medium
G_m; ε_m - electric constants of the membrane
G_e; ε_e - electric constants of the external solution.

Equation 1. :

$$N = 4\pi\varepsilon_0\varepsilon_e r_2^3 E^2 \left\{ \left(\frac{C_1}{C_2} - \frac{B_1}{B_2}\right)\frac{\omega/\omega_{C1}}{1+(\omega/\omega_{C1})^2} + \left(\frac{B_1}{B_2} - \frac{A_1}{A_2}\right)\frac{\omega/\omega_{C2}}{1+(\omega/\omega_{C2})^2} \right\}$$

where

$$A_1 = \varepsilon_m(\varepsilon_e-\varepsilon_i) + \frac{d}{r_2}(\varepsilon_i-\varepsilon_m)(2\varepsilon_m+\varepsilon_e) \qquad ; \qquad d = r_2-r_1$$

$$A_2 = -\varepsilon_m(\varepsilon_i+2\varepsilon_e) - \frac{2d}{r_2}(\varepsilon_i-\varepsilon_m)(\varepsilon_e-\varepsilon_m)$$

$$B_1 = -G_m(\varepsilon_e-\varepsilon_i) - \varepsilon_m(G_e-G_i) - \frac{d}{r_2}((G_i-G_m)(2\varepsilon_m+\varepsilon_e)+(2G_m+G_e)(\varepsilon_i-\varepsilon_m))$$

$$B_2 = G_m(\varepsilon_i+2\varepsilon_e) + \varepsilon_m(G_i+2G_e) + \frac{2d}{r_2}((\varepsilon_i-\varepsilon_m)(G_e-G_m)+(\varepsilon_e-\varepsilon_m)(G_i-G_m))$$

$$C_1 = -G_m(G_e-G_i) - \frac{d}{r_2}(G_i-G_m)(2G_m+G_e)$$

$$C_2 = G_m(G_i+2G_e) + \frac{2d}{r_2}(G_i-G_m)(G_e-G_m)$$

Equation 2 : $\omega_{c1} = C_2/\varepsilon_0 B_2$; Equation 2a: $\omega_{c2} = -B_2/\varepsilon_0 A_2$

229

related to the ß-dispersion range. When dielectric constants and conductivities were used as in animal or plant cells, the characteristic frequency (ω_{c1}) was strongly affected by the electric properities of the membrane (charging processes perpendcular as well as in the plane of the membrane). The second characteristic angular frequency (ω_{c2}) was related to the electric properties of the internal and external media. These properties of the function will be discussed in greater detail below. The analytical terms for the characteristic angular frequencies are represented in Eqns. 2 and 2 a.

The rotation phenomenon, described here, is first of all based on conductivity effects. To complete the theoretical part of this paper, the dipole effect, too, should be briefly explained. Born, in 1920, published a theoretical paper in which the molecular dipole movement in rotating high-frequency fields was calculated. At angular frequencies in the GHz-range, dipoles (e.g., water dipoles) could not follow the field vector fast enough. Therefore, a torque acted per unit volume and leads to rotation of all the liquid. This effect was experimentally verified by Lertes (1921) and was found to be related to the γ-dispersion range. Therefore, theoretically, rotation of a single-shell sphere could be observed in three frequency ranges.

On principle, rotation effects could also be possible in the α-dispersion frequency range. Those effects, however, cannot be calculated by the model presented here, since for such consideration attention should be given to the effects of membrane foldings, double layers, and to other constellations.

3. Experimental technique to measure electrorotation of cells

Depicted in Fig. 3 is a diagram of the electronic setup to generate the rotating electric field. In the easiest way the sinus waves of a high-frequency generator are split up and inversed to get four outputs each with a phase shift of 90°. To avoid transduction of a DC field, the generator is coupled to the electrodes by capacitors. A voltage of about 10 V is necessary for the experiments, depending on electrode distance. From the electronic point of view, it is also possible to apply simple square-topped pulses instead of sinus waves (Pilwat and Zimmermann 1983, Fuhr et al. 1984).

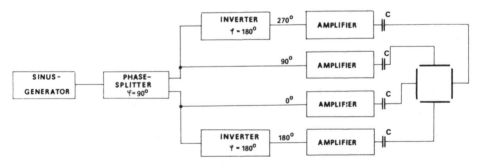

Fig. 3: Basic circuit diagram of the setup for a 4-phase generator.

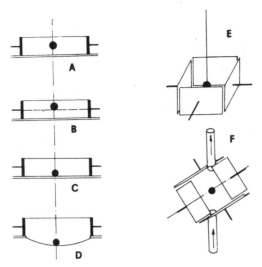

Fig. 4: Various possible setups for measurement of electrorotation of
particles:
(A) horizontal chamber for slowly sedimentating cells
(B) horizontal chamber with density boundary
(C) horizontal chamber with bottom made of hydrophobic material
     (i.e. low friction)
(D) chamber with free-hanging droplet and positioning of particle
     on lower surface
(E) chamber for observation of macroscopic bodies hanging from
     filament
(F) vertical chamber allowing lifting of particle by fluid suction
     through vertical tubes.

The torque resulting from the outside rotating electric field leads
to spin of cells, with its angular velocity depending on the frictional
forces. A small rotating sphere (Re < 1) indicates a frictional torque
(N_f), depending on the radius (r), viscosity (η), and angular
velocity (ω_c), as follows (see Berker 1963):

$$N_f = 8\pi\ \eta\ r^3\ \omega_c \qquad\qquad [3]$$

Significant spin of cells cannot be observed unless the cells move
without additional friction. Several possible ways to realize this
condition are given in Fig. 4. In all cases the frictional force given
by Eqn. (3) is so large that the cell spin is by many orders slower than
the rotation frequency of the applied field. Therefore, it can be
measured easily by microscopic observation.

A chamber consisting of four electrodes made of stainless steel or
platinum and mounted on a glass slide was used in the simplest way. The
distance between electrodes was 2 to 5 mm, the depth of the chamber being
0.7 to 3 mm. A chamber of that kind could be closed by a cover slip and
mounted under a microscope. Fig. 4A illustrates the setup to measure the
rotation of cells during sedimentation. Hence, cells can thus be
continuously observed over an extended period of time. The diagram in
Fig. 4F shows a vertical chamber with observation through a horizontal

microscope. This setup is closed on both sides. The chamber will be filled by two vertical tubes, enabling at the same time, adustment of the sedimenting cell by careful suction of the medium.

Larger cells can be placed on the border between two liquids of different densities (e.g. sucrose-mannitol solutions). We used this method mostly in investigations of plant cells and protoplasts. Conductivities of these two solutions have to be adjusted. In some cases cells are able to rotate even after sedimentation (Fig. 4C), especially if the bottom of the chamber consists of hydrophobic material. However, in such case the experiment will be strongly affected by small impurities on the surface. Many cells are stable enough to resist on the water-air surface of a hanging droplet (Fig. 4D). In this case, measurement is possible up to the point where surfactants (proteins etc.) occupied the surface, building up a film and, therefore, avoiding further rotation. This phenomenon occured usually after a definite time and did not lead to continuous moderation of rotation but to its abrupt block. Therefore, the result of measurement is not so strongly affected.

Even electrorotation of macroscopic bodies is possible as we demonstrated in previous publications (Fuhr et al. 1985, Fuhr 1985). In such case, objects can be best positioned with lowest friction by letting them hang on from a long filament (Fig. 4E).

4. General aspects of electrorotation experiments

The behaviour of animal cells in rotating electric fields can be described by the model demonstrated in Chapter 2 and, accordingly by Eqn. 1. A multi-shell model has to be applied, on principle, to plant cells and protoplasts, in view of the presence in them of a large central vacuole. Yet, more important, the most effective contribution to the torque of these cells has always been made by the plasma membrane, resp. by the plasma membrane as a complex with the cell wall. Hence, even plant cells can be treated with good approximation, using the equation derived for a single-shell model.

While the factor r^3 is contained in both Eqns. 1 and 3, it will be cancelled if the angular velocity of the spinning cell is calculated for the case of stationary rotation, stating $N = N_f$. The real angular velocity of the cells, consequently, did not depend primarily on the cell radius. Addtional though lower dependence, however, was given by occurence of the cell radius ($r = r_1 \simeq r_2$) in the part of Equ. 1 enclosed in braces. The resulting dependence of the torque on the cell radius is depicted in Fig. 8.

Furthermore, Eqn. 1 indicates the torque (N) as being dependent on the square of the field strength (E). This relation has been experimentally checked for a large number of cells and even marcroscopic bodies (Arnold and Zimmermann 1982a, Glaser et al. 1983, Fuhr et al. 1985, Fuhr 1985). To unify electrorotation experiments carried out with different field strength, we used a parameter (R), called "rotation" of cells, dividing the measured angular velocity (ω_c) by the square of applied field strength (E):

$$R = \omega_c/E^2 \qquad\qquad [4]$$

This parameter is usually in the order of 10^{-8} to 10^{-7} rad.m^2.s^{-1}.V^{-2} ("Rotation", therefore, has the same significance for electrorotation experiments as "mobility" has for the measurements of cell electrophoresis).

In other figures (Figs. 5,6,7,8) the function of Eqn. 1 is plotted with various parameters. In these cases, a torque number (N') is used.

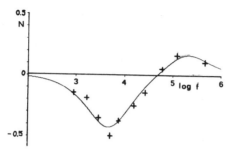

Fig. 5: Electrorotation (R) of a liquid filled glass sphere (r = 18.5 mm) as a function of the frequency (f). G_i = 1.2 mS/m, G_m = 10^{-12} S/m, G_e = 0.35 mS/m, ε_i = 40, ε_m = 4.5, ε_e = 70, d = 0.5 mm.

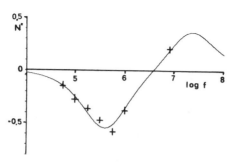

Fig. 6: Electrorotation of a single erythrocyte (r = 2.9 μm) washed in isotonic 30 mM NaCl-sucrose solution and resuspended in NaCl solution, G_e = 0.022 S/m in conductivity. (The following values were predicted: ε_e = 80, ε_i = 50.) The regression coefficient of the curve is 0.94.

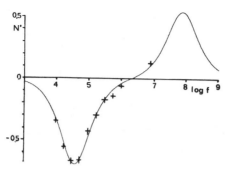

Fig. 7: Electrorotation of a single protoplast of Kalanchoe daigremontiana (r = 22.3 μm) in a solution of G_e = 5.2 mS/m. The points were fitted by a curve of Eqn. 1 with the following parameters: c_m = 1.83 mF/m², G_i = 0.813 S/m. The regression coefficient of the curve is 0.988.

233

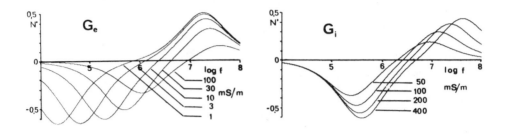

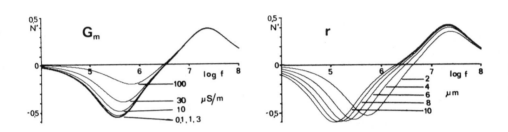

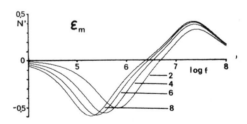

Fig. 8: The torque number (N') for rotating cells as a function of
frequency (f in Hz) for single-shell cell model.

G_e = variation in external conductivity
G_i = variation in internal conductivity
G_m = variation in membrane conductivity
r   = variation in cell radius
ε_m = variation in membrane dielectric constant (resp.
specific membrane capacity)
In spite of these modified parameters, the following fixed
values were considered: G_m = 0.1 µS/m, G_e = 20 mS/m,
G_i = 0.2 S/m, ε_m = 4, r = 3 µm.

This dimensionless number is identical with that part of the formula (Eqn. 1) which is enclosed in braces. In all these calculations and plots we used the frequency f = ω/2π in Hz. Figs, 5, 6 and 7 are examples in which experimental points from single cells are plotted together with a fitted curve corresponding to Eqn. 1. Fitted maximum negative rotation at the first characteristic frequency, (ω_{c1} resp. f_{c1}), was related to the corresponding torque number (N'_{c1}) to standardise the rotation measured. All experimental values were subsequently standardized by this factor.

At first the theoretical considerations were checked experimentally on liquid filled glass spheres. Two extrema of the torque (N) with different polarities were expected and measured (Fig. 5). By using a small inside conductivity and a mixture of alcohol and water as external solution, the second characteristic frequency shifts toward lower frequencies (f_{c2} = 163 kHz). These data are in good agreement with the function predicted by Eqn. 1.

Experiments with cells also showed an icreasing spin with frequencies rising towards a maximum f_{c1}=450 kHz for the erythrocyte (Fig. 6) and f_{c1} = 37.5 kHz for the protoplast (Fig. 7). In this frequency range, the direction of spin was opposite to that of the rotating field. This was in accordance with the negative value of the torque resulting from Eqn. 1. At frequencies of 3.8 MHz or 1.9 MHz, the curves crossed the abscissa, and rotation of cells in the same direction as the electric field was consequently observed at f = 8 MHz.

These examples stand for many experiments on various cells indicating, on principle, the same behaviour. The expected maximum of the curve at the second characteristic frequency, (ω_{c2} resp. f_{c2}), between 10^7 and 10^8 Hz, cannot be measured yet for technical reasons.

Dependence of such curves on variation in experimental conditions is depicted in Fig. 8. The characteristic frequency, f_{c1}, was shifted significantly by variation cell radius (r) as well as by change of membrane capacity (resp. ϵ_m). External conductivity (G_e) had a large effect on the position of the first characteristic frequency. In contrast to this, membrane conductivity (G_m), as well as internal conductivity (G_i), affected the characteristic frequency, f_{c1}, only to a lower extent. These parameters, on the other hand, were of large influence on rotation proper, i.e. on the value of N_{c1}.

The membrane of the cell was fully short-circuited in terms of capacitance at the second characteristic frequency (ω_{c2} resp. f_{c2}). In this case, membrane conductivity (G_m), therefore, had an impact neither on f_{c2} nor on the value of N'_{c2}. The effects of membrane capacity (resp. ϵ_m) and cell radius (r) were small as well. The second characteristic frequency as well as the torque were mostly determined, at this point, by internal conductivity, (G_i).

Electrorotation at such high frequencies, as mentioned earlier, was not measured for technical reasons. In Chapter 6, however, we shall describe the possibility of using the experimentally measurable point, where curves cross the abscissa, (f_0), to calculate values of internal conductivity of the cell.

5. Evaluation of membrane conductivity changes

As indicated in Fig. 8, increase in membrane conductivity led to an upward shift of the first characteristic frequency (f_{c1}) and, at the same time, to decrease in rotation. As pointed out, electrorotation of

cells could consequently be used as an indicator for integrity of the
membrane as an ion barrier (Glaser et al. 1983). Recently, Arnold et al.
(1985) used this method to measure the impact of ionophores on
permeability of thylacoid membranes.

Fig 9 demonstrates that both effects, i.e., shift of the first
characteristic frequency (f_{c1}), as well as change of torque numbers at
this frequency (N'_{c1}), reached constant degrees always at low values of
membrane conductivity. In low-conductivity outside solutions (Fig 9,
curve 5) this occured at $G_m < 10^{-7}$ S/m. In solutions of higher
conductivity, however, the value was shifted to membrane conductivities
of 10^{-5} to 10^{-4} S/m. In other words, the region of membrane
conductivity measureable by electrorotation was limited and depended
strongly on the conductivity of the outside solution.

To illustrate the practicability of this method, we shall demonstrate
experiments undertaken to study the dynamics of membrane pores produced
by electric breadown in human erythrocytes (Fig.10). For that purpose,
we measured first, rotation (R) as a function of frequency (f) using for
control cells which were not treated by the electric field. The first
characteristic frequency was determined from these measurements, as in
the case demonstrated in Fig. 6, ($f_{ci} = 320$ kHz). The membrane
conductivity of these cells is known as $G_m < 10^{-6}$ S/m. Rotation of
cells treated by an electric field were subsequently measured
particularly at this frequency. The torque number of the cells at this
frequency was calculated by Eqn. 1 as a function of
membrane activity. We found

$$R_{rel} = \frac{R}{R_{max}} = \frac{N'}{N'_{c1}} \qquad [5]$$

under the condition that friction of cells did not change as a function
of G_m during the experiment (i.e. shape should remain constant). In
the resulting function R_{rel} (G_m) (Fig. 11), different from function
N'_{c1} (G_m) in Fig 9, due consideration was given to the condition that
measurement of field-treated cells, indicating increased membrane
conductivity, was not conducted at a frequency which was in any case
equal to the first characteristic frequency. Fig. 11 indicates that
there is no difference, no matter if minimum conductivity of the control
cells (i.e. for the case of $R_{max} = 1$) is taken as 10^{-6} S/m or lower.

This example shows how modification of membrane conductivity as a
function of ion permeability can be evaluated by the method of
electrorotation. The advantage of ths method is derived from the
possibility that single cells can be checked in a suspension. It has
even been possible to evaluate this way the percentage of cells
perforated by the applied field impulse. It should be mentioned, on the
other hand, that the values determined in this experiment in fact
depended on several predictions. The reliability of this method will be
discussed below.

6. Determination of relative membrane capacity and internal conductivity

Arnold et al. (1982) found earlier that specific membrane capacity
(C_m) could be calculated from the value measured or from the first
characteristic frequency (f_{c1}). Internal and external conductivities
as well as cell radius (r) should additionally be taken into
consideration for that purpose. Membrane conductivity, (G_m), was
considered as being negligibly small:

$$c_m = \frac{G_i \; G_e}{f_{c1} \; \pi \; r \; (G_i + 2G_e)} \qquad [6]$$

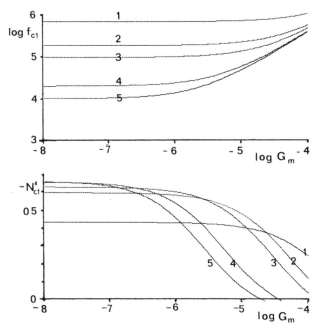

Fig. 9: First characteristic frequency (f_{c1}) and torque number (N') as functions of membrane conductivity (G_m in S/m) for conditions of various external conductivities (G_e): 1 = 50 mS/m, 2 = 10 mS/m, 3 = 5 mS/m, 4 = 1 mS/m, 5 = 0.5 mS/m. Additional parameters are: r = 3 μm, ε_m = 4, G_i = 0.2 S/m).

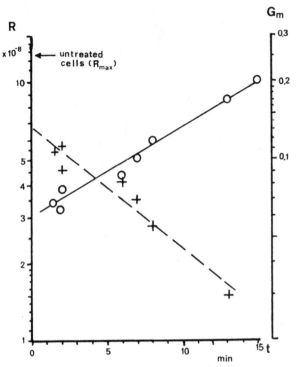

Fig. 10: Rotation (R) of human erythrocytes treated with a field pulse
(10^6 V/m, 0.2 s) in isotonic 30 mM NaCl-sucrose solution
and stored for definite time (t) at 4°C in 0.8 mM NaCl-sucrose
solution (+ ----- +). Rotation has been measured at fixed
frequency (f = 320 kHz). Measured rotation has been converted
by means of curve in Fig. 10 into membrane conductivity
o———————o, G_m in mS/m). Result indicates subsequent opening of
produced pores, following electric breakdown.

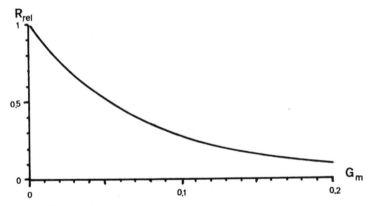

Fig. 11: Relative rotation
($R_{rel} = R/R_{max}$) as function of membrane conductivity (G_m in
mS/m) for human erythrocytes, measured
under following conditions: G_e = 20 mS/m, G_i = 155 mS/m,
f = 320 kHz, c_m = 4.42 mF/m^2.

Our model, as given in Chapter 2 (Eqn. 1), may be expected to include this equation as a special case for $G_m \to 0$. This equation, however, does not explicitly include c_m, but instead of this the dielectric constant of membrane (ε_m) and membrane thickness, (d):

$$c_m = \frac{\varepsilon_o\,\varepsilon_m}{d}$$

[7]

Using this relation, it is possible to derive the following function of c_m which, however, also contains the parameters ε_m and d:

$$C_m = \frac{1}{f_{cl}\pi}\left[\frac{G_i\,G_e}{r(G_i + 2G_e) + 2d(\varepsilon/\varepsilon_m - 1)(G_e + G_i)} - \frac{G_i + G_e}{2d\,(G_i + 2G_e)}\,G_m\right]$$

[8]

(In this case $\varepsilon_i = \varepsilon_e = \varepsilon$) It may be easily seen that for the vanishing membrane conductivity ($G_m = 0$) this equation will be reduced nearly to the Eqn. 6, despite an additional term in the denominator. Use of real values of the parameters revealed that this term was small in relation to term $r(G_i + 2G_e)$.

A second measuring parameter is required for independent calculation of relative membrane capacity and internal conductivity. We have pointed out earlier that we used for that purpose the intersection of the curve with the abscissa (f_o), as shown in Fig. 8, with the position of this point depending on various parameters. For practical use, it is possible to determine this value by measuring the rotations (R_1, R_2) at frequencies (f_1, f_2) on both sides as closely as possible to the assumed frequency, f_o. Parameter f_o can then be calculated, using the following logrithmic fitting:

$$\lg f_o = \frac{N_1\,\lg f_2 - N_2\,\lg f_1}{N_1 - N_2}$$

[9]

Eqn. 1 allows to deduce the following function:

$$f_o = \left[\frac{C_2^2\,(A_2 B_1 - A_1 B_2) + B_2^2\,(B_1 C_2 - B_2 C_1)}{4\pi^2\varepsilon_o^2[A_2^2(B_2 C_1 - B_1 C_2) + B_2^2\,(A_1 B_2 - A_2 B_1)]}\right]^{\frac{1}{2}}$$

[10]

parameters A_1, A_2, B_1, B_2, C_1, C_2 are the same as in Eqn. 1.)
By using now Eqn. 8 to gether with Eqn. 10, specific membrane capacity, (c_m), and internal conductivity, (G_i), can be iteratively calculated on the basis of the values measured of f_{cl}, f_o as well as r, G_e, and ε_e. Yet, to that end, additional definite assumptions should be made of parameters d, ε_i, and G_m.

Some values representative of various cells are given in Table 1.

Table 1   Specific membrane capacity (c_m), and internal conductivity (G_i) of various cells calculated from measured values of f_{cl} and f_o

	G_e [mS/m]	r [µm]	c_m [mF/m^2]	G_i [mS/m]
Human erythrocytes	22	2.9	3.8±0.3	155±14
Human thrombocytes	12	0.9	6.3±0.8	108± 8
Kalanchoe daigremontiana (protoplasts)	5.2	22.3	1.8	813
	5.2	22.0	2.5	2633
	5.2	12.9	1.5	2030
	5.2	22.6	0.9	2121
	12	21.9	2.4	1420
Avena sativa (protoplasts)	25	25.8	3.6	752

Membrane conductivity was assumed to be $G_m = 10^{-7}$ in all cases. Mean values measured from six to eight cells are given in the table for erthrocytes and platelets; sizes were more or less homogeneous. In contrast to this, however, the protoplast values were recorded from measurements of single cells and exhibited large divergency. This was actually attributable to the variability of individual cells rather than to the method, as demonstrated by the good fitting of values in Fig. 9.

The parameters thus calculated depended on the accuracy of the values measured (f_o, f_{c1}, r, G_e) as well as on the assumed parameters (d, G_m). To evaluate these inaccuracies, for example, we varied within reasonable limits the parameters for the case of erythrocyte measurements and calculated a sensitive factor (s) as follows:

$$ s\frac{1}{2} = \frac{\bar{X}_2}{\bar{X}_1}\frac{\partial X_1}{\partial X_2} \tag{11}$$

This factor revealed the extent to which parameter X_2 was affected by variation of X_1. This number is valid, in fact, only at special bounds because it is based on non-linear relations.

The most sensitve dependence on resulting specific capacities was related to the cell radius which could be measured with high accuracy only for large cells (e.g. protoplasts) but not for erythrocytes and platelets, as may be seen from Table 2. The accuracy of G_i, however, was not so strongly influenced. Deviation of cell shape from the ideal sphere, as postulated in the model, was not taken into consideration in this calculation. Especially the influence of the flat form of erthrocytes may produce an additional effect.

Table 2   Sensitivity factors for various parameters of curve for evaluation of erythrocyte parameters (see Tab. 1).

PARAMETER:	REGION OF VARIATION:			s for c_m	s for G_i
r	2	...	3.5 μm	−1.02	$< 10^{-3}$
f_{c1}	0.35	...	0.5 MHz	−1.66	−1.99
f_o	3	...	4.5 MHz	+0.70	+2.85
G_e	20	...	30 mS/m	+0.45	−0.95
G_m	10^{-7}	...	10^{-5} S/m	−0.07	−0.09
d	5	...	9 nm	$< 10^{-3}$	$< 10^{-3}$

External conductivity of the solution can be measured with high accuracy, but possible variations produced by ion extrusion of suspended cells in the course of the experiment should be considered in such case. For this, it is necessary to measure the conductivity of the cell suspension at various junctures during the experiment. Change in G_e will primarily affect the accuracy of the calculated value of G_i, as may be seen from Table 2.

The unmeasured parameters, which have to be assumed, are without significant effect on the calculated values of c_m and G_i. Remarkable modification of results has to be expected only for perforated cells where $G_m > 10^{-5}$ S/m. Further decrease in G_m has no effect. Membrane thickness (d) is absolutely insignificant if changed within reasonable limits. In this case, according to Eqn. 6, the value of c_m changes in correspondence to d.

7. Discussion

In this paper it is demonstrated that electrorotation is a useful method to study membrane properties of single cells. This method allows one to indicate differences in the behaviour of individual cells, in contrast to impedance measurements on cell suspensions, where only average values of all cells in the suspension are measured. This makes electrorotation useful for checking various cell physiological processes and testing drug interactions. In this way the change of the membrane conductivity produced by ionophores has been measured (Glaser et al. 1983, Arnold et al. 1985).

In order to evaluate the membrane capacity (c_m), the membrane conductivity (G_m), and the internal conductivity (G_i) of the cells independently, it is necessary, however, to measure more than one characteristic parameter of electrorotation. The first characteristic frequency (f_{c1}), and additionally the frequency at the intersection point of the curve with the abscissa (f_o) makes it possible to calculate c_m and G_i assuming ($G_m < 10^{-6}$ S/m as demonstrated in Chapter 6. The change in membrane conductivity can best be measured following the change of the rotation speed (resp. the "rotation" R). This value, however, strongly depends upon the friction of the rotating cells. The rotation force can be calculated with sufficient accuracy only in the case of exactly spherically shaped cells that present a smooth surface and which hang freely in suspension. Those conditions do not occur in most cases. The Rotation (R), therefore, is better used as a relative parameter in cases where the kinetics of the process to be analysed does not by itself affect cell shape.

Additionally, it has to be considered that even the position of the first characteristic frequency (f_{c1}) can be shifted by the deviation of the cell shape from the geometry of a sphere. This is to be considered especially in the case of erythrocytes, which present discoidal (normocytes) or creneted (echinocytes) shapes. After switching on the electric field, the erythrocytes orientate quickly in such a way that the largest dipole is directed toward the plane of the rotating field. The normocyte therefore rotates as a flat disc. It is possible that the nonspherical shape of erythrocytes shifts the parameters, calculated from the measured values f_{c1} and f_o.

As pointed out aleady in previous publications (Arnold and Zimmermann 1982a,b. Glaser et al. 1983), the specific membrane capacity calculated from electrorotation experiments is usually lower than 10 mF/m^2, as measured by the impedance method (see e.g. Schwan 1963). Arnold and Zimmerman (1982a) found for plant protoplasts c_m = 4.8 ± 0.7 mF/m^2. They considered this to result from the superposition of the two membranes (plasmalemma and tonoplast) surrounding the vacuole. In the case of thylacoid membranes, which actually are single membranes, they found a larger value of 9.3 ± 0.7 mF/m^2, which, however, is smaller than the value 17 ... 20 mF/m^2, found by DeGroth et al. (1980a,b) by another method.

The specific membrane capacity of all cells has been found lower than 10 mF/m^2 even in our experiments. In the case of erythrocytes, this may be attributed to some degree by their non-spherical shape. Experiments on macroscopic model bodies, however, indicate that this maximally results in a shift by a factor <2. In the case of Kalanchoe protoplasts, obviously, the specific capacity is diminished by an incomplete removal of the cell wall during preparation, and in a complicated manner influenced by both membranes. In the case of Avena sativa protoplasts a higher specific capacity has been found.

Electrorotation is a new method and its application to various problems depends on its further technical completion. A weighty limitation is given by the necessity to use external solutions of low conductivity. By increasing the external ionic strength, i.e. increasing G_e, the values f_{c1} and f_{o1} shift to higher frequencies, the rotation speed decreases to some extent (see Fig. 8), and the applied electric field leads to an increasing current and, consequently, to a warming of the system. This can be overcome by developing an electronic setup which allows to perform measurements at higher frequencies, and to shorten the time of measurement, i.e. the time in which the electric field is applied. Arnold and Zimmermann (1985) have used a direction-reversal compensation technique which allowed them to find the first characteristic frequency in one single measurement.

Acknowledgement

For constructing the electronic devices for electrorotation we are very grateful to Jan Gimsa. Measurements on thrombocytes and those in Fig. 10 were carried out by the students Marcel Egger and Jutta Engel. We thank Thorsten Muller for careful measurements on macroscopic bodies and plant protoplasts. The assistance of Mrs. Jutta Donath is gratefully acknowledged.

References

Arnold W M, Wendt B, Zimmermann U and Korenstein R 1985 Biochim. Biophys. Acta $\underline{813}$ 117

Arnold W M and Zimmermann U 1982a Zeitschr. Naturf. $\underline{37c}$ 908

Arnold W M amd Zimmermann U 1982b Naturwiss. $\underline{69}$ 297

Berker R 1963 Handbuch der Physik VIII/2 (Berlin) 217

Born M 1920 Zeitschr. f. Physik $\underline{1}$ 221

DeGrooth B G, Van Gorkom H J and Meiburg R F 1980a Biochim. Biophys. Acta $\underline{589}$ 299

DeGrooth B G, Van Gorkom H J and Meiburg R F 1980b FEBS Letters $\underline{113}$ 21

Fuhr G. 1985 über die Rotation dielektrischer Körper in rotierenden Feldern. Dissertation, Humboldt University, Berlin

Fuhr G, Glaser R and Hagedorn R 1985 Biophysical J. (in press)

Fuhr G, Hagedorn R and Göring H 1984 studia biophysica $\underline{102}$ 221

Glaser R, Fuhr G and Gimsa J 1983 studia biophysica $\underline{96}$ 11

Hagedorn R and Fuhr G 1984 studia biophysica $\underline{102}$ 229

Hertz H 1881 Wiener Annalen der Physik $\underline{13}$ 266

Heydweiller A 1897 Verhandlungen der Deutsch. Physik. Ges. $\underline{16}$ 32

Holzapfel C, Vienken J and Zimmermann U 1982 J. Membrane Biol. $\underline{67}$ 13

Hub HH, Ringsdorf H and Zimmermann U 1982 Angew. Chemie $\underline{21}$ 134

Lampa A 1906 Wiener Berichte $\underline{115}$ 1659

Lertes P 1921a Zeitschr. f. Physik $\underline{4}$ 315

Lertes P 1921b Zeitschr. f. Physik $\underline{6}$ 56

Pilwat G and Zimmermann U 1983 Bioelectrochem. Bioenergetics $\underline{10}$ 155

Sauer F A and Schlögel W 1985 in: Chiabrera A, Nicolini C and Schwan H.P. (Eds.) (New York: Plenum Publ.) (in press)

Schwan H P 1963 Physical Techniques in Biol. Research $\underline{6}$ 323

RECENT STUDIES ON ELECTROKINETIC POTENTIALS IN HUMANS

Eugene Findl*, Howard Gutermann**, and Robert J. Kurtz

BioResearch, Inc.
315 Smith Street
Farmingdale, NY   11735

1.   INTRODUCTION

The field of bioelectrochemistry is primarily the study of the interaction between ions, cells and fields. More specifically, it is the science dealing with cellular membranes and the forces causing ions to become displaced on the surface of or through the channels in membranes. There are a number of forces that can displace ions, ranging from electric, magnetic and electromagnetic fields to chemical gradients, to mechanical forces. It is the mechanical force aspect that this article addresses.

There are several ways that mechanical forces cause bioelectrochemical effects or are generated by bioelectrochemical processes. The original basis for believing electric fields would be beneficial for bone regeneration was based upon the premise that mechanical stressing of dried bone generates piezoelectric potentials and without such stressing, bone decalcifies. More likely however, is that bone stress generates bioelectrokinetic potentials, or if you prefer, streaming potentials, rather than piezoelectric. In a similar fashion, it is standard dogma that mechanical force generated by muscle contraction is caused by chemically induced, electrochemically transmitted pulses to myoneural cells. Such electrochemical signals are said to be the cause of the electrocardiogram (ECG) and the electromyogram (EMG) potentials. As Findl et al (1977, 1982) have demonstrated, it is more probable that the ECG and EMG potentials are bioelectrokinetic rather than myoneural.

Bioelectrokinetic potentials are quite simply the electric potentials generated by the mechanical displacement of ions on the surface of cells. Such displacement can be caused by the motion of an ionic fluid over the surface of a cell, or the motion of a cell in an ionic fluid. All that is required is that the ionic fluid and cell move relative to each other.

Present Address

*   15730 Hartsook Street, Encino, CA   91436
**   31 Rochester Road, Newton, MA   02158

Thus, any motion, from contraction of the heart to bending of a finger, to the flow of blood in the cardiovascular system will generate a bioelectrokinetic, i.e., streaming potential.

2. EAG Background

Findl et al. (1980, 1984) and Conti et al. (1982) have described some of the factors involved in measuring the bioelectrokinetic potentials generated by blood flow in the arterial system. The technique used to measure arterial streaming potentials is known as electroarteriography or by the acronym EAG. Basically, the technique involves the measurement of the streaming potential generated by the pulsatile flow of blood in an artery. By comparing the bioelectrokinetic potential waveforms of a patient being examined with those of normal and diseased EAG's of others, the patients degree of artherosclerosis can be evaluated. With proper placement of the EAG electrodes, the location of blockages in arteries can also be determined. In addition, by measuring the time differences between the peak of the cardiogram QRS signal (measured simultaneously), one can obtain information regarding hardening of the arteries, i.e. arteriosclerosis. Thus, both athero and arteriosclerosis can be evaluated by the EAG technique.

Figure 1 illustrates the factors we look at in evaluating the status of a normal EAG pattern. In patients with no symptoms of arterial problems, the EAG pattern closely follows the pressure wave characteristics of an artery. If the artery starts to get blocked by atherosclerotic plaques, the EAG pattern starts to flatten out and the reversed blood flow portion (below the base line) is eliminated. With severe atherosclerosis, the EAG pattern is little more than a low magnitude hump type signal.

For arteriosclerosis evaluation, the T2 time period is used. As an artery hardens, the T2 time decreases. Figure 2 illustrates typical T2 times for about 30 patients, principally male, ranging in age from 20 to 65. The dorsalis pedis artery was used for this measurement and all readings were normalized to account for differences in the height of patients.

3. Recent Studies

Before getting into the new feature of the EAG, it is appropriate to describe the latest EAG electrode arrangement and method of measurement. Figure 3 illustrates the design and placement of same over a wrist artery. Prior to testing, the skin is first cleaned with an alcohol/water solution. As shown on Figure 3, the electrode consists of three, in line silver coated metallic discs. Common mode signal rejection is facilitated by elimination of potentials common to both end electrodes and the center electrode. EAG potentials, less the common mode signals, are amplified and fed to appropriate readout devices such as an oscilloscope and strip chart recorder. Tac Gel R conductive adhesive is used to coat each electrode. The disc electrodes are attached to three contact springs to insure good skin contact. The springs in turn are attached to brass snaps used as electrical contacts. Electrical connection to the amplifier is made via a connector attached to a Velcro strap. The strap is used to hold the

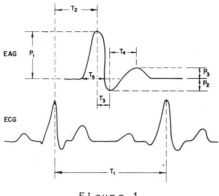

Figure 1

Representation of EAG and ECG signals defining
various timing (T) and peaks amplitude (P)
data points

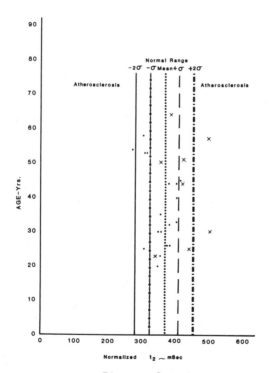

Figure 2

Normalized T$_2$ times of Dorsalis Pedis
Arteries of 30 Patients

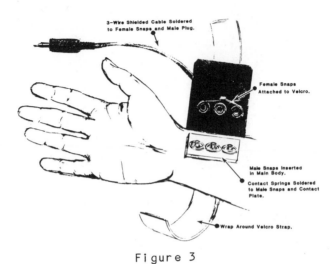

Figure 3

Latest EAG Electrode Design

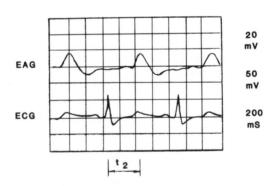

Left Dorsal Pedis Artery – Patient: RJ

$$[t_2 = 59 \text{ msec.}]$$

Figure 4

Typical EAG Waveform Using Latest
Electrode Design

electrode in place. Last, but not least, a small pressure cuff
is placed over the electrode and inflated to 40-60 mm Hg, to
further ensure good contact with the skin.

Typical results obtained using this electrode and
procedure are shown on Figure 4. The upper waveform is the EAG
and the lower is the ECG. Based upon the EAG waveform, and the
time difference between the EAG and ECG peaks, the patient does
not appear to have any arterial problems in the arterial system
upstream of the measurement point, which in this case was the
left dorsalis pedis artery (top of the left foot).

While conducting tests on a series of patients to verify
the usefullness of the three in line style of electrode, a new
feature of the EAG technique was noted. Simply put, it was
noted that the EAG signal changed rather remarkably when the
patient fully relaxed and fell asleep. A sequence of EAG
patterns with such a patient is shown on Figure 5. Shown
thereon are a sequence of 10 EAG waveforms taken on the right
dorsalis pedis of a 37 year old asymptomatic male. The
pressure cuff was maintained at 42 mm Hg. Each waveform was
taken approximately 1 minute after the preceding one. Thus,
9 minutes elapsed between first and last. Figure 5a represents
the typical normal EAG pattern as the patient started to doze
off. His heart rate was about 71/minute. A minute later, as
shown in Figure 5b, the pattern doubled in magnitude and the T2
timing decreased. Heart rate remained about 72/minute. On
Figure 5c, the amplitude scale changed from 20 to 50 mv/div.
Heart rate increased to 81/min. A decided variation in
amplitude from heart beat to heart beat can be seen. On Figure
5d, amplitude increased by about 75% and the heart rate
slightly increased. As shown on Figure 5e, amplitude varied
considerably from beat to beat while the heart rate stayed at
82/min.   On Figure 5f, heart rate decreased to about 72/min.
and the EAG waveform stabilized. The next waveform, Figure 5g,
was taken at a scan speed of 200 msec. to expand the waveform.
Amplitude had again increased, while heart rate remained
constant. Figure 5h shows a sharp decrease in amplitude while
heart rate remained constant. Figure 5i shows that the EAG
signal had drastically decreased to zero, even after a 10 fold
increase in signal amplification. On the last of this
sequence, Figure 5j, a most interesting fact to be noted is
that the signal polarity reversed. Heart rate also increased
to 76/min.

Another example of this reversal phenomena is shown on
Figure 6. Here we have the left wrist radial artery EAG of a
30 year old male volunteer. The initial EAG signal is shown on
Figure 6a. In this case, the signal polarity was set such that
the normal EAG peak was negative so that the T2 time could be
more easily measured. As can be seen in going from Figure
6a-6d, as the patient dozed off, the signal amplitude and
polarity changed in the brief elapsed time between recordings.

At first, this seemingly anomalous EAG pattern behavior
was quite puzzling. After a few days of cogitation, the basis
for the seemingly anamalous behavior was deduced to be due to
the closed flow loop aspects of the cardiovascular system.
Such closed loop systems have been described by Findl (1982) in
terms of an electrical analog, such as shown on Figure 7.
Utilizing this type of ladder network analog, it is possible to
demonstrate that the streaming potential generated in a closed
loop is dependent upon both the locally generated streaming
potential and more remotely generated potentials. As the
locally measured potential decreases, as during a flow decrease

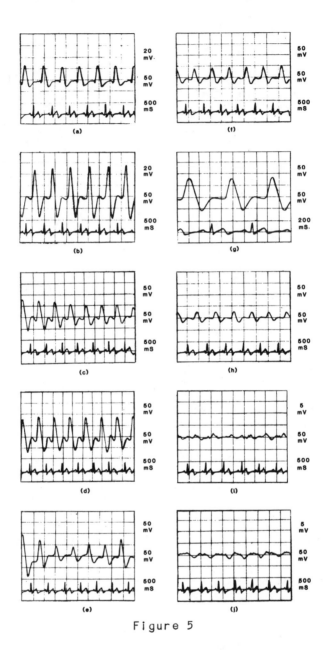

Figure 5

Sequence of 10 EAG Waveforms on Right Dorsalis
Pedis of a Patient During Awake/Sleep Period

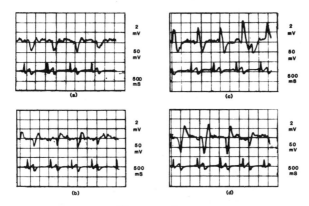

Figure 6

Left Wrist Radial Artery EAG Sequence
During Awake/Sleep Period

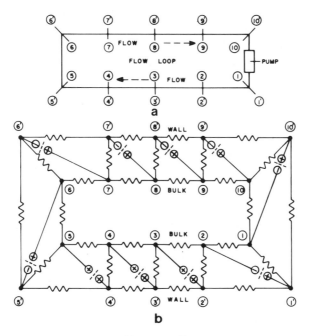

Figure 7

Basic Closed Loop Analog

in an artery, locally generated potentials, or rather locally measured potentials, become more significantly influenced by remotely generated potentials. In the extreme, the measured potential may be almost entirely due to remotely generated potentials.

By far the largest magnitude electrokinetic signal generated in the cardiovascular system is the ECG generated by the heart. This signal is about 100 times larger than the EAG signal and is measurable anywhere in the body. However, it is readily descriminated against by means of the common mode signal rejection technique previously noted. With the ECG signal rejected, the largest magnitude EAG signals are expected in the major blood vessels having the highest blood velocities and shear rates. One would expect the ascending and descending aorta, abdominal aorta, iliacs, carotids, etc., to be large magnitude signal contributors compared to the peripheral arteries. Therefore, unless the flow rate and the resultant blood/blood vessel interfacial shear in a peripheral artery is of sufficient magnitude, the EAG electrodes will detect potentials from more distant signal sources.

Without going into any mathematical or ladder network analysis, let us simply state that because of the closed loop aspect of the cardiovascular system and the fact that Kirchoff's Laws related to electrical circuits are operative, the summation of the potentials going around the loop must be zero. Therefore, signal polarities and magnitudes that are measured, are dependent upon the entire flow loop, not just the local measurement section. Thus, EAG signal changes noted when a patient's peripheral blood flow diminishes significantly are a simple manifestation of Kirchoff's Laws. The phenomenon noted may have significant diagnostic value in determining blood flow characteristics in the major arteries previously noted, particularly those of arteries in the abdomen and chest cavity. Proof of this remains to be determined.

4. EAG/Streaming Potential Theory

EAG signals, measured at the skin surface, are the result of two steps; (a) generation of the signal within the artery due to the flow of blood and (b) propagation of the signal from the artery to the skin surface. The waveform produced at the artery depends on closed loop hemodynamic and electrochemical factors. Its alteration in transmission to the surface is a function of the electrical conductivity of the tissue and the geometry of the blood vessel and tissue. The relation of the EAG signal to hemodynamic parameters requires the evaluation of both processes.

The classical mathematical relationship describing streaming potentials in tubular, open flow loops under laminar flow conditions is generally attributed to Helmholtz, although a number of other electrochemical investigators preceded him.

$$E_S = L' \cdot D \cdot \zeta \cdot \Delta P / 4 \pi \cdot \mu \cdot k \cdot L^* \qquad (1)$$

where E_S = the streaming potential, L' = spacing of measurement electrodes, D = dielectric constant of the double layer, ζ = the interfacial zeta potential, ΔP = pressure drop of flowing fluid measured over length L^*, μ = the viscosity of the double layer electrolyte, k = the conductivity of the double layer electrolyte and L^* = the length of tubing over which the pressure drop is measured. Since double layer values of D, and k are not readily available or determinable, it is common practice to use the properties of the bulk electrolyte for calculations. This unfortunately results in the zeta potential being more or less a fudge factor to make the measured value of E turn out to be mathematically correct. Regardless of the mathematical inaccuracies of the Helmholtz equation, it does provide insight into streaming potential generation.

A similar and more general equation for streaming potentials was developed by Findl. This relationship relates streaming potential to flow velocity, rather than a pressure drop. (See Kurtz, Findl, Kurtz & Stormo, 1976).

$$E_s = 0.4 D \cdot \zeta \cdot L' \cdot U^{1.75} \cdot \rho^{1.75} / \, \Pi \cdot \mu^{0.75} \cdot d^{1.25} \cdot k \qquad (2)$$

where U = the electrolyte flow velocity, ρ = the electrolyte density and d = diameter of tubing in the measurement section, with all other symbols as before. This equation appears to be valid in both laminar and turbulent flow regimes, at least to Reynolds numbers up to 10 for smooth bore tubing.

The factor of the degree of tubing smoothness has largely been neglected in the literature. Such neglect is unfortunate. It is a crucial factor in streaming potential generation.

In order for a streaming potential to be generated, the interfacial double layer must be perturbed by the flow of the electrolyte. From both hydrodynamic and electrochemical viewpoints, if the double layer lies wholly within the hydrodynamic shear plane, no such perturbation can occur. In essence, unless the diffuse portion of the double layer extends out for hundreds of angstroms and the Reynolds number of the fluid is 10^5 or greater, streaming potentials in ideal smooth bore tubing will be negligible. Yet, we are well aware that streaming potentials can be measured under considerably less "severe" conditions.

This seeming contradiction can be resolved if one simply realizes that real world flow conduits are not ideally smooth. Figure 8 illustrates, in a somewhat exaggerated form, the role a surface plays in streaming potential generation. Quite simply, the surface roughness juts out into the flow stream, resulting in a portion of the double layer being perturbed by the flow. To the best of our knowledge, no one has derived a mathematical relationship for the effect of surface roughness on the magnitude of the streaming potential.

Schlicting (1960) took another approach to modifying Helmholtz's equation for use in both the laminar and turbulent flow region.

$$E_s = D \cdot \zeta \cdot T_W / \, 2 \, \pi \cdot k \cdot d \cdot \mu \qquad (3)$$

where T_W = the wall shear stress, which is proportional to fluid velocity and other symbols are as before.

In the case of electronically and/or electrochemically conductive flow conduits, a correction factor must be added to all equations for streaming potential. This correction factor is needed to account for potential decreases due to Ohm's Law shunting effects. Thus, a more general form of the Schlichting equation takes the form

NOTE
⊕ - DISPLACED DOUBLE LAYER IONS

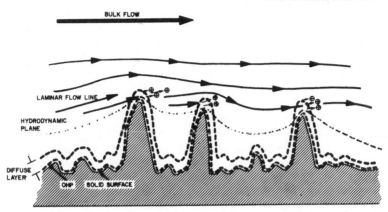

Figure 8

Illustration of the Effect of Surface Roughness
on Generation of Streaming Potentials

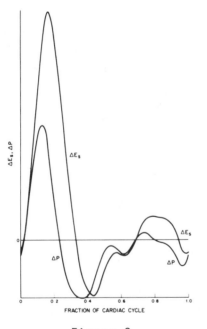

Figure 9

EAG and Pressure Gradient Waveforms

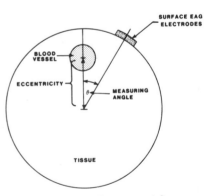

Figure 10

Crosssection of Idealized
Limb

$$E_s = (D \cdot \zeta \cdot T_w \, / \, 2\pi \cdot k \cdot d \cdot u) \, (1/1+G) \qquad\qquad (4)$$

where G is a tubing wall conductivity factor. For an insulating flow conduit, G approaches zero; for a conductive flow conduit, G approaches large values.

Using Schlichting's equation the streaming potential may be evaluated for any flow pattern in which the wall shear stress can be found. A fairly good first approximation to the velocity profiles occurring in arteries may be obtained using the theory for pulsatile flow through rigid tubes developed by Womersley (1955). This theory allows the velocity profiles within the artery to be computed given the pressure gradient waveform in the form of a Fourier series. As a sample calculation, a pressure gradient waveform given in this form by MacDonald (1974) for a canine femoral artery was used. The velocity profile and wall shear stress were calculated using Womersley's equations and the streaming potential was assumed to be proportional to the wall shear stress as indicated by equation (4).

The resultant waveform for the streaming potential is shown on Figure 9 with the measured pressure gradient waveform shown for comparison. Scale factors are not shown for E on this figure because of the uncertainties in the values of the other constants in equation (4). However, it is clear that the generated streaming potential waveform closely follows the shape of the pressure gradient curve. The slight lag seen in Figure 9 between the two curves is the result of viscous damping of the pressure wave.

Transmission of the EAG signal from the artery to the skin surface was modeled by assuming that the blood vessel was a cylindrical tube located eccentrically within a cylinder of muscular tissue (the limb), with the axes of the two cylinders being parallel. This configuration is shown in Figure 10, with EAG surface electrodes placed on the skin surface on a line parallel to the blood vessel. The streaming potential generated within the artery was assumed to be constant per unit length of artery and uniform around the arterial circumference.

Evaluation of the effect of the periodicity of the streaming potential on the tissue and surface voltage profiles indicates that the surface voltage will have a waveform identical with that of the artery because of the very low capacitance of the surrounding tissue. This result is generally true in electrophysiological problems, as previously shown by Plonsey and Heppner (1967). Therefore, combining this result with Figure 9, it is evident that the surface EAG signal measured will be a direct indication of both the shear stress at the arterial wall and the pressure gradient waveform. Changes in the normal EAG profile will therefore be indicative of hemodynamic changes that are known to occur in atherosclerotic regions. Variations in local wall shear stress have also been implicated in atherogenesis.

The attenuation of the EAG signal in its transmission to the skin surface is a complicated function of the limb radius, blood vessel radius, vessel eccentricity, and the angle between the measuring electrodes and the blood vessel. This function was evaluated for a time-invariant streaming potential per unit length of artery by solving Laplace's equation $\nabla^2 \cdot_V = 0$ in the

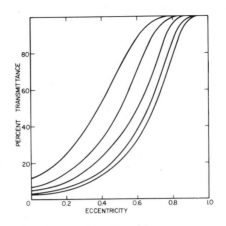

Figure 11

Variation of EAG transmittance with vessel
eccentricity. From right to left, blood vessel
radii are 1, 2, 5, 10 and 20% of idealized limb
radius. Measuring angle is zero degrees.

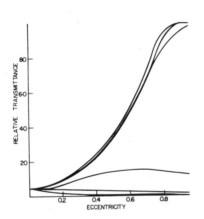

Figure 12

Effect of measuring angle on EAG transmittance.
From top to bottom, measuring angle is 0, 1, 5, 10,
45, 90 and 180 °. Blood vessel radius is 5% of
limb radius

region between the two cylinders, assuming a constant axial
voltage gradient at the arterial surface and a voltage gradient
of zero at the skin surface.
    Transmittance of the EAG signal is plotted in Figures 11
and 12. Figure 11 shows the transmittance as a function of

eccentricity, normalized to the limb radius with the blood
vessel radius as a parameter. Experimentally, it is found that
for arteries close to the surface (eccentricity 0.8 or more)
the EAG signal-to-noise ratio is about 25:50. Therefore, for
vessels more centrally located within the limb or vessels
located within larger tissue masses (i.e., the torso),
transmittance may be so low as to inhibit accurate analysis of
the EAG profile, unless the anamalous behavior noted previously
can be used for this purpose.

Figure 12 illustrates the variation in transmittance with
eccentricity for various measuring angles. It can be seen that
fairly small (less than 10 degrees) errors in electrode
placement will not produce significant degradation of the
signal.

5. Summary

Recent studies have indicated a new feature of the EAG
technique. This feature relates to the possibility of being
able to measure blood flow parameters in arteries distant from
the point at which the measurement is made. In essence, it may
be possible to determine blood flow parameters in deeply buried
major arteries that are not now accessible to any of the
presently available techniques, such as impedance
plethysmography and ultrasound.

References

1. Conti, J.C., Strope, E.R., Griffiths, C.L., Findl, E.,
Kurtz, R.J., 1982, Non Invasive Assessment of the
Cardiovascular System, ed. D. Dietrich, (John Wright, PSG,
Inc., Boston), pp 313-316.

2. Findl, E., Kurtz, R.J., 1977, Electrochemical Studies of
Biological Systems, ed. D.T. Sawyer, (Am. Chem. Soc. Symp.
Series No. 38, Washington), pp 3-14.

3. Findl, E., Kurtz, R.J., Strope, E.R., Conti, J.C., 1980
Proc. 15th Ann. Mtg. AAMI, San Francisco, p 46.

4. Findl, E., 1982, Modern Aspects of Electrochemistry, Vol.
14, ed's J.O.M. Bockris, B.E. Conway, R.E. White, (Plenum
Press, New York), pp 509-555.

5. Findl, E., Conti, J.C., Strope, E.R., Milch, P.O.
Griffiths, C.L., Kurtz, R.J., 1984, Bioelectrochem.
Bioenergetics, 12, 3-14.

6. Kurtz, R.J., Findl, E., Kurtz, A.B., Stormo, L., 1976, J.
Colloid Interface Sci., 57, 28.

7. Macdonald, D.A., 1974, Blood Flow in Arteries, (Williams and
Wilkins, Baltimore), pp 146-173.

8. Plonsey, R. Heppner, A., 1967, Bull. Math. Biophys. 29,

9. Schlichting, H., 1960, Boundary Layer Theory, (McGraw
Hill, New York), pp 502-517.

10. Womersley, J.R., 1955, J. Physio., 127, 553-563.

EFFECTS OF PH BEFORE AND AFTER DEEP-FREEZE STORAGE

AT -80° C ON THE ZETA POTENTIALS OF BONES

William K. Elwood, Stephen D. Smith, and H.E. McKean

Departments of Anatomy and Statistics
University of Kentucky
A.B. Chandler Medical Center
Lexington, KY   40536-0084

INTRODUCTION

During the investigation of electroosmosis with bone in vitro (Elwood and Smith, 1983, 1984c), we became concerned about the effects of storage and storage media on the electrokinetic properties of bone. When changes in the electrokinetic characteristics of bone in vitro were monitored by measuring streaming potentials and calculating zeta(ζ)-potentials it was found that storage media could affect the ζ-potentials of bone stored overnight at 4° C or for several weeks at -80° C (Elwood and Smith 1984a, 1984b). Changes in the electrokinetic properties of the bone were considered to be reflective of its chemical and physical state, and hence of its possible suitability as a bone bank specimen. The changes were also considered to be a likely result of ion adsorption to or exchange between the bone surface and the "fixed" layer of electrolyte solution adjacent to its surface, which includes the double layer discussed in other contributions to this symposium.

Since the work of others indicated that collagen and apatite have isoelectric points (Young 1963; Kambara et al. 1978; Somasundaran 1968; Larsen 1966) a surface of exposed bone matrix (a combination chiefly of apatite and collagen) might also be expected to exhibit a discrete isoelectric point (i.e., a zero zeta-potential point). If bone matrix exhibits such an isoelectric point then ionic exchange with and adsorption to the matrix should be affected accordingly by the pH of the storage or measurement medium.

Streaming potential measurements were used by Kambara et al. (1978). to determine the isoelectric points of human dental enamel (pH 3.8 and 4.7 for enamel of permanent and deciduous teeth, respectively). Although enamel differs from bone in several important respects including ratio of inorganic to organic content, composition of the organic components, and absence of cells from enamel, we thought that an isoelectric point for bone might be demonstrated by an experimental approach similar to that used by Kambara et al.

The following experiments reported here were performed to determine the effects of pH on the potentials of bone in vitro and to determine whether an isoelectric point could be shown for bone.

MATERIALS AND METHODS

Femurs, tibias and humeri were removed from dogs shortly after their sacrifice. The dogs had been used by the Department of Surgery or the Department of Anesthesiology for experimental purposes, and were obtained at the close of their acute experiments. The ends of the bones were removed by means of a Stryker saw, and the marrow was removed by inserting a glass rod through the marrow cavity. Cleansing was completed by scrubbing with a test-tube brush. A length of rubber tube was slipped over each end of a prepared bone so as to hold platinum wire electrodes between the tubing and the bone. One end of each electrode was positioned in the lumen of the marrow space at the end of the bone while the free end was connected to an oscilloscope (502A Dual Beam Oscilloscope, Tektronix Inc., Portland, Oregon). The free end of each rubber tube was placed in a flask containing the fluid used to measure the streaming potential. One of the rubber tubes was passed through a peristaltic pump. When the pump was in operation, a pulse of fluid was propelled through the tubes and marrow cavity. The potential that developed between the ends of the bone with each pulse of fluid was recorded on polaroid film from the image on the oscilloscope screen. A Millar pressure gauge inserted into the tubing on the up-stream side of the bone registered simultaneous pressure changes. The rate of flow was kept at 3.12 ml per pulse throughout the experiment. In order to establish the validity of measurements made in a closed-loop system (vs an opened-loop system) of flowing fluid, some preliminary determinations were made with an opened-loop pattern. These were made with only one end of the rubber tubing in a beaker containing the buffer (used to obtain streaming potential measurements) and the other end in a separate beaker to collect the fluid after its passage through the tubing and bone. No significant differences were observed between the resultant signals obtained with the opened and closed loops, so the more convenient closed-loop configuration was used. Other measurements made as controls indicated that potentials generated by the rubber tubing (expansible) did not contribute significantly to the potentials measured. Specific conductivity was determined with a conductivity bridge (Model RC-16B2, Industrial Instruments, Inc., Cedar Grove, NJ) for each solution used for streaming potential measurements. Zeta potentials were calculated by using the equation

$$\zeta = \frac{4\pi\eta\lambda S}{DP}$$ where ζ = zeta-potential, η = viscosity coefficient;

λ = specific conductance, S = streaming potential, D = dielectric constant, and P = pressure difference between the two ends of the bone. [0.01 was used for viscosity coefficient, 80 for the dielectric constant, and 0.006 kg/cm^2 (an average of measured values) for pressure.] We recognize that the zeta potentials we calcualted are not the strictly-defined zeta potentials one would obtain from a well-defined and homogeneous material's surface. In fact, they are a sort of "compound" or "average" potential (given the heterogeneous nature of the endosteal surface); but we think that they are at least useful as figures of merit, if not as absolute standards of reference.

Fluids Used to Determine Streaming Potentials

A solution of 0.01 M N-2-hydroxyethylpiperazine-N[1]-2-ethanesulfonic acid (HEPES) in 5.2% glucose brought to pH 7.4 with NaOH was used to determine the streaming potentials of each bone before and after the subjection of bone to a pH sequence (see paragraph below). Buffered

HEPES thereby served as a reference solution to monitor change in electrokinetic properties of bone as a consequence of exposure to a pH sequence.

One pH sequence was composed of a series of universal buffer (univ. buffer) solutions, pH 3, 4, 5, 6, 7, 8, 9 and 10, prepared by adjusting with 0.2 N NaOH the pH of a solution consisting of 0.0286 M each of citric acid, potassium dihydrogen phosphate, boric acid, and barbital, at room temperature (23° C) (Data for Biochemical Research, 1969). A corresponding pH sequence composed of a composite of three different buffer systems (citric acid-disodium hydrogen phosphate, monosodium dihydrogen phosphate-disodium hydrogen phosphate, and potassium chloride-boric acid) was prepared and streaming potentials measured using these solutions were compared to those using the universal series. Only the results with the universal series are presented in this paper. (A comprehensive treatment of the entire study has been submitted for publication elsewhere, Elwood and Smith, '85).

Experimental Procedure

Twelve pairs of bones (femurs, tibias and humeri) from four dogs were used. Each pair of bones was arbitrarily divided into an "experimental" and a "control" bone. First, streaming potentials of the twelve experimental bones were determined with HEPES (pH 7.4). Second, streaming potentials were determined for six of the bones using the universal buffer in an ascending pH sequence (i.e., pH 3, 4, 5, 6, 7, 8, 9 and 10) and for the remaining six bones using a descending sequence (i.e., pH 10, 9, 8, 7, 6, 5, 4, and 3). Third, streaming potentials were again determined with buffered HEPES. No prestorage streaming potentials were determined for control bones. They remained in buffered HEPES at room temperature while prestorage potential measurements were made of their mates. In this way, both experimental and control bones remained at room temperature for the same length of time before storage. Both experimental and control bones were then immersed and stored in HEPES at -80° C. Bones from two of the dogs were stored for 3 weeks; those from the other two were stored for 8 weeks. Following storage, the bones were thawed and the determinations of streaming potentials were repeated in the same order as before storage for each bone (i.e., HEPES → universal buffer pH 3 to 10 → HEPES, or HEPES → univ. buffer pH 10 to 3 → HEPES). Streaming potentials for the control bones were measured in the same pH sequence of univ. buffers as their mates. The control bones thus served to determine whether prestorage subjection of bones to a pH sequence may have altered their storage stability and/or electrokinetic characteristics. All bones were marked so that the same end of the bone was facing the direction of fluid flow for the later streaming potential determinations as for the first.

An estimation of differences in ζ-potentials attributable to differences in specific conductivity was made by arbitrarily using the ζ-potentials calculated with mean streaming potentials of bones at pH 3 as a reference for the ascending pH sequence and at pH 10 for the descending pH sequence.

Statistical Analyses

An SAS package (File Management and Statistical Analyses by SAS. SAS User's Guide, 1982, Cary, N.C.) on an IBM main-frame computer was used to analyze the data. Factors analyzed included storage time (3 vs 8 wks), direction of pH sequence (pH 3 to 10 vs pH 10 to 3), changes in potentials with storage (pre- vs poststorage), and effects of prestorage treatment on poststorage values (experimental vs control).

RESULTS

An analysis of the data showed no consistent or significant difference between the 3- and 8-week storage results; the results of the two storage periods were therefore pooled. Figs. 1 and 2 show the pre- and poststorage, and control results. Plotting ζ-potentials vs pH formed trend lines that were essentially linear. (Potentials determined with HEPES were not included in trend lines.)

Monitoring of Univ. Buffer Trend Lines

Controls. A single common trend line was suitable statistically for all control values (Compare Figs. 1 and 2). The equation of the common line was: $y = 0.240 + 0.072$ pH, where y = predicted potential.

Prestorage. A single trend line statistically served all prestorage data whose common equation was $y = 0.242 + 0.047$ pH (y = predicted potential). (Compare Figs. 1 and 2.) Furthermore, since prestorage and control potentials did not differ significantly a single line could be fitted by both sets of data with a common equation of $y = 0.241 + 0.054$ pH.

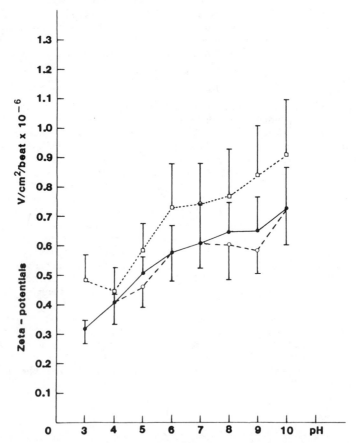

Fig. 1.   Zeta-potentials of bone in vitro calculated from streaming potentials determined in an ascending pH sequence of universal buffer. ●——● before storage at -80° C. o - - -o after storage. ◻···◻ after storage controls.

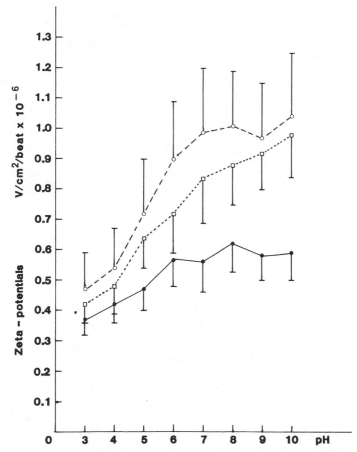

Fig. 2. Zeta-potentials of bone in vitro calculated
from streaming potentials measured in a descending pH sequence of universal buffer.
●——● before storage at -80° C. ○- - -○ after
storage. □ ···· □ after storage controls.

Poststorage. Whether the pre- and poststorage trend curves were
statistically different depended on the pH sequence. No significant difference was found between pre- and poststorage measurements made during
pH sequences 3 to 10 (Fig. 1); however, with the pH sequence 10 to 3, the
poststorage trend line was considerably different from the prestorage one
(P < 0.04) (Fig. 2).

Contribution of Differences in Specific Conductivity to ζ-potential
Differences. A comparison of estimated values vs measured ones indicated that changes in ζ-potentials from one pH to another in the univ.
buffer sequence could be attributed for the most part to differences in
specific conductivity of the fluids. See Fig. 3.

A comparison of Fig. 3 with Figs. 1 and 2 showed that variations
away from the zero "base" line corresponded to similar deviations of the
corresponding trend lines. (If an observed value is identical to an
estimated value, it lies on the "zero" line.)

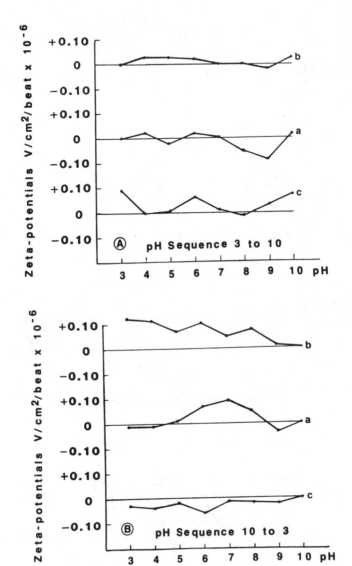

Fig. 3 (Parts A and B). Estimation of zeta-potential differences between one pH and another not accounted for by differences in specific conductivity. Zeta potentials calculated with mean streaming potentials at pH 3 for the ascending pH sequence or at pH 10 for the descending pH sequence were arbitrarily used as zero values with the exception that, in the case of control, "c", pH sequence 3 to 10 where pH 4 was used for 0 value. 0 = point where estimated zeta-potential equals the observed potential. Specific conductivity was the only variable. Deviation to the + side means that the estimated value was less than the observed one. A = pH sequence 3 to 10, B = pH sequence 10 to 3, b = before storage at -80° C, a = after storage at -80° C, and c = control after storage at 80° C.

No significant difference occurred between ζ-potentials measured in HEPES before vs after a pH sequence when measurements were made either pre- or poststorage. However, analyses showed that potentials obtained in HEPES before storage were significantly higher than those obtained after storage, whether comparison was made with potentials from the same bone ($p < 0.03$) or from its control mate ($p < 0.05$). Comparison of potentials of controls and their experimental mates after storage showed no significant difference.

DISCUSSION

Zeta-potentials were determined for bone in vitro at pH 3, 4, 5, 6, 7, 8, 9 and 10 with univ. buffer in either an ascending or a descending pH sequence before, and again after, storage at $-80°$ C. Failure to find a point of zero potential within the pH range used indicated that no isoelectric point was demonstrable with the bone as prepared by us. Anderson and Eriksson (1970) have suggested that if the isoelectric point of bone hydroxyapatite and that of bone collagen differed, there would be no true isoelectric point for bone. Since the surfaces of the bone preparations we used for ζ-potential calculations included even more than collagen and hydroxyapatite (the chief components of extracellular bone matrix), e.g., cells and adherent soft tissue, the likelihood of obtaining an isoelectric point would be expected to be further reduced.

Covariate analyses of the data indicated that three factors, i.e., pH sequence, storage, and prestorage treatment, had an interrelated effect on the observed potentials. When the pH sequence 10 to 3 was used, the poststorage potentials differed significantly from the prestorage ones, whereas no significant difference occurred between poststorage and prestorage values obtained with the pH sequence of 3 to 10. We do not know whether the same results would have been obtained if we had performed the second series of measurements immediately following the first set without an intervening storage period. Such an experiment would have indicated whether storage was an important factor in conjunction with the pH sequence. It may be relevant in this respect to note that the pre- and post-pH sequence ζ-potentials obtained in HEPES before storage were not significantly different, nor were the corresponding measurements made with HEPES after storage. One might have expected the prestorage values made after the pH 10 to 3 sequence to differ from those made before the pH 10 to 3 sequence, and likewise to differ from measurements obtained before and after the pH 3 to 10 sequence.

It is curious that all the prestorage values obtained with HEPES were significantly greater than all poststorage ones determined with HEPES, which strongly suggests a storage effect. Our own previous studies had indicated that potential values dropped considerably, but not significantly at the $p < 0.05$ level when bone was stored in HEPES at $-80°$ C (Elwood and Smith, 1984b). In contrast, a corresponding poststorage reduction in potentials was not evident when measurements were made within the univ. buffer pH series.

Estimates of ζ-potentials compared with the observed data indicated that within each pH sequence category (e.g., prestorage pH 3 to 10 experimental, prestorage pH 10 to 3 experimental, poststorage pH 3 to 10 control, poststorage pH 10 to 3 control, poststorage pH 3 to 10 experimental, and poststorage pH 10 to 3 experimental), any differences observed between values determined at one pH vs another could be attributed to differences in specific conductivity. Hence, change in the electrokinetic status of the bone surface would occur as a shift of the entire pH trend line and be evident as either a rather uniform increase or decrease in ζ-potentials. This circumstance occurred with the pre- vs post storage, pH 10 to 3 sequence of potentials and resulted in an overall increase in ζ-potentials. (Fig. 2)

The heterogeneity of the endosteal bone surface, the sequence of solutions to which the surface was exposed, the solubility of surface components, the elution of substances from the endosteal surface, and ionic exchange between the buffers and the bone surface are all factors which could affect ζ-potentials. Time also may be an important consideration. If we assume a corollary status between the study of Neuman and Neuman (1958) regarding the kinetics of exchange of labeled phosphate with hydroxyapatite and the possible kinetics of our experiment, we would envision four compartments involved in an interaction between the buffer and bone. These four compartments would be as follows: first, the bulk of the buffer passing through the marrow cavity; second, a diffusion layer (non-fixed potential gradient) adjacent to the surface; third, a layer strongly adsorbed (fixed) to the surface; and fourth, an intrabone pool deep to the fixed layer. Exchange of phosphate between the corresponding first and second compartments in the Neuman experiment was very rapid (a matter of minutes); between the second and third compartments it was slower (measured in hours); and between the third and fourth compartments it was slowest of all.

The observations of Neuman and Neuman suggest that since the time required to measure streaming potentials is very short (less than a minute), there would be insufficient time for a significant interaction except between the bulk of the buffer and the diffusion layer. Since the streaming potentials reflect principally the movement between the diffusion and the fixed layer, one would suspect that the potentials obtained reflected a fixed layer that remained fairly constant in character and a diffusion compartment that changed with each change of the buffer solution. When the bones were transferred from the acidic or basic end of a pH sequence into HEPES for storage, and when the bones were later thawed, a much greater length of time was available for an exchange between the diffusion and fixed compartments. If the exchange between the diffusion and fixed compartments were sufficient to change the character of the fixed layer, then a second series of streaming potential measurements (i.e. poststorage) would differ from the first (i.e. prestorage). Although the diffusion compartment would be affected in a similar manner with each change of buffer solution, the fixed layer would no longer be the same.

In summary, differences in steaming potential measurements of endosteal bone in vitro with univ. buffer at pH 3, 4, 5, 6, 7, 8, 9 and 10 could be correlated chiefly to differences in specific conductivity of the measuring solutions. Potentials measured during a pH sequence of 10 to 3 after storage at -80° C differed significantly from those obtained during the same sequence prior to storage; however, no differences occurred between pre- and poststorage ζ-potentials when measurements were made during pH sequences of 3 to 10. No isoelectric point was observed for bone.

ACKNOWLEDGMENTS

This investigation was supported by Biomedical Research Support Grant #RR05374 from the Biomedical Research Support Branch, Division of Research Facilities and Resources, NIH.

The authors wish to thank Dr. James O'Reilly, Department of Chemistry, University of Kentucky, for his help in determining conductances and for the use of his Conductivity Bridge.

REFERENCES

Anderson, J. C., and Eriksson, C., 1970, Piezoelectric properties of dry and wet bone, Nature, 227:491.

Elwood, W. K., and Smith, S. D., 1983, Electroosmosis in compact bone, J. Bioelec., 2:37.

Elwood, W. K., and Smith, S. D., 1984a, Effects of freeze-thawing, storage medium, and incubation on the streaming potentials of bone in vitro, J. Bioelec., 3:67.

Elwood, W. K., and Smith, S. D., 1984b, Effects of refrigerated (4° C) and deep-freeze (-80° C) storage in buffered HEPES, pH 7.4 on the zeta-potentials of bone, J. Bioelec., 3:385.

Elwood, W. K., and Smith, S. D., 1984c, Electroosmosis in compact bone in vitro, J. Bioelec., 3:409.

Elwood, W. K., Smith, S. D., and McKean, H. E., 1985, Effects of pH on the zeta-potentials of bone in vitro, J. Bioelec. Submitted for publication.

Kambara, M., Asai, T., Kumasaki, M., and Konishi, K., 1978, An electrochemical study on the human dental enamel with special reference to isoelectric point, J. Dent. Res., 57:306.

Larsen, S., 1966, Solubility of hydroxyapatite, Nature, 212:605.

Neuman, W. F., and Neuman, M. W., 1958, "The Chemical Dynamics of Bone Mineral", University of Chicago Press, Chicago.

Somasundaran, P., 1968, Zeta potential of apatite in aqueous solutions and its change during equilibration, J. Colloid and Interface Sc., 27:659.

Young, E. G., 1963, Occurrence, classification, preparation and analysis of proteins, in: "Comprehensive Biochemistry", Vol. 7, M. Florkin and E. H. Stotz, eds., Elsevier Publishing Co., New York.

____, 1969, "Data for Biochemical Research", R. M. C. Dawson, D. C. Elliot, W. H. Elliot and K. M. Jones, eds., 2nd ed. p. 485. Oxford University Press, New York.

ELECTROCHEMICAL AND MACROMOLECULAR INTERACTIONS AT RED BLOOD CELL SURFACE

Kung-ming Jan and Shlomoh Simchon

Departments of Physiology and Medicine, Columbia University
College of Physicians and Surgeons, New York, NY 10032

ABSTRACT

Red blood cells aggregate to form rouleaux in the presence of macro-molecules. Neutral and charged macromolecules were used to induce RBC aggregation quantified by microscopic observation. Variations of cell surface potential were achieved by the removal of RBC surface charge with neuraminidase treatment or by changing the ionic composition of the fluid medium. RBC aggregation by neutral polymer dextran is enhanced by removal of RBC surface charge and decreased by reduction of ionic strength. RBC aggregation by heparin requires the presence of sialic acids at cell surface and enhanced by reduction of ionic strength. It is concluded that the surface charge of RBCs plays a significant role in cell-to-cell interactions.

INTRODUCTION

Red blood cells (RBCs) aggregate to form rouleaux in the presence of high molecular weight mocromolecules, e.g. fibrinogen (Fåhraeus, 1929) and dextrans (Thorsén and Hint, 1950). This type of cell-to-cell inter-action is one of the major determinants of erythrocyte sedimentation rate (Fåhraeus, 1929) and low-shear blood viscosity (Chien et al., 1966). The mechanism of RBC aggregation has been extensively studied and is postu-lated to be due to macromolecular bridging between adjacent cell surfaces (Chien, 1975). According to this theory, in order for RBCs to aggregate, the macromolecular bridging energy has to overcome the disaggregating energies which include the electrostatic repulsion between the charged surfaces (Jan and Chien, 1973a), the membrane strain energy (Skalak et al., 1981) and the work done by mechanical shearing (Chien et al., 1977).

Human RBCs are negatively charged, primarily from the ionogenic carboxyl groups of sialic acid at cell surface (Cook et al., 1961, Eylar et al., 1962). The surface sialic acids can be removed by enzymatic treatment with neuraminidase (Jan and Chien, 1973a). Studying the aggregations of normal and neuraminidase-treated (N-treated) RBCs in macromolecular suspension, Jan and Chien (1973a) suggested that the surface charge of RBCs inhibits or prevents RBC aggregation. For a given surface charge density, the macromolecular bridging force is a function of the number of bridging macromolecule per cell pair, the number of adsorption bonds per molecule, and the adsorption force per bond (Chien, 1975). The adsorption of macromolecules onto RBC surface when RBCs form rouleaux has also been demonstrated (Brook, 1973; Chien et al., 1977). The results are in agreement with the model of RBC aggregation proposed from the macromolecular bridging theory.

The purpose of the present study was to compare the nature of cell-to-cell interaction induced by neutral macromocules, dextran (Dx), and by negatively charged macromolecules, heparin. Experiments were designed to investigate the role of surface electrical potentials in RBC aggregation by variations of the charge and ionic environment at cell surfaces.

MATERIALS AND METHODS

Cells

Fresh blood samples from healthy human subjects were drawn into heparinized syringes. After centrifugation and the removal of plasma and buffy coat, the RBCs were washed three times with a saline solution. The saline solution is composed of 0.15 M NaCl and contains 0.125 g/dl human serum albumin to prevent cell crenation and hemolysis (Ponder, 1948) and 12 mM Tris (hydroxymethyl) amino-methane buffer at pH = 7.4.

To prepare the charge depleted RBCs, neuraminidase from Vibrio Cholerae (Calbiochem-Behring Corp., La Jolla, CA) was added to suspensions of washed RBCs in saline (hematocrit = 45 percent), at a concentration of 0.05 unit/ml. The suspension was incubated at $37°C$ under constant shaking for 60 min and subsequently washed three times with saline. Charge depletion of greater than 90 percent reduction of the cell electrophoretic mobility was confirmed with the use of a cylindrical microelectrophoretic apparatus (Rank Bros., Cambridge, England) at $25°C$.

Suspending Medium

The macromolecules used to induce RBC aggregation were Dx 70 (Pharmacia Labs, Uppsula, Sweden; $M_w = 63,000$, $M_n = 41,500$) and heparin (Sigma Chemical Co., St. Louis, MI). The macromolecules were dissolved in 150 mM NaCl solution with concentrations up to 10 g/dl. In order to study the effect of variations in ionic strength on RBC aggregation, suspending media were also prepared in iso-osmotic mixtures of 150 mM NaCl and 300 mM sucrose in various ratios.

Quantification of RBC aggregation

The washed RBCs were suspended in the solutions containing macromolecules at hematocrit = 1 percent. The quantification of RBC aggregation was by direct microscopic observations in a hemocytometer chamber (Jan and Chien, 1973a). The microscopic aggregation index (MAI), which gives the average number of RBCs in each aggregation unit, was calculated as the ratio of the number of RBCs per unit area on the photomicrograph to the number of aggregation units in the same area.

Determination of adsorption isotherms

Adsorption isotherms of Dx 70 on RBC surfaces were determined by the anthrone reagent (Semple, 1957) according to the methods described by

Chien et al. (1977). To determine the adsorption isotherms of heparin, the heparin was conjugated with ^{125}I-Bolton-Hunter reagent (ICN Bio-medicals, Inc., Irvine, CA) at pH = 8.5 and purified by gel filtration using Sephadex-G 10. After dialysis and lyophylization, the radioactive heparin was dissolved in the saline at concentrations up to 10 g/dl. Each solution contained a trace amount of ^{51}Cr-human albumin (Squibb Diagnostics, New Brunswick, NJ) for trapping correction (Chien et al., 1977). After adding RBCs and centrifugation, the adsorption isotherms were determined counting by the radioactivities of the supernatant and the packed RBCs. The quantity of macromolecules adsorbed per RBC was calculated according to Chien et al. (1977).

Intrinsic viscosities of macromolecules vs. ionic strength

Saline was mixed with distilled water in various ratios to obtain NaCl solutions of various ionic strengths. Dx 70 and heparin were dissolved in each of these NaCl solutions in concentrations from 0.25 to 5.0 g/dl. The viscosity of each solution was measured by a co-axial cylinder viscometer (Chien et al., 1966) at 37°C. Since the viscosity of each solution is independent of the shear rate, values obtained at 5.2 and 0.52 sec^{-1} were averaged. The intrinsic viscosities of Dx 70 and heparin were calculated according to the methods described by Jan (1985).

RESULTS

Normal and neuraminidase (N-treated) RBCs suspended in the saline solution were monodispersed. The addition of Dx 70 or heparin to RBC suspensions caused aggregation of RBCs. The degree of RBC aggregation was a function of the macromolecular concentration, the integrity of surface charge, and the ionic strength of the suspending medium (Fig. 1). With Dx70, normal RBCs at normal ionic strength (I = 150 mM) showed char-acteristics aggregation and disaggregation phases, with an optimum con-centration for aggregation at 4 g/dl. The N-treated RBCs not only showed a greater aggregation than normal RBCs, but also did not exhibit the dis-aggregation phase seen in normal RBCs. A decrease in ionic strength caused a reduction in aggregation of normal RBCs in dextran, but had no effect on the aggregation of N-treated RBCs. With heparin, normal RBCs showed appreciable aggregation at concentration of 1 g/dl. An increase in heparin concentration caused a progressive increase in RBC aggrega-tion, and there was no disaggregation phase of the normal RBCs at concen-trations of heparin up to 10 g/dl. The N-treated RBCs, however, showed no aggregation in heparin. A decrease in ionic strength resulted in an increase in aggregation of normal RBCs in heparin. The N-treated RBCs, however, showed no aggregation despite the reduction in ionic strength.

The surface adsorptions of dextran and heparin molecules rose with the bulk concentrations (Fig. 2). With Dx 70, the adsorption isotherms showed a smooth curve up to a concentration of approximately 5 g/dl; the adsorption curve for N-treated RBCs was not significantly different from that for normal RBCs. When the bulk concentration of dextran was above 5 g/dl, secondary adsorption curves were found for both types of RBCs. The surface adsorption of dextran in the secondary adsorption region was significantly higher for normal than N-treated RBCs. With heparin, there was significant difference in adsorption isotherms between the two types of RBCs. The surface adsorption of heparin was significantly higher for N-treated RBCs than for normal RBCs.

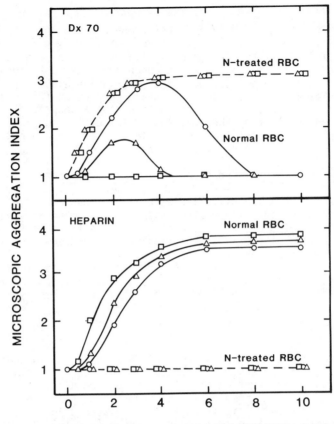

Fig. 1. Effects of ionic strength (I) on aggregations of
normal RBCs (solid lines) and N-treated RBCs
(dashed lines) in NaCl solution plus Dx 70 (upper
panel) and heparin (lower panel). The ionic
strengths were I = 150 mM (open circle), 100 mM
(open triangle), and 50 mM (open square). The
total osmolarity was 300 mOsm, balanced by
sucrose. The figure gives the relation between
the microscopic aggregation index and macromolecular concentration at various I.

     The effects of varying ionic strength on the intrinsic viscosities
($/\eta/$) of Dx 70 and heparin were illustrated in Fig. 3. The intrinsic
viscosity of Dx 70 was constant ($/\eta/$ = 0.26 dl/g) over a wide range of
ionic strength. Heparin, on the other hand, showed a marked increase in
intrinsic viscosity as the ionic strength of solution was reduced. The
intrinsic viscosity of heparin was 0.14 dl/g at I = 150 mM and it increased to 0.31 dl/g as the ionic strength approached zero.

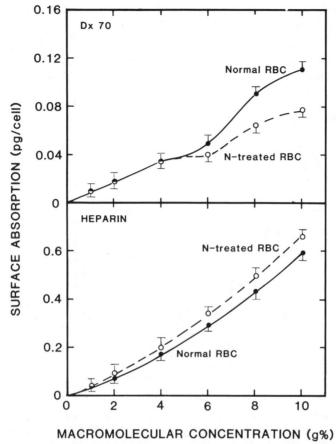

Fig. 2. Adsorption isotherms of Dx 70 (upper panel) and
heparin (lower panel) on the surfaces of normal
(solid lines) and N-treated RBCs, (dashed
lines). Abscissa is the bulk concentration of
macromolecules. The vertical bars represent SEM.

DISCUSSION

RBC aggregation in macromolecular suspension is primarily due to
macromolecular bridging between cell surfaces. The attractive force is
provided by short-range adsorption forces of the macromolecules onto the
cell surface. These forces include the electrostatic attraction, hydrogen bonding, and van der Waals forces. For dextran molecules, the nature
of the adsorption forces onto RBC surfaces has been postulated to be
hydrogen bonding and van der Waals forces (Jan, 1979). For heparin
molecules, which are highly negatively charged macromolecules, the nature
of surface adsorption onto RBC surfaces has not been well understood.
The surface adsorption of macromolecules is in a dynamic process (LaMer
and Healy, 1963), so that at equilibrium state, the rate of adsorption
is equal to that of desorption. The adsorbed macromolecules consist of
adsorbed segments and extended segments at the surface (Silverberg,

1962). When the extended segments of the adsorbed macromolecules are
sufficient in number and length, they can be further adsorbed on the
surface of the adjacent cells and result in aggregation of cells. In the
present investigation, the adsorption isotherms of Dx 70 and heparin
(Fig. 2) showed continuous curves of surface adsorption at the concentration ranges studied. These findings are in support of the macromolecular

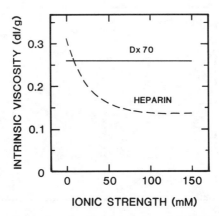

Fig. 3. Effects of ionic strength on the intrinsic viscosities of Dx
70 and heparin in NaCl solution. The figure gives the
relation between the intrinsic viscosity and the ionic
strength.

bridging theory of RBC aggregation. With Dx 70, the adsorption isotherms
show secondary curves at bulk concentrations above 5 g/dl and the surface
adsorption of dextran was greater on normal RBCs than on N-treated RBCs.
These findings may suggest that there is an increase in availability of
adsorption sites on normal RBCs because of disaggregations of RBCs. With
heparin, N-treated RBCs showed a greater surface adsorption of heparin
than normal RBCs, suggesting that heparin adsorption is probably at the
sites of non-ionic region of RBC surfaces and that removal of sialic
acids results in an increase in the surface area for heparin adsorption.

    The major opposing force against the macromolecular bridging force
in RBC aggregation is a long-range electrostatic repulsion between
adjacent cell surfaces. The zeta potential of RBCs in saline calculated
from Helmholtz-Smoluchowski equation is -15 mV at 25°C (Jan nd Chien,
1973a). Neuraminidase treatment results in a marked reduction of the
zeta potential of RBCs to a level of less than -2 mV (Jan and Chien,
1973a). According to the theory of electrical double layer (Overbeek,

1952), the negatively charged cell surface is surrounded by an ionic
atmosphere composed of counter-ions and co-ions in the suspending medium.
The ultrastructure of the intercellular relation in the rouleaux has
shown that the RBC surface may be assumed to be an infinitally large
plane (Jan and Chien, 1973a). For this flat plate model, the potential
distribution in the diffuse double layer (ψ) is a function of the
surface potential (ψ_0), the distance normal to the surface (x), and
the ionic composition of the solution. The potential profile can be
approximated as:

$$\psi = \psi_0 e^{-\kappa x} \tag{1}$$

is the Debye-Huckel function:

$$\kappa = (8 \pi z^2 F^2 c / DRT)^{1/2} \tag{2}$$

where c is the concentration of electrolytes in moles per unit volume, z
is the valency of ion, D is the dielectric constant of medium, and F, R
and T are the Faraday constant, the gas constant and the absolute
temperature, respectively. The potential decreases almost exponentially
from the surface with a slope κ. The reciprocal of κ is customarily
called the "thickness" of the diffuse double layer. The neuraminidase
treatment of RBCs results in a decrease of ψ_0, whereas reduction in
the ionic strength of the suspending medium causes an increase in ψ_0
due to a reduction in the screening of surface charge by counterions and
an increase in "thickness" of the double layer according to equation (2)
(Jan and Chien, 1973b).

The repulsive force between adjacent RBC surfaces results from the
overlap of the two electrical double layers. For the parallel plate
model with separation of d, the potential energy of repulsion per unit
area (E_R) is:

$$E_R = (64 \ cRT/\kappa) \ \gamma^2 \ ^{-\kappa d} \tag{3}$$

where $\gamma = \tanh (zF\psi_0/4RT)$. With neutral macromolecules, dextran,
findings on the aggregation of normal and N-treated RBCs (Fig. 1) suggest
that the surface charge of RBCs inhibits RBC aggregation. A decrease in
ionic strength causes a reduction in aggregation of normal RBCs in
dextran, but has no effect on the aggregation of N-treated RBCs. At low
ionic strength, the electrostatic repulsive interaction between normal
RBCs is increased due to an increase in the surface potential and an
increase in the "thickness" of the double layer. At I = 150 mM, the
disaggregation of RBCs at high concentration of dextran is attributed to
an increase in the surface electrostatic repulsion. Brooks (1973) first
proposed that as a result of volume exclusion by macromolecules, the
effective ionic strength of the suspending medium decreases, and this
effect would be most pronounced at cell surface where macromolecular
concentration is increased. Indeed, the zeta potential of RBCs is increased by the presence of dextrans (Jan and Chien, 1973a). The increases in ψ_0 and the "thickness" of electrical double layer result in
an enhancement of the electrostatic repulsive interaction according to
equations (2) and (3). This theory explains satisfactorily the mechanism
of disaggregation of RBCs by high concentrations of dextran. However,
the possibility that the alignment of sialic acids at cell surface may
be altered in response to the presence of dextran (Chien, 1980) can not
be excluded.

Heparin-induced RBC aggregation requires the presence of the negatively charged sialic acids at RBC surfaces (Fig. 1). Heparin molecules

are also negatively charged. Enhancement of the charge interaction by lowering the ionic strength causes an increase in aggregation of normal RBCs. These findings led to a hypothesis that heparin molecules bind with RBC surface sialic acids through a short-range electrostatic attraction between charged molecules of the same sign but of dissimilar charge densities (Jan, 1980). Studies on the adsorption isotherms, however, suggest that heparin adsorption onto RBC surfaces increased by removal of sialic acids. It appears that in heparin-induced RBC aggregation, heparin molecules require surface charges to cause RBC aggregation but do not bind with the surface charge. In the presence of surface charge, the conformation or molecular length of heparin is probably altered in favor of bridging adjacent cell surfaces. It is likely that the adsorbed heparin molecules are more extended in the negative potential environment at the charged surface. The mechanisms may be related to the counterion distribution in the electrical double layer or to the altered dielectric constant in the region (Jan, 1985). When the surface changes are removed, the extended heparin molecules resume their natural conformation.

The molecular dimension of macromolecules in the solution can be estimated by determination of the intrinsic viscosity of the macromolecular solution (Houwink, 1952). As the axial ratio and/or the hydrodynamic volume of the macromolecules increase, the intrinsic viscosity is increasingly greater (Tanford, 1961). The constancy of intrinsic viscosity of Dx 70 over a wide range of ionic strength (Fig. 3) indicates that its molecular dimension is not a function of the surrounding ions. The ionized sulfate and carboxyl groups of heparin molecule, however, may establish diffuse double layers and result in mutual repulsions among the intramolecular ionized groups. Low ionic strength causes an increase in this intramolecular repulsive force and results in an expansion of the molecule. At high ionic strength, the diffuse double layers are compressed and the molecules then exist in a more compact form (Overbeek and Bungenberg de Jong, 1952). Such compression-expansion effects of ionic strength on the macromolecular dimension of heparin are illustrated by the increase in intrinsic viscosity as the ionic strength of the solution is reduced (Fig. 3). Therefore, in a system consisting of RBCs suspended in a solution of charged macromolecules, variations in the ionic strength of the solution may change not only the repulsive force between the cells but also the dimension of the macromolecule. The repulsive forces between charged surfaces tend to prevent RBC aggregation, while the effectiveness of a macromolecule in bridging RBCs is a function of the macromolecular length. The degree of RBC aggregation by heparin in various ionic strengths thus represent the net effect of these two factors. At low ionic strength, the increased effectiveness of heparin in inducing RBC aggregation (Fig. 1) may indicate the predominance of the important effect of molecular expansion over that of the increased electrostatic repulsion between RBC surfaces. Because the molecular dimension of Dx 70 is not a function of the ionic strength (Fig. 3), the diminished aggregation of normal RBCs at low ionic strength (Fig. 1) is only due to an increased repulsive force between cells. Therefore, the aggregation of N-treated RBCs in Dx 70 is independent of the ionic strength of solution.

ACKNOWLEDGMENT

The authors would like to thank Dr. Shu Chien for helpful discussion. This work was supported by NHLBI Research Grants HL-16851 and Training Grant HL-07114.

REFERENCES

Brooks, D. E., 1973, The effect of neutral polymers on the electrokinetic potential of cells and other charged particles. II. A model for the effect of adsorbed polymer on the diffuse double layer. J. Colloid Interface Sci., 43:687.

Brooks, D. E., Charalambous, J., and Janzen, J., 1978, The molecular mechanism of erythrocyte aggregation, in: "Proc. Third Internat. Congr. on Biorheology", Y. C. Fung and J. G. Pinto, ed.), LaJolla, CA., pp. 91.

Chien, S., Usami, S., Taylor, H. M., Lundberg, J. L., and Gregersen, M. I., 1966, Effects of hematocrit and plasma proteins on human blood rheology at low shear rates. J. Appl. Physiol., 21:81.

Chien, S., 1975, Biophysical Behavior of red cells in suspensions, in: "The Red Blood Cell", 2nd ed.; D. MacN. Surgenor, ed., Academic Press New York; Vol. 2, pp. 1031.

Chien, S., Simchon, S., Abbott, R. E., and Jan, K.-M., 1977, Surface adsorption of dextrans on human red cell membrane. J. Colloid Interface Sci., 62:461.

Chien, S., Sung, A. L., Kim, S., Burke, A. M., and Usami, S., 1977, Determination of aggregation force in rouleaux by fluid mechanical technique. Microvasc. Res., 13:327.

Chien, S., 1980, Aggregation of red blood cells: An electrochemical and colloid chemical problem, in: "Advances in chemistry series, no. 188 Bioelectrochemistry: Ions, surface, membrane", M. Blank, ed., American Chemical Society, 3.

Cook, G. M. W., Heard, D. H., and Seaman, G. V. F., 1961, Sialic acids and the electrokinetic change of the human erythrocyte. Nature, 191:44.

Eylar, E. H., Madoff, M. A., Brody, O. V., and Oncley, J. L., 1962, The contribution of sialic acid to the surface charge of the erythrocyte. J. Biol. Chem., 237:1992.

Fåhraeus, R., 1929, The suspension stability of blood. Physiol. Rev., 9:241.

Houwink, R., 1952, The formation and structure of macromolecules. Macro-molecular sols without electrolyte character, in: "Colloid Science", H.R., Kruyt, ed., Elsevier Publishing Co., Amsterdam, Houston, New York, London, Vol. II, Chapter II and VI.

Jan, K.-M., and Chien, S., 1973a, Role of surface electric charge in red blood cell interactions. J. Gen. Physiol., 61:638.

Jan, K.-M., and Chien, S., 1973b, Influence of ionic composition of fluid medium on red cell aggregation. J. Gen. Physiol., 61:655.

Jan, K.-M., 1979, Red cell interactions in macromolecular suspension. Biorheology., 16:137.

Jan, K.-M., 1979, Role of hydrogen bonding in red cell aggregation. J. Cell Physiol., 101:49.

Jan, K.-M., 1980, Electrochemical basis of heparin-induced red blood cells aggregation, in: "Advances in chemistry series, no. 188 Bioelectrochemistry: ions, surfaces, membranes", M. Blank, ed., American Chemical Society, 143.

Jan, K.-M., Usami, S., and Chien, S., 1982, The disaggregation effect of dextran 40 on red cell aggregation in macromolecular suspensions. Biorheol., 19:543.

Jan, K.-M., 1985, Roles of surface electrochemistry and macromolecular adsorption in heparin-induced red blood cell aggregation. Biorheol., in press.

LaMer, V. K., and Healy, T. W., 1963, Adsorption-flocculation reaction of macromolecules at the solid-liquid interface. Rev. Pure Appl. Chem. 13:112.

Overbeek, J. Th. G., 1952, Electrochemistry of the double layer. Electrokinetic phenomena. Interaction between colloidal particles. Kinetics of flocculation, in: "Colloid Science", H.R. Kruyt, ed., Elsevier Publishing Co., Amsterdam, Houston, New York, London, vol. II, Chap. IV-VI.

Overbeek, J. Th. G., and Bungenberg de Jong, H. G., 1952, Sols of macromolecular colloids with electrolytic nature, in: "Colloid Science", H.R. Kruyt, ed., Elsevier Publishing Co., Amsterdam, Houston, New York, London; vol. II, Chap. VIII.

Ponder, E., 1948, "Hemolysis and related phenomena", Grune and Stratton, New York, pp. 398.

Semple, R. E., 1957, An accurate method for the estimation of low concentration of dextran in plasma. Cana. J. Biochem. Physiol., 34:383.

Silverberg, A., 1962, The adsorption of flexible macromolecules. Part I. The isolated macromolecule at a plane interface. J. Phys. Chem., 66:1872.

Skalak, R., Zarda, P. R., Jan, K.- M., and Chien, S., 1981, Mechanics of rouleau formation. Biophysical J., 35:771.

Tanford, C., 1961, "Physical chemistry of macromolecules", John Wiley and Sons, Inc., New York, London Sidney, pp. 710.

Thorsén, G., and Hint, J., 1950, Aggregation, sedimentation and intra-vascular sludging of erythrocytes. Acta Chir. Scand. (Suppl.), 154:1.

THE MYOSIN FILAMENT; CHARGE AMPLIFICATION AND CHARGE CONDENSATION

G.F. Elliott, E.M. Bartels and R.A. Hughes

Open University, Oxford Research Unit, Foxcombe Hall
Berkeley Road, Boars Hill, Oxford, OX1 5HR, England

Current models for the molecular mechanism of muscular contraction are
derived largely from structural and biochemical information. Experimental
evidence has traditionally been obtained as answers to questions like "where
are the various contractile proteins located, what are the physical and
chemical properties of the isolated, purified proteins in solution, how do
the kinetics of muscle chemistry correlate with structural changes in working
muscles?" A detailed picture of many essential events in muscle contraction
has been obtained in this manner and has been incorporated into current views
of the sliding filament hypothesis, see for example the monographs of A.F.
Huxley (1980) and Bagshaw (1982).

A direct physical approach to any working mechanism of microscopic
dimensions should include (1) the distribution of matter, and importantly
(2) the distribution of electric charge. Some understanding of these two
distributions would provide an essential step in defining a working mechanism
in physical terms. While the experimental paradigm in the biological
sciences usually includes substantial effort to define the distribution of
matter in systems which may have been simplified, in solution for example,
the distribution of electric charge is often overlooked or ignored in kinetic
schemes, largely because of the technical difficulties in obtaining a reliable index of charge patterns at high spatial and temporal resolution. Even
in contracting muscle, where finite changes in the distribution of contractile material occur on a measureable time scale, very little effort has been
made to analyse the mechanism in the physical terms of charge density and
distribution.

Collins and Edwards (1971) showed that Donnan potentials could be
observed from vertebrate striated muscle (using 3 M KCl electrodes) after the
membrane had been rendered porous by glycerol treatment, and that these potentials could be used to calculate the fixed charge on the contractile
proteins. After this pioneering work we combined the same approach with
X-ray and light diffraction (Elliott et al 1978, Naylor et al 1985) to
derive the linear charges on the thick (myosin-containing) and thin (actin-
containing) filaments in the muscle. The theoretical and practical basis of
the method of filament-charge measurement is discussed in Elliott and Bartels
(1982), Elliott et al (1984) and Naylor (1982).

Using light microscopy to locate the microelectrode tip, we also demonstrated (Bartels and Elliott 1980, 1981, 1985) different Donnan potentials

in the A- and I-bands of glycerinated vertebrate striated muscle (and in mechanically skinned barnacle muscle). In rigor (the absence of ATP) the negative A-band potential is about twice the I-band potential. In relaxing solution (containing ATP) the A-band potential in the glycerinated muscle falls to the same as the I-band potential. This potential fall is also observed in a solution containing pyrophosphate (PP_i) but not one containing ADP or AMP-PNP (Bartels and Elliott 1983), so that it appears to be caused by the pyrophosphate moiety of the ATP molecule. In a contracting solution (Ca^{++} and ATP present) both A-band and I-band potentials exceed resting values in the early phase of tension development, and then fall below resting values during the recovery phase (experiments made with slow muscle fibres, mechanically or chemically skinned, Bartels and Elliott 1984a and b).

There are some differences in the behaviour of the A- and I-band potentials in relaxing solutions between muscles that have been glycerinated and those that have been skinned mechanically or chemically. We have shown (Bartels and Elliott 1982, 1985) that these may arise from the presence in skinned muscles of a parallel system, probably the sarcoplasmic reticulum, which charges up as the thick filaments discharge, and which is not present (or not functional) in glycerinated muscle or in skinned muscle treated with specific detergents to remove the sarcoplasmic reticulum.

The A-band potential change on the addition of ATP is a dramatic effect. In collaboration with Dr. Peter Cooke we have developed threads of purified myosin, the major protein component of a muscle A-band, and have made Donnan measurements on such threads (Bartels et al 1985). Figure 1 shows the results of a typical experiment, the myosin molecular charge falls sharply with the addition of small amounts of ATP, the fall is essentially over by 0.5 mM [ATP]. This is about twice the concentration of myosin molecules in the threads, so that the effect may be stoichiometric, with one or two ATP-binding sites per molecule. The potential (and charge) decrease coincides with the rise of [$MgATP^{--}$] in the solution, and the charge in this experiment falls from about 75e (electronic charges) molecule^{-1} in the absence of ATP to about 50e in the presence of 0.5 mM [$MgATP^{--}$], so the change is too large for 2 ATP molecules (~ 4-8e). In a relaxing solution (containing ATP) the ligand is known to be bound, and to be released when the muscle goes into the contraction cycle and into (ATP-free) rigor also; the consequent change due to this alone would be in the reverse direction to our observations. The implication is that the charge effect must be amplified in amount and reversed in sign, and our working hypothesis is that the ion-binding properties throughout the protein are modified as a result of some initiating event.

McLachlan and Karn (1982) have shown that in a myosin rod (the α-helical tail of the myosin molecule) there are 38 repeat sequences of 28 amino acids, each sequence showing a similar and pronounced charge pattern. The charge change would, on this basis, be about one extra electronic charge for each of these 38 sequences. Alternatively, the change might be confined to the S-2 region of the myosin molecule, the first twelve of these sequences. In this case the charge change per sequence would be about 3e, which looks interestingly as though it might be one phosphate ion. The hypothesis is that the binding of ATP/PP_i to myosin (the major effect in muscle relaxation) causes the release of negative ions from the myosin rod, or at least from part of it. Alternatively, positive ions might be bound rather than negative ions released. For a possible mechanism, we favour the ideas of Saroff (Loeb and Saroff 1964), that ion-binding to proteins is by hydrogen bonding onto networks of charged side chains clustered along the polypeptide chain. In myosin such clusters (Saroff sites) may be set up between the myosin rods in the filament or between the two polypeptide chains in one myosin rod (Figure 2). Mechanical stress transmitted through the filament backbone might cause small alterations in the structure of the Saroff sites (Figure 3) giving rise to the co-operativity in effects which we have observed in muscle A-bands

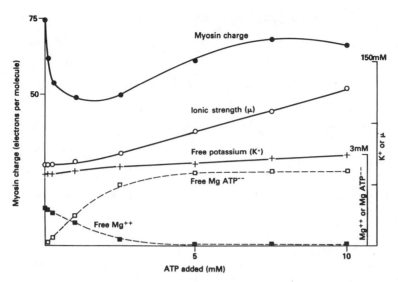

Fig. 1. An ATP titration experiment on threads containing
filaments of purified myosin shows that the molecular
charge reduction occurs sharply at low values of
added ATP, and is maximal by about 0.5 mM [ATP].
The levels of several of the solution constituents are
plotted; the fall in charge coincides with the rise in
[MgATP^{--}]. The subsequent increase in charge is probably a non-specific ionic-strength effect, since it
coincides with the rise in μ and we have observed
similar effects in both rigor and relaxing solutions.

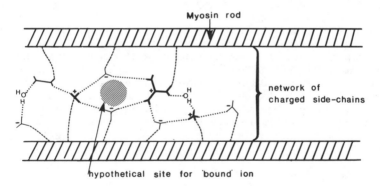

Fig. 2. A hypothetical Saroff site, a network of charged side-
chains between two myosin rods. Such networks are
apparent from the work of McLachlan and Karn (1982),
although the form of this one is taken from tobacco
mosaic virus coat protein. Notice that the vertical
scale between the rods has been much exaggerated to
show the network, in all probability a myosin site
is much more compact than appears in this diagramatic
representation.

A natural question arises; if the charge reduction that we have observed
is initiated by ATP/PP$_i$ binding to a specific site, where is that site on the

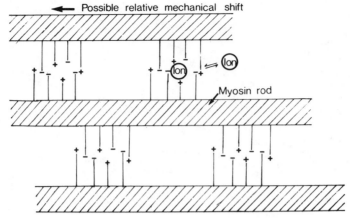

Fig. 3. In a myosin filament, with many identical Saroff
sites between adjacent myosin rods, the binding
or release of an ion at one site may cause a relative mechanical shift of the two rods, affecting
the adjacent sites and giving rise to co-operativity of the ion-binding or release.

myosin molecule? Since the effect is observed in myosin rod gels as well as
in whole myosin gels (Bartels et al 1985) it would seem that the effect
cannot be caused by ATP/PP$_i$ binding at the primary ATP-ase site on the globular myosin head (S-1). There must, then, be some further ATP-affected site
on the α-helical backbone of the myosin rod. Such secondary ATP sites have
been observed by a number of previous workers, including Harrington and
Himmelfarb (1972), whose studies revealed the presence of one or two binding
sites in the rod segment of myosin. They investigated the association/
dissociation of myosin filaments at pH 8, and remarked that these rod sites
exert a profound effect on the stability of the filaments at low substrate
concentrations. They did not, however, find substantially different behaviour between ATP, ADP, AMP-PNP or PP$_i$. In our experiments, at pH 7 where the
filaments are associated, ATP and PP$_i$ do cause the charge decrease, ADP and
AMP-PNP do not do so. There seem to be some parallels between our experiments and those of Harrington and Himmelfarb, but the details remain to be
explained.

A second question should also be asked. Does the striking decrease in
negative charge in the presence of ATP/PP$_i$ stem from the <u>release of negative
ions from the protein</u>, or from the <u>binding of positive ions to the protein</u>
when the ATP/PP$_i$ initiates some Saroff-type mechanism? Here the clues are
contradictory. Experiments in which the anion type is varied (T.D. Bridgman,
E.M. Bartels and G.F. Elliott, paper in preparation) show that there are
differences between the filament charges measured in different anions. This
could imply that anion binding in rigor is the important effect, and that
anion release occurs on ATP/PP$_i$ addition. On the other hand, Lewis and
Saroff (1957) showed that myosin binds about 20 K$^+$ ions at pH 7 and [K$^+$] =
0.1 M, and showed moreover that actomyosin (that is, myosin linked to actin
in the rigor configuration) bound less K$^+$ ions under similar conditions.
This work suggests that changes in cation binding might be important, but
unfortunately Lewis and Saroff did not investigate the effect of ATP upon K$^+$
binding. This second question could probably best be answered by equilibrium
dialysis experiments using radioactive isotopes, and we intend to carry these
out as soon as possible.

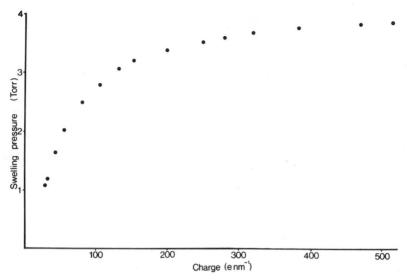

Fig. 4. The swelling pressure in Torr as a function of surface charge, calculated as explained in the text
for filaments of charge radius 15 nm, at a Debye
length of 1.0 nm. Charge saturation begins at about
100 e nm^{-1} and is advanced by about 200 e nm^{-1}.

We have demonstrated that charge changes can be observed in muscle in
situations which are of direct physiological importance. Do these charge
changes give rise to effects which are relevant to the function of muscle?
The charge changes might affect the electrostatic swelling pressure within
the muscle filament lattice, and thus change the transverse force balance
that must exist in that lattice (Elliott 1968, April 1975, Millman and Nickel
1980). However, Millman and Nickel (1980) first drew attention to the
charge-saturation effect for muscle, that the calculated electrostatic repulsive force between the filaments does not increase beyond a certain level
with filament charge; the alternative formulations of Elliott and Bartels
(1982) and Naylor (1982) have also shown this effect.

Charge saturation is related to counter-ion charge condensation (Manning
1969, 1978). Manning has shown that for those molecules that may be approximated as a line charge, there is a maximum molecular charge which can affect
the field produced by the molecule, because if the molecule charge exceeds
this critical maximum, the excess charge is neutralised by the condensation of
counter ions in the region of the line charge. The critical charge, in this
line charge approximation, is one electronic charge per Bjerrum length, and
this length, at 25°C in an aqueous environment, is about 0.7 nm.

The Manning analysis applies to line charges, and has been used successfully to explain experimental effects in systems of small polyion radius such
as DNA molecules (e.g. Manning 1978). It is not immediately clear how the
analysis may be applied to systems of much larger polyion radius, such as
muscle thick filaments. Lampert (1983, see also Lampert and Crandall 1980),
has considered this problem theoretically, and gives numerical tables which
can be used to estimate the maximum effective charge and confinement radius
for an isolated charged cylinder (of finite radius) in an electrolyte. An
alternative approach is to derive the swelling pressure saturation as the
filament charge increases, and this is shown in Figure 4, calculated using
the computer programme described in Elliott and Bartels (1982), based on

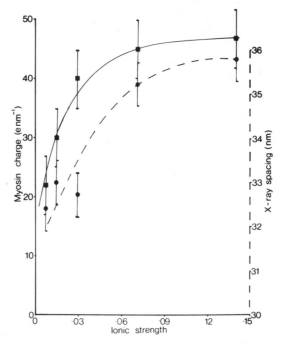

Fig. 5. X-ray-spacing (d_{10}), round points, and myosin filament charge, square points, measured in rigor solution of similar composition obtained by serial
dilution of a standard salt solution to different
ionic strengths (for details of the solutions see
Naylor et al 1985). All were at pH 7, phosphate
buffer, and all X-ray spacings were taken at sarcomere lengths of 3.8 μm. Experimental errors are
shown. The saturation in myosin filament charge is
a real effect, for higher ionic strengths the filament charge falls, but the X-ray spacing is then very
difficult to measure because the reflections become
very diffuse. The correlation between X-ray spacing
and filament charge is clear (curves fitted by eye).

Alexandrowicz and Katchalsky (1963). Figure 4 shows that for filaments of
15 nm charge radius, at a Debye length of 1.0 nm, the charge saturation
occurs at between 100 and 200 e nm^{-1}. This agrees with the 100 e nm^{-1}
figure given by Millman (1985, this volume). 100 e nm^{-1} on the surface a
15 nm radius filament is about 1 e nm^{-2} of filament surface area, so that the
individual charges under these circumstances are on the average separated by
rather more than the Bjerrum length, which seems intuitively reasonable.

In view of the agreement between us and Millman (1985), and the intuitive reasonableness of the order of magnitude, it might be hoped that the
maximum effective charge derived from Lampert (1983) would also agree. Unfortunately this is not so, and calculation from this paper suggests a maximum effective charge of 14 e nm^{-1} for cylinders of radius 15 nm at a Debye
length of 1.0 nm. The table in Lampert's paper which leads to this conclusion is derived by numerical integration, so that we are unable to go further
into this discrepancy.

It is possible, however, to demonstrate experimentally that the X-ray

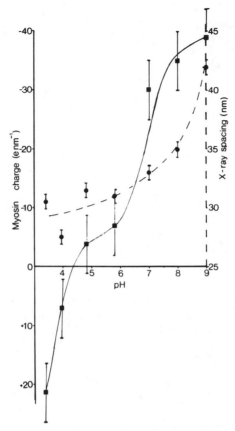

Fig. 6. X-ray spacing (d_{10}), round points, and myosin filament charge, square points, in rigor solutions all
at ionic strength ~ 0.014 M, and of different pH's.
All X-ray spacings were taken at a sarcomere length
of 3.8 μm except for those at pH 8 and 9, which are
at 3.45 μm (we are indebted to Mr. R.J. Ward for these
two measurements). At pH's below 6.0, citrate buffer
was used, at pH 7 and 8, phosphate buffer and at pH
9, borate buffer. Most of the data are from Naylor et
at 1985, with the addition of subsequent experiments
at pH 8 and 9. Experimental errors are shown. The
general correlation between charge and spacing is
clear. The curves are fitted by eye, and we have not
chosen to suggest a minimum in the X-ray spacing between pH 4 and 5, since the measurements are not convincing although a minimum would be expected at
around the isoelectric point, pH 4.4.

spacing between the muscle filaments is sensitive to the myosin charge. This
is most easily seen at long sarcomere lengths, where the situation is not
complicated by actin-myosin interactions. The results for ionic strength
variation are plotted in Figure 5, and for pH variation in Figure 6. The
data for these Figures are largely from the results of Naylor et al (1985),
and the Figures are plotted for rigor muscle at a sarcomere length of 3.8 μm,
which is zero overlap between the thick and thin filaments. These Figures
show clearly that the interfilament repulsive forces are not charge saturated
in the experimental regime in rigor, and that they certainly will not be so
at the lower filament charges seen in relaxed muscle.

Our contracting muscle results, in conjunction with the measurements on relaxed and rigor muscles, would fit naturally into Harrington's (1971) model where the force is generated by some ionic-led phase change within the S-2 link between the myosin head (S-1) and the filament backbone. They would also fit into models where the basic event is ionic swelling in the filament lattice, converted to longitudinal force by the myosin heads acting as dragging anchors (e.g. Elliott, Rome and Spencer 1970 modified as suggested in the discussion following Elliott et al 1978). In either case we may recall the words of Albert Szent-Györgyi (1941), "I was always led in research by my conviction that the primitive, basic functions of living matter are brought about by ions, ions being the only powerful tools which life found in the sea-water where it originated. Contraction is one of the basic primitive functions".

ACKNOWLEDGEMENTS

We are indebted to Ms. K. Jennison, Mr. R.J. Ward and Dr's Peter Cooke, Barry Millman and Carl Moos for enthusiasm and constructive criticism. This work was aided by a grant from the Science and Engineering Research Council.

REFERENCES

Alexandrowicz, Z., and Katchalsky, A., 1960, Colligative properties of poly-electrolyte solutions in excess of salt, J. Polymer. Sci. A., 1:3231.

April, E.W., 1975, Liquid crystalline characteristics of the thick filament lattice of striated muscle, Nature, 257:139.

Bagshaw, C.R., 1982, "Muscle contraction", Chapman and Hall, London and New York.

Bartels, E.M. and Elliott, G.F., 1980, Donnan potential measurements in the A- and I-bands of cross-striated muscles and calculation of the fixed charge on the contractile proteins, J. Musc. Res. Cell Motility, 1: 452.

Bartels, E.M., and Elliott, G.F., 1981, Donnan potentials from the A- and I-bands of skeletal muscle, relaxed and in rigor, J. Physiol., 317:85P.

Bartels, E.M. and Elliott, G.F., 1982, Donnan potentials in rat muscle: Difference between skinning and glycerination, J. Physiol., 327:72P.

Bartels, E.M. and Elliott, G.F., 1983, Donnan potentials in glycerinated rabbit skeletal muscle: the effect of nucleotides and of pyrophosphate, J. Physiol., 343:32P.

Bartels, E.M. and Elliott, G.F., 1984a, Changes in the Donnan potentials from the A- and the I-bands of contracting skeletal muscles, Acta. Physiol. Scand., 121(3):A20.

Bartels, E.M. and Elliott, G.F., 1985, Donnan potentials from the A- and I-bands of glycerinated and chemically skinned muscles, relaxed and in rigor, Biophys. J., 48:61.

Bartels, E.M. and Elliott, G.F., 1984b, Donnan potentials from contracting muscle, J. Musc. Res. Cell Motility, 5:227.

Bartels, E.M. and Elliott, G.F., 1985, Donnan potentials from the A- and I-bands of glycerinated and chemically skinned muscles, relaxed and in rigor, Biophys. J., in press.

Bartels, E.M., Cooke, P.H., Elliott, G.F. and Hughes, R.A., 1985, Donnan potential changes in rabbit muscle A-bands are associated with myosin, J Physiol., 358:80P.

Collins, E.W. and Edwards, C., 1971, Role of Donnan equilibrium in the resting potentials in glycerol-extracted muscle, Am. J. Physiol., 221:1130

Elliott, G.F., 1968, Force-balances and stability in hexagonally-packed poly electrolyte systems, J. Theoret. Biol., 21:71.

Elliott, G.F., Rome, E.M., and Spencer, M., 1970, A type of contraction hypothesis applicable to all muscles, Nature, 226:417.

Elliott, G.F., Naylor, G.R.S. and Woolgar, A.E. 1978, Measurements of the
electric charge on the contractile proteins in glycerinated rabbit
psoas using microelectrode and diffraction effects, In:"Ions in
Macromolecular and Biological Systems", D.H. Everett and B. Vincent,
eds. Scientechnica Press, Bristol.

Elliott, G.F. and Bartels, E.M. 1982, Donnan potential measurements in extended hexagonal polyelectrolyte gels such as muscle, Biophys. J., 38:
195.

Elliott, G.F., Bartels, E.M., Cooke, P.H. and Jennison, K., 1984, Evidence
for a simple Donnan equilibrium under physiological conditions,
Biophys. J., 45:487.

Harrington, W.F., 1971, A mechanochemical mechanism for muscle contraction,
Proc. Natnl. Acad. Sci., 68:685.

Harrington, W.F. and Himmelfarb, S., 1972, Effect of adenosine di- and tri-
phosphates on the stability of synthetic myosin filaments, Biochemistry, 11:2945.

Huxley, A.F., 1980, "Reflections on muscle", Liverpool University Press,
Liverpool.

Lampert, M.A., 1983, Maximum effective charge and confinement radius for the
Coulomb condensate for an isolated charged sphere or cylinder in an
electrolyte, Chem. Phys. Letters, 96:475.

Lampert, M.A. and Crandall, R.S., 1980, Non-linear Poisson-Boltzmann theory
for polyelectrolyte solutions: the counterion condensate round a line
charge as a delta-function, Chem. Phys. Letters, 72:481.

Lewis, M.S. and Saroff, H.A., 1957, The binding of ions to the muscle proteins, J. Am. Chem. Soc., 79:2112.

Loeb, G.I. and Saroff, H.A., 1964, Chloride and hydrogen-ion binding to
ribonuclease, Biochemistry, 3:1819.

McLachlan, A.D. and Karn, J., 1982, Periodic charge distributions in the
myosin rod amino acid sequence match cross-bridge spacings in muscle,
Nature, 299:226.

Manning, G.S., 1969, Limiting laws and counterion condensation in polyelectrolyte solutions, J. Chem. Phys., 51:924.

Manning, G.S., 1978, The molecular theory of polyelectrolyte solutions with
applications to the electrostatic properties of macromolecules, Ann.
Rev. Biophys., 11:179.

Millman, B.M. and Nickel, B.G., 1980, Electrostatic forces in muscle and
cylindrical gel systems, Biophys. J., 32:49.

Millman, B.M., 1985, Long range forces in cylindrical systems: muscle and
virus gels, This volume.

Naylor, G.R.S., 1982, On the average electrostatic potential between the filaments in striated muscle and its relation to a simple Donnan potential, Biophys. J., 38:201.

Naylor, G.R.S., Bartels, E.M., Bridgman, T.D. and Elliott, G.F., 1985, Donnan
potentials in rabbit psoas muscle in rigor, Biophys. J., in press.

Naylor, G.R.S., Bartels, E.M., Bridgman, T.D. and Elliott, G.F., 1985, Donnan
potentials in rabbit psoas muscle in rigor, Biophys. J., 48:47.

Szent-Györgyi, A., 1941, Discussion, Studies from the Institute of Medical
Chemistry, University of Szeged, 1:69.

DONNAN POTENTIALS GENERATED BY THE SURFACE CHARGE ON MUSCLE FILAMENTS

Ernest W. April and Robert A. Aldoroty*

College of Physicians & Surgeons, Columbia University

New York, New York 10032

Abstract. Electrochemical potentials (measured with microelectrodes) in the liquid-crystalline lattices of striated muscle
result from Donnan equilibriums established by the fixed charge
associated with the myosin rods and actin filaments in the ionic
cytosol. From these electrochemical potentials and the unit-cell
volumes (measured by x-ray diffraction) together with Donnan
theory, the effective linear charge densities are derived for the
respective myosin rods and actin filaments. Electrochemical potentials also were measured under conditions that varied the parameters of the Donnan equation. Modeling predicted Donnan potentials
for the same conditons. The predicted Donnan potentials correlate
well with the measured electrochemical potentials for both the
A-band and I-band liquid-crystalline lattices.

INTRODUCTION

Macromolecular order in many biological systems has been ascribed to
a balance between intermolecular forces (Bernal and Fankuchen, 1941;
Elliott, 1968; Brenner and McQuarrie, 1973; Millman et al., 1984). The
crystalline characteristics and deformability of the array of myosin
rods, which comprises the highly ordered and optically anisotropic
A-band of striated muscle, has led to the conceptualization of this
structure as a smectic B_1 liquid crystal (Elliott and Rome, 1969;
April, 1975a, 1975b; Hawkins and April, 1981, 1983a, 1983b). In the
crayfish contractile system to be discussed (Figure 1), the myosin
rods (each 4.5 um long and 18 nm wide) align in a p6mm lattice with an
approximate 69 nm interaxial spacing. The I-band of actin filaments (each

---

*Dr Aldoroty's address: The Mt. Sinai Medical Center, New York, NY.

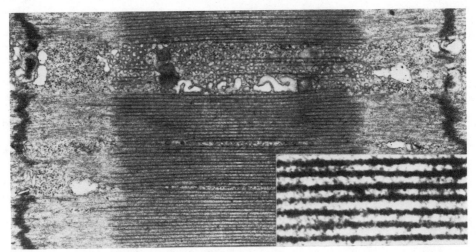

Fig. 1. Electron micrograph of a longitudinal section of a crayfish
striated muscle fiber (magnification approximately 15000x). The
field spans one sarcomere from Z-line to Z-line. The oriented
myosin rods in the A-band lattice are highly ordered. The oriented actin filaments in the I-band are less ordered, but become
highly ordered where they interdigitate the A-band. The inset
shows detail of the myosin rod lattice in the region where there
is no filament overlap (magnification approximately 90000x).
(Electron micrographs courtesy of Dr A. Eastwood.)

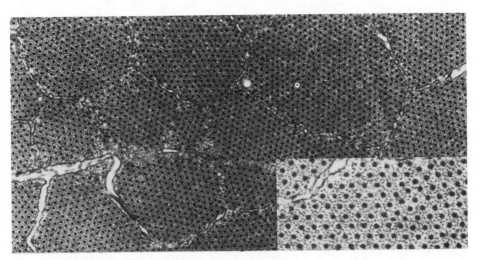

Fig. 2. Electron micrograph of a cross section through the highly
ordered A-band of crayfish striated muscle in a region where
there is filament overlap (magnification approximately 25000x).
The inset demonstrates the hexagonal arrangement of myosin rods
and the interstitial organization of the interdigitating actin
filaments (magnification approximately 50000x). (Electron micrographs courtesy of Dr A. Eastwood.)

3.6 um long and 8 nm wide) appears to be a nematic liquid crystal. The actin filaments interdigitate both ends of the A-band lattice to lie at interstitial sites, forming a 6:1 unit-cell ratio (Figure 2). Each A-band and its two associated half-I-bands form the sarcomere; the amount of actin interdigitation is a function of the sarcomere length (Figure 3). The alternation of A-bands and I-bands along the muscle fiber that gives it a striated appearance microscopically (see Figure 4).

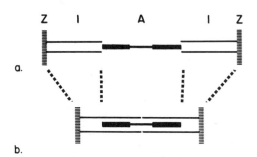

Fig. 3. Schematic representations of the relationships between the myosin and actin filaments of crayfish muscle. The actin filaments are 3.6 um long and the myosin filaments are 4.5 um long. Condition a: the fiber is stretched to a sarcomere length of 11.9 um where there is no filament overlap. Condition b: The fiber is allowed to shorten to a sarcomere length of 7.4 um where filament overlap is maximum.

It is plausable that intermolecular forces are responsible for the order and stability of the A-bands and that the repulsive component of the necessary force-balance results from the electrostatic surface charge on the myosin rods and the interdigitating actin filaments. The electrochemical properties of the A-band lattice initially were investigated by Rome (1968). She found that changes in pH, ionic strength, and divalent cation concentration affected the interfilament spacing. From this she inferred that the strength of the electrical double layer was changing as a function of surface charge or ionic screening. These results were confirmed in our laboratory (April et al., 1971, 1972).

Thus, electrostatic repulsive forces arising from the electical double layer appeared to be a dominant factor in macromolecular spacing and the stability of the A-band of striated muscle. Elliott (1968) derived a plausable balance between London-Van der Waals' forces and long-range electrostatic forces for a hexagonally-packed polyelectrolyte system similar to the muscle A-band lattice. This concept subsequently received theoretical support (Brenner and McQuarrie, 1973) and continues to receive experimental support (April, 1975a, 1975b; Millman and Nickel, 1980; Millman et al., 1983, 1984).

THEORY AND METHODOLOGY

Crayfish myosin rods and actin filaments have isoelectric points at aproximately pH 4.4 (April et al., 1972, April, 1978). Therefore at physiological pH, the myosin rods and actin filaments carry a net negative charge which is a reasonable source of repulsive force. However, only direct measurement of these charges could resolve the hypothesis that the A-band lattice is a liquid crystal in which electrostatic repulsive forces are instrumental in maintaining order and stability. These surface charges on the rods and filaments should generate electrochemical potentials of Donnan origin. Such potentials arise from an equilibrium condition in which charged structural elements in one phase of a two-phase ionic system cause an unequal distribution of diffusible ions (Overbeek, 1956). The lattice of charged myofilaments in an ionic solution satisfies these criteria. Indeed, Elliott (1973) suggested that the A-band of striated muscle may be regarded as a single phase, the dimensions of which are described by the volume that the myosin rods occupy, and that this phase would exhibit Donnan and osmotic properties.

The Donnan potential is a function of fixed charge concentration and mobile ion concentration. According to Benedeck and Villars (1979), the Donnan potential can be expressed as

$$E = \frac{kT}{e} \ln\left(\sqrt{1 + \frac{ZC_z}{4C^2}} - \frac{ZC_z}{2C}\right) \qquad (1)$$

where E is the Donnan Potential, k is the Boltzmann constant, T is absolute temperature, e is the charge on an electron, Z is the charge along the filaments, C_z is the filament concentration, and C is the external mobile ion concentration. A useful form of the Donnan equation, presented by Elliott and his colleagues (1978), is

$$ZC_z = 2C \sinh \left(\frac{Ee}{kT} \right) \qquad (2)$$

Hence, by measuring the potential (E) and lattice parameters (to obtain C_z), it should be possible to calculate the magnitude of the effective surface charge (Z) on the myosin rods and actin filaments.

The methodology utilized crayfish long-tonic muscle fibers in the relaxed condition and with the surface membrane removed (April et al., 1971; Reuben et al., 1971). In addition, all internal membranous structures, a possible source of electrochemical potential, were removed by treatment with Triton X-100 (Aldoroty and April, 1984). Sarcomere length was measured by light diffraction; fiber diameter was measured by optical microscopy; and myosin lattice spacing was measured by low-angle x-ray diffraction (April et al., 1971; Aldoroty and April, 1984). From the lattice spacing $(d_{1,0})$, the unit-cell area (A) of the lattice is obtained by the formulation

$$A = \frac{(d_{1,0})^2}{\sin 60^\circ} \qquad (3)$$

The product of the unit-cell area and the length of myosin rod (4.5 um) provides the A-band unit-cell volume. It has been shown (Hawkins and April, 1981, 1983a) that the relative unit-cell volumes (determined by x-ray diffraction measurements) correlate well with the relative muscle fiber areas (determined by light microscopy) for muscle fibers under the same experimental conditions. Hence, it is possible to calculate absolute unit-cell volumes of the A-band and I-band with a minimum of x-ray diffraction measurements. From the absolute unit-cell volume, the filament concentration is

$$C_z = \frac{n}{NV} \qquad (4)$$

where C_z is the myosin rod or actin filament concentration, n is the number of filaments per unit cell (myosin: 1; actin: 6), N is Avagadro's number, and V is the absolute volume of the unit cell.

While measurement of electrochemical potentials from vertebrate muscle have been reported (Elliott et al., 1978; Bartels et al., 1980, 1982; Elliott and Bartels, 1981, Naylor, 1981), it was difficult in those preparations to determine optically whether the electrode was in an A-band or I-band. The use of crayfish striated muscle, in which the sarcomere is four times as large, obviates this problem. Electrochemical potentials (Figure 4) were measured in crayfish muscle using microelectrodes filled with 3 mol/l KCl (0.3 um tip diameters with resistances of 10 to 15 MOhm) and standard electrophysiological technique (Aldoroty and April, 1984).

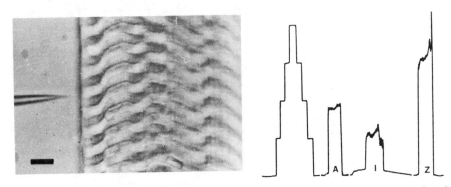

Fig. 4. a: Micrograph of a microelectrode tip adjacent to an A-band of a crayfish skinned muscle fiber. The polarizing optics render the A-band birefringent. (The marker bar represents 10 um.) b: Typical recordings from an A-band, I-band and Z-line. (Each step on the calibration trace on the left represents 1 mV.)

The technique for recording electrochemical potentials was validated using gels of various agar concentrations, electrodes of various resistances, and salt solutions of various species and various concentrations both inside and outside the electrode (Aldoroty and April, 1984). These experiments demonstrated absence of significant variation in measured potentials under conditions that would change the magnitude of the artifactual diffusion potentials (functions of ionic mobilities and transference numbers). It must be concluded from these experiments that diffusion potentials do not significantly contribute to the measured electrochemical potential in this system (Aldoroty and April, 1984).

Determination of charge density. To calculate the charge per unit cell, the charge per myosin rod, and the charge per actin filament, initial measurements of the electrochemical potential were made at an extended sarcomere length (Figure 3a) at which all of the actin filaments are withdrawn from the A-band lattice. Under this condition, the electro-chemical potential measured from the A-band is produced only by the charge associated with the myosin rods and, due to the lattice structure, the charge per myosin rod is equivalent to the charge per unit cell. By applying equation 2, a charge (Z) of 3.6×10^5 e$^-$/myosin rod is obtained (Aldoroty et al., 1985). Then the myosin rod linear charge density (distributed along the 4.5 um length of the rod) is 6.6×10^4 e$^-$/um. The electrochemical potential measured from the I-band is produced only by charge associated the actin filaments. By applying equation 2, a charge (Z) of 2.4×10^4 e$^-$/actin filament is obtained. The actin filament linear charge density (distributed along the 3.6 um filament length) is 6.8×10^3 e$^-$/um (Aldoroty et al., 1985). Thus, the myosin rods represent the major contribution to the magnitude of the measured electrochemical potential, yet the contribution from the actin filaments is significant.

A number of electrochemical potentials were measured at a sarcomere length of 9.7 um. This particular sarcomere length represents a point at which approximately half of the A-band is overlapped by interdigitating actin filaments (a condition between Figure 3a and 3b). From the electro-chemical potential at this sarcomere length, a charge within the unit cell of 2.8×10^5 e$^-$ is calculated using Donnan theory. Alternatively, from the geometry of the unit cell (a 1:6 myosin to actin ratio), the charge per unit cell at any sarcomere length may be calculated from the appropriate contributions of linear charge density from the individual filaments. The fixed charge contribution of each filament to the A-band unit cell is

$$ZC_z = Z_m Cz_m + Z_a Cz_a \tag{5}$$

where, for myosin, $Z_m = l_m(6.6 \times 10^4 \, \text{e}^-/\text{um})$,
while, for actin, $Z_a = 2[l_a - 0.5(l_s - l_m - l_z)](6.8 \times 10^4 \, \text{e}^-/\text{um})$, and
where l_m is the length of the myosin rod (4.5 um), l_a is the length of the actin filament (3.6 um), l_z is the length of the Z-line (0.2 um) and l_s is the sarcomere length (9.7 um). This calculation also predicted a unit-cell charge of 2.8×10^5 e$^-$ at this sarcomere length.

Effects of filament overlap on potential. In a similar manner, an A-band electrochemical potential can be measured at any degree of filament overlap. Sarcomere length was varied by adjusting with a micromanipulator the amount of stretch on the muscle fiber. Hence, the amount of actin filament interdigitation into the myosin-rod lattice is altered thereby varying the amount of charge in the unit cell. The electrochemical potentials measured from the A-bands and the I-bands are plotted against the sarcomere length in Figure 5.

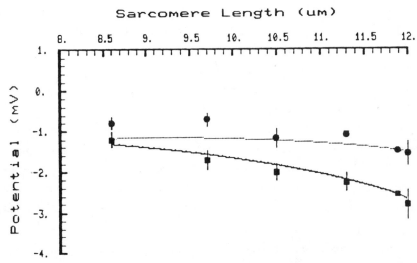

Fig. 5. Potential as a function of sarcomere length. The squares (■) represent electrochemical potentials measured from A-bands; the circles (●) represent electrochemical potentials measured from I-bands and the bars indicate SD. The curves are calculated from an algorithm based on Donnan theory. The lower curve is the predicted Donnan potential for the A-band; the upper curve is the precicted Donnan potential for the I-band.

The Donnan potential also can be predicted for any sarcomere length. Because the distance between the filaments also varies in a predictable manner with sarcomere length (April and Wong, 1976), the filament concentration (C_z) varies as well as the total charge (Z) in the unit cell. This behavior is described by the nonconstant-volume relationship in skinned fibers and is a linear function of sarcomere length (April and Wong, 1976). Using the nonconstant-volume relationship, the linear charge densities of the myosin rod and actin filament, and equations 2 and 5, the curves in Figure 5 are calculated. Over a range of sarcomere lengths,

the electrochemical potentials measured from the A-band closely agree
with the calculated Donnan potentials; there is moderate agreement
between the I-band potentials (Aldoroty and April, 1985).

Effects of lattice volume on potential. The electrochemical potential can be measured at any lattice volume. Lattice volume, and hence
filament concentration can be perturbed independently of other variables
by appling external osmotic compression to skinned fibers. Osmotic
pressure was generated by various concentrations of polyvinylpyrrolidone
(molecular mass: 360,000 g/mol), which is a non-ionic polymer that is
impermeant to both the A-band lattice and the I-band lattice. The electrochemical potentials measured from the A-bands and I-bands are plotted
against the unit-cell volume in Figure 6.

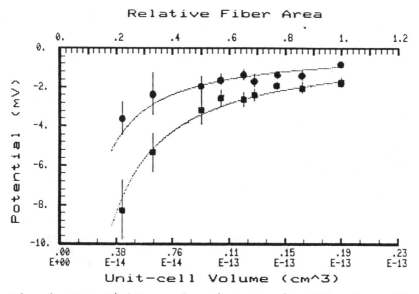

Fig. 6. Potential as a function of unit-cell volume. The
squares (■) represent electrochemical potentials
measured from A-bands; the circles (●) represent
electrochemical potentials measured from I-bands; the
bars indicate SD. The curves represents the predictions of Donnan potential in the A-band and I-band as
a function of filament concentration calculated from
an algorithm based on Donnan theory.

The Donnan potential also can be predicted for any degree of lattice
compression. From the unit-cell volume, the filament concentration (C_z)
is obtained. Equation 2 can be used to predict the magnitudes of the

Donnan potentials over a range of filament concentrations. These calculations form the curve in Figure 6. Over a range of decreasing lattice volume, the measured A-band and I-band electrochemical potentials agree with the predicted Donnan potentials (Aldoroty and April, 1982).

Effects of ion concentration on potential. External ion concentration also can be varied. The precise external ion concentrations were determined by using computer programs based on the Henderson-Hasselbach equation with the appropriate dissociation constants of the included salts. The electrochemical potentials measured from the A-bands and I-bands are plotted against cation concentration in Figure 7.

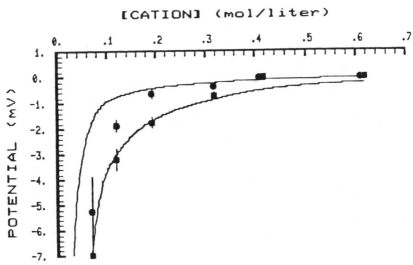

Fig. 7. Potential as a function of cation concentration. The squares (■) represent electrochemical potentials measured from A-bands; the circles (●) represent electrochemical potentials measured from I-bands and the bars bars indicate SD. The curves represent the predictions of Donnan potential in the A-band and I-band as functions of ionic concentration and are calculated from an algorithm based on Donnan theory.

The Donnan potential also can be predicted as a function of external ion concentration (C). Variation of external ion concentration also results in a change in filament concentration (C_z) (April et al., 1972). Equation 2 can be used to predict the Donnan potential. These calculations provide the curves in Figure 7. The measured A-band electrochemical potentials agree with the predicted Donnan potentials over the range of ionic concentrations investigated; there is moderate agreement between I-band potentials (Aldoroty, 1984).

Effects of pH on potential. Finally, electrochemical potentials can be measured as a function of filament charge. The fixed charge on the myofilaments is perturbed by varying the pH of the bathing solution. The electrochemical potentials measured from the A-bands and I-bands are plotted against pH in Figure 8 which is essentially a titration curve (Aldoroty, 1984).

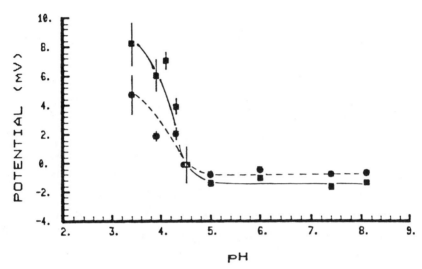

Fig. 8. Potential as a function of pH. The squares (■) represent electrochemical potentials measured from A-bands;
the circles (●) represent electrochemical potentials
measured from A-bands; the bars indicate SD. The
curves are fit to the data by hand.

In all of the previous experiments, with the exception of those in which pH was varied, the variables perturbed in the Donnan equation could be determined by independent measurements or calculations. Hence, the applicability of Donnan theory is validated in each instance. While pH is known to affect filament charge (Z), the manner in which it does so is not precisely known for the myosin rods and actin filaments. Change in pH also results in changes in filament concentration (C_z) and in the external ion concentration (C). Figure 8 assumes the shape of a titration curve and describes the effect of hydrogen ion concentration on the myosin rods and actin filaments. The isoelectric point, occuring at 0 mV and indicating the absence of a Donnan potential, corresponds to a pH of 4.5. This A-band isoelectric point closely agrees with the pH (4.4) at which the myosin filaments are in closest approximation as determined by x-ray diffraction (April et al., 1972; April, 1978). Therefore, Donnan theory can be used to calculate the charge (Z) on the filaments as a function of pH. The relationship between pH and charge correlates well with the relationship between pH and interfilament spacing. This supports the concept that electrical double layers associated with the myosin rods and actin filaments produce long-range electrostatic repulsive forces which, in turn, contribute to the force-balance that orders and stabilizes the A-band of striated muscle.

CONCLUDING REMARKS

Measurement of electrochemical potentials from polyelectrolyte systems provides a method for determining the surface charge of the population of the macromolecular components. The ability to vary the parameters in a Donnan equation and predict a Donnan potential that closely corresponds to the measured electrochemical potentials argues stongly for the applicability of Donnan theory to such an experimental system. The demonstrated success of the model implies that the microelectrode accurately measures Donnan potentials generated by the electrical double layers associated with the myosin rods and the actin filaments. The existance of measurable electrical double layers associated with the arrays of contractile filaments in striated muscle substantiates a source of long-range electrostatic repulsive force that has been implicated in the order and stability of macromolecular liquid-crystalline systems.

ACKNOWLEDGEMENT

Supported in part by a grant from the National Institutes of Health, 5R01-AM15876, and the Equitable Life Assurance Society of the United States through the Insurance Medical Scientist Scholarship Fund.

REFERENCES

Aldoroty, R.A. and E.W. April. 1982. Microelectrode measurement of A-band Donnan potentials as a function of relative volume. Biophys. J. 37:121a.
Aldoroty, R.A. 1984. "Donnan Potentials from Striated Muscle" Ph.D. Thesis. Columbia University, New York, NY.

Aldoroty, R.A. and E.W. April. 1984. Donnan potentials from striated
muscle liquid crystals. A-band and I-band measurements. Biophys. J.
46:369-379.

Aldoroty, R.A., N.B. Garty, and E.W. April. 1985. Donnan potentials from
striated muscle liquid crystals. Sarcomere length dependence. Biophys.
J. 47:89-96.

April, E.W. 1975a. Liquid crystalline charateristics of the thick
filament lattice of striated muscle. Nature. 257:139-141.

April, E.W. 1975b. The myofilament lattice: Studies on isolated fibers
IV. Lattice equilibria in striated muscle. J. Mechanochem. Cell
Motility. 3:111-121.

April, E.W. 1978. Liquid crystalline contractile apparatus in striated
muscle. In: "Mesomorphic Order in Polymers and Polymerization in
Liquid-Crystalline Media", American Chemical Society Symposium Series,
No. 74. Ed: A. Blumstein, American Chemical Society, pp 248-255.

April, E.W., P.W. Brandt and G.F. Elliott. 1971. The myofilament lattice:
Studies on isolated fibers I. The constancy of the unit-cell volume
with variation in sarcomere length in a lattice in which the thin-to-
thick myofilament ratio is 6:1. J. Cell Biol. 51:72-82.

April, E.W., P.W. Brandt and G.F. Elliott. 1972. The myofilament lattice
Studies on isolated fibers II. The effects of osmotic strength, ionic
concentration and pH upon the unit-cell. J. Cell Biol. 53:53-65.

April E.W. and D. Wong. 1976. Non-isovolumic behavior of the unit cell
of skinned striated muscle fibers. J. Mol. Biol. 101:107-114.

Bartels, E.M., T.D. Bridgman and G.F. Elliott. 1980. A study of the
electrical changes on the filaments in striated muscle. J. Muscle Res.
Cell Motility 1: 194.

Bartels, E.M. and G.F. Elliott. 1982. Donnan potentials in rat muscle
differences between skinning and glycerination. J. Physiol. (London).
327:72-73P.

Benedek, G.B. and F.M.H. Villars."Physics with Illustrative Examples
of Medicine and Biology, Volume 3: Electricity and Magnetism." Addison
Wesley. California. 1979. Chapter 3.

Bernal, J.D. and I. Fankuchen. 1941. X-ray crystalographic studies on
plant virus preparations. J. Gen. Physiol. 25:111-165.

Brenner, S.L. and D.A. McQuarrie. 1973. Force balances in systems of
cylindrical polyelectrolytes. Biophys. J. 13:301-331.

Elliott, G.F.. 1968. Force-balance and stability in hexagonally-packed
polyelectrolyte systems. J. Theor. Biol. 21:71-87.

Elliott, G.F. 1973. Donnan and osmotic effects in muscle fibers without membranes. J. Mechanochem. Cell Motility. 2:83-89.

Elliott G.F. and E.M. Bartels. 1981. Donnan potential measurements in
extended hexagonal polyelectrolyte gels such as muscle. Biophys. J.
38:195-199.

Elliott, G.F., G.R.S. Naylor and A.E. Woolgar. 1978. Measurements of
the electric charge on the contractile proteins in glycerinated rabbit
psoas using microelectrode and diffraction effects. In: "Ions in
Macromolecular Biological Systems", Colston Papers No. 29, Ed: D.H.
Everett and B. Vincent. Scientific Press, England, pp 329-339.

Elliott, G.F. and E.M. Rome. 1969. Liquid crystalline aspects of muscle
fibers. Molec. Cryst. Liq. Cryst. 5:647-650.

Hawkins, R.J. and E.W. April. 1981. X-ray measurements of the bulk
modulus of the myofilament liquid crystal in striated muscle. Molec.
Cryst. Liq. Cryst. 75:211-216.

Hawkins, R.J. and E.W. April. 1983a. The planar deformation behavior
of skinned striated muscle fibers. Molec. Cryst. Liq. Cryst. 101:
315-328.

Hawkins, R.J. and E.W. April. 1983b. Liquid crystals in living tissues.
In: "Advances in Liquid Crystals.", Academic Press, New York. Vol 6,
pp 243-264.

Millman, R.M., T.C. Irving, B.G. Nickel, and M.E. Loosley-Millman. 1984. Interrod forces in aqueous gels of tobacco mosaic virus. Biophys. J. 45:551-556.

Millman, B.M. and B.G. Nickel. 1980. Electrostatic forces in muscle and cylindrical gel systems. Biophys. J. 32:49-63.

Millman, B.M,. K. Wakabayashi, and T. Racey. 1983. Lateral forces in the filament lattice of vertebrate striated muscle. Biophys. J. 41:259-267.

Naylor, G.R.S. 1981. On the average electrostatic potential between the filaments in striated muscle and it relation to a simple Donnan potential. Biophys. J. 38:201-204.

Overbeek, J. ThG. 1956. The Donnan equilibrium. Prog. Biophys. Biophys. Chem. 6:57-84.

Reuben, J.P., P.W. Brandt, M. Berman and H. Grundfest. 1971. Regulation of tension in the skinned crayfish muscle fiber. I. Contraction and relaxation in the absence of Ca (pCa>9). J. Gen. Physiol. 57:385-407.

Rome, E. 1968. X-ray diffraction studies of the filament lattice of Striated muscle in various bathing media. J. Molec. Biol. 37:331-344.

LONG-RANGE ELECTROSTATIC FORCES IN CYLINDRICAL SYSTEMS:

MUSCLE AND VIRUS GELS

Barry M. Millman

Biophysics Interdepartmental Group
Physics Department, University of Guelph
Guelph, Ontario, N1G 2W1, Canada

More than forty years ago it was suggested (Bernal and Fankuchen, 1941) that in certain aqueous gels, formed from cylindrical biological structures, stability results from a balance of long-range forces – particularly electrostatic repulsive and van der Waals attractive forces. Calculations based on this assumption were generally consistent with observed changes in equilibrium spacings in hexagonal, aqueous gels of tobacco mosaic virus (TMV) as the ionic strength and pH were varied.

Some years later, G. F. Elliott (1968) proposed that a similar balance of long-range forces could provide stability in the A-band filament lattice of vertebrate cross-striated muscle. Comparison of Elliott's calculated forces with x-ray diffraction data on lattice spacings, particularly those of Rome (1967, 1968), showed that such a balance of long-range forces was feasible in muscle as well. In the following years, both the theory and the experimental data (for both muscle and TMV systems) were elaborated by several individuals and qualitative agreement between theory and experiment was demonstrated, but some problems emerged in quantitative comparisons (e.g. Shear, 1970; Miller and Woodhead-Galloway, 1971; Brenner and McQuarrie, 1973; April, 1975; Parsegian and Brenner, 1976; Morel and Gingold, 1979). Furthermore, the data on which these comparisons were based was limited in that only a single balance point could be recorded for any experimental condition (e.g. of ionic strength and pH). Where a discrepancy occurred between theory and experiment, it was not possible to assign it specifically to an error in calculation of either the attractive or repulsive forces nor to the basic assumptions.

This limitation was essentially removed in 1976 when the osmotic stress technique was developed and applied (initially) to lipid multibilayer systems (LeNeveu et al., 1976). With this technique, a known external pressure is applied to a regular system through the osmotic pressure generated by large, uncharged and inert macromolecules (e.g. dextran, polyvinylpyrrolidone) which, because of their size, are unable to enter the aqueous spaces in the structure being studied. The net pressure (or force) in a lattice could now be studied over a whole range of lattice spacings. Since, at equilibrium, the external pressure was balanced by an internal repulsive lattice pressure, graphs of lattice pressure as a function of lattice spacing could be obtained under a variety of conditions through varying the ionic strength and pH of the external solution or the sarcomere length and physiological state in muscle.

In my laboratory, we have been studying the relationship between external pressure and lattice spacing in hexagonal systems of charged cylindrical rods in ionic solutions: in particular aqueous gels of TMV, lattices formed by muscle thin filaments and the A-band filament lattice of striated muscle from which the cell membrane has been removed (Millman, 1981; Millman and Bell, 1983; Irving and Millman, 1983; Millman, et al., 1983, 1984). As well, we have refined the calculation of electrostatic pressure using both linearized and non-linearized solutions to the Poisson-Boltzmann equation, and compared these calculations to the observed data (Millman and Nickel, 1980). In the case of TMV gels, and to a somewhat lesser extent the thin filament lattice of muscle, the experimental systems provide a reasonable approximation to the assumptions used in the electrostatic calculations: i.e. smooth, incompressible cylinders with a fixed charge spread uniformly over their surfaces. In the case of the muscle A-band lattice, these assumptions do not apply so well. The A-band lattice is a double array of thick and thin filaments, and the major component (the thick filaments) are not smooth, but have projections which stick out a considerable distance from the backbone surface in a regular (screw) pattern along the filament length (Squire, 1981). Part, at least, of the charge is on these projections and, moreover, they are capable of shifting their position radially (and probably axially and azimuthally as well) in response to changes in the osmotic pressure or the physiological state of the muscle (i.e. from relaxation to contraction or rigor). Because the TMV gels (and to a lesser extent the thin filament lattices) are well-characterized and approximate well to the theoretical model, we have used data from these two systems to assess the electrostatic calculations. The electrostatic calculations have then been used to calculate expected pressures in the muscle A-band lattice. We have attempted to explain departures from the expected (calculated) curves by relating these to structural changes in the filament lattice, particularly changes related to the thick filaments.

TMV GELS

Our results with aqueous gels of TMV have been reported in detail elsewhere (Millman et al., 1984). Purified virus particles, when concentrated, form microcrystalline gels, which give sharp, powder-like reflections by small-angle x-ray diffraction. From these reflections, the lattice spacings (i.e. the interaxial distances of the viral particles) can be measured. Addition to the external solution of polymer molecules (dextran) that are large enough so that they do not enter the lattice spaces, causes the application of an osmotic pressure to the lattice, shrinking it until the net internal lattice pressure equals the applied external pressure. The amount of shrinking depends on the pH and ionic strength of the external solution, which affect the rod charge and ionic shielding respectively.

Data at pH 6 and ionic strength of 0.1 M (Fig. 1, filled circles) show that the logarithm of lattice pressure decreases almost linearly with increases in the lattice spacing. Calculations of electrostatic pressure, based on the known viral charge and (charge) diameter of 14 electrons/nm and 18 nm respectively along with the ionic concentrations in the external solution, give a curve (dashed line) in very good agreement with the data.

As the ionic strength was increased (to 0.9 M) or decreased (to 0.01 M), the data and the calculated electrostatic pressure curves were shifted similarly (Fig. 1). In general, the slope of the experimental pressure/spacing curve changed much as predicted (Table 1a). At an ionic strength of 0.9 M, however, the change was considerably less than predicted; presumably because at ionic strengths above 0.1 M the assumptions in the Poisson-Boltzmann theory are no longer appropriate. At the lower ionic strength (0.01 M), there was a good match between the slopes, though the whole experimental curve was shifted in a direction which suggested a slightly decreased charge (or charge diameter). On average, the experimental slopes were slightly less steep than the calculated ones (Table 1a).

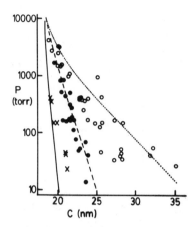

Fig. 1. Interaxial spacing (C) as a function of externally
applied osmotic pressure (P) for tobacco mosaic
virus gels in sodium phosphate buffer at pH 6.0,
compared with calculated electrostatic pressure
curves at ionic strengths of 0.9 M (X and solid
line), 0.1 M (● and dashed line), and 0.01 M
(○ and dotted line).

In comparison with the changes in the pressure/spacing curve with ionic strength, changes with pH (over the range from 5.0 to 7.2), indicated no change in slope and only a very small change in position of the curves, consistent with a decrease in rod charge as the pH of the external solution was lowered (Millman et al., 1984). Overall, in the case of TMV gels, the calculated electrostatic pressure agreed reasonably well with the experimental data, particularly around normal physiological conditions: pH near 7 and ionic strength near 0.1 M.

THE MUSCLE THIN FILAMENT LATTICE

In several types of muscles where thick filaments are absent over considerable lengths of the muscle (or sarcomere), the thin filaments form a limited or poorly-developed hexagonal lattice. Thus, in the smooth "catch" muscles of molluscs (Lowy and Vibert, 1967) or in vertebrate smooth muscle (Elliott and Lowy, 1968), a diffuse reflection, corresponding to a lattice spacing of 12 to 15 nm, is seen in the equatorial x-ray diffraction pattern which can be identified with the thin filaments. A similar

Table 1. Slope (with standard errors) of calculated and experimental curves for $\log_{10}$ pressure as a function of lattice (interaxial) spacing

ionic strength (mM)	slope of calc. electrostatic pressure curve (nm^{-1})	slope of experimental curve (nm^{-1})	ratio of slopes calc./exptl.
(a) Tobacco mosaic virus in phosphate buffer at pH 6.0 (from Millman et al., 1984)			
870	-1.56	-0.58 ± 0.05	2.69
98	-0.52	-0.50 ± 0.05	1.04
48	-0.36	-0.24 ± 0.09	1.50
10	-0.17	-0.17 ± 0.02	1.00
(b) Thin filament lattice of demembranated muscles in buffered 100 mM KCl solutions: pH 7.0. (1) rabbit taenia coli (2) ABRM of Mytilus (3) rabbit psoas with thick filaments removed (4) I-band, frog semitendinosus, stretched beyond filament overlap			
(1) 116	-0.62	-0.39 ± 0.04	1.59
(2) 171	-0.74	-0.35 ± 0.06	2.11
(3) 122	-0.62	-0.37 ± 0.07	1.68
(4) 159	-0.71	-0.30 ± 0.03	2.37
(c) A-band lattice of relaxed frog sartorius muscle			
108	-0.34	-0.22 ± 0.02	1.56

reflection is seen in striated muscle stretched to long sarcomere lengths (Rome, 1972; Millman and Bennett, 1976) and in striated muscle from which the thick filaments have been dissolved (Millman and Bell, 1983).

Like TMV gels, the thin filament lattice is compressed by external osmotic pressure and the relationship between pressure and lattice spacing is roughly logarithmic (Fig. 2, 3). Four skinned muscle preparations have been examined: a vertebrate smooth muscle - taenia coli from rabbit; a smooth molluscan "catch" muscle from the mussel - the anterior byssus retractor (ABRM); frog semitendinosus muscle stretched to a sarcomere length where thick and thin filaments no longer overlap (i.e. beyond 3.6 μm), and rabbit psoas muscle from which the thick filaments had been dissolved with a solution of high ionic strength (buffered 0.6 M KCl). The latter two preparations are vertebrate cross-striated muscles where normally the thin filament x-ray reflection is masked by the A-band lattice reflections except under special conditions such as those described. (More details of the A-band structure are given later.)

Results from the two smooth muscles are shown in Fig. 2, along with

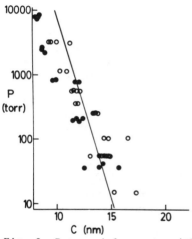

Fig. 2. Interaxial spacing (C) as a function of externally applied osmotic pressure (P) for thin filament lattice from skinned muscles: rabbit taenia coli (●), Mytilus byssus retractor (○). Line is the electrostatic pressure curve for a filament diameter of 9 nm, charge of 15 e/nm and ionic strength of 140 mM.

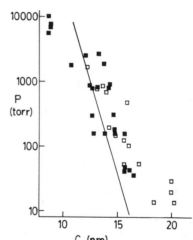

Fig. 3. As Fig. 2 but for rabbit psoas muscle with thick filaments dissolved (■) and frog semitendinosus stretched beyond filament overlap (□). Line as in Fig. 2 but for a filament diameter of 10 nm.

the electrostatic pressure curve calculated for an hexagonal lattice of cylindrical rods with surface charge of 15 electrons/nm and a charge diameter of 9 nm. The charge is the net charge estimated for thin filaments (Bartels and Elliott, 1985) and the charge diameter is approximately that determined in several recent studies (e.g. Egelman and Padron, 1984).

The data from the two striated muscle preparations are shown in Fig. 3 along with the electrostatic pressure curve calculated as in Fig. 2, except that a charge diameter of 10 nm was used. The data from the striated muscles lie at slightly greater spacings (for the same external pressure) than those from the smooth muscles. This may be because of the presence of troponin which, in striated muscle is found on the surface of the thin filaments additional to the actin and tropomyosin (see below). In the smooth muscles, only actin and tropomyosin are found. The troponin would be expected to increase the effective diameter of the thin filaments, thus shifting the pressure/spacing curves to slightly greater spacings.

In all four cases, the experimental data lie close to the calculated electrostatic pressure curves, although in every case the slope of the experimental curve is less steep than that of the theoretical curve (Table 1b). Thus, the electrostatic calculations appear to underestimate the electrostatic forces at any specific filament separation.

ELECTROSTATIC FORCES IN HEXAGONAL GELS: CONCLUSIONS

What can we conclude about electrostatic forces from the above studies of hexagonally-packed, cylindrical rods in ionic solutions? In the electrostatic calculations, there were no fitting constants: all parameters were derived either directly from the experimental conditions (e.g. the ionic strength of the external solution) or determined from other experiments (i.e. rod charge and diameter). Thus the agreement between theory and experiment is unlikely to be fortuitous. If, in the regime of our experiments, the repulsive forces arise from electrostatic repulsion between the rods (and it is difficult to imagine what other repulsive forces could be contributing significantly), then some specific conclusions follow.

Firstly, calculations based on the Poisson-Boltzmann equation (e.g. Millman and Nickel, 1980) can provide a good first approximation to the electrostatic repulsive forces in such gels. Because of screening by the free ions in the solution, the electrostatic forces fall off rapidly as one moves away from the charged surface and the magnitude of the slope of the pressure/spacing curve will decrease with decreasing ionic strength (Fig. 1). The calculated magnitude of the slope, however, is often too great by a factor of almost two (Table 1). A similar discrepancy between the calculated and experimental slopes has been found in the case of charged phospholipid multibilayers (Lis et al., 1981; Loosley-Millman et al., 1982), where it appeared to result from strong gradients in solvent polarization near the charged surfaces (see e.g. Gruen and Marcelja, 1983). This is likely the cause of at least some of the discrepancy in the hexagonal systems and thus represents a factor that needs to be considered in electrostatic calculations.

Secondly, for a given charge position on the cylindrical rods (i.e. charge diameter), there is a very limited range over which electrostatic forces can be detected, whatever the magnitude of the surface charge. In general, changes in either charge or charge diameter will cause a parallel shift of the electrostatic pressure/spacing curve, but will not change the slope of the curve. Increasing surface charge beyond a certain level (about 100 electrons/nm) produces only a very small shift in the curve (Fig. 4). Thus, in a given ionic solution, there is an upper limit to the pressure which can be found at any specific distance from a charged surface. Furthermore, there is a lower limit to the interrod spacing. Ultimately, this is the point where the rods come in contact, but in practise, approach of the rods will be limited at a somewhat greater separation because of hydration forces between the rods. If the rods have a hydration layer similar to that of phospholipid multibilayers (e.g. Rand and Parsegian, 1984), the rods can only approach to within about 2 nm of each other, limiting still further the range over which electrostatic forces affect the lattice. These limits are illustrated in Fig. 4. Only in a small, approximately triangular region of the pressure/spacing curve can electrostatic forces be the dominant repulsive force. The conclusion derived from Fig. 4 is that under a given ionic condition, changes in rod charge produce relatively small changes in electrostatic pressure over a very limited range of lattice spacings. Changes in electrostatic forces over a much greater range of lattice spacings are possible if there are changes in the position of the charge (i.e. in the charge diameter). For this reason, in cylindrical systems electrostatic forces are much more sensitive to the position of the charge than to the amount of charge (Millman and Nickel, 1980).

THE A-BAND LATTICE OF VERTEBRATE STRIATED MUSCLE

The A-band of vertebrate striated muscle is an hexagonal lattice formed from two types of protein filaments: thick filaments found at the corners of the lattice with thin filaments lying at the trigonal points between the thick filaments. Each thin filament is a double-stranded helix of globular actin monomers with rod-like tropomyosin molecules in the grooves of the helix, together forming a roughly cylindrical filament 9 to 10 nm in diameter. In striated (but not smooth) muscle, a pair of troponin molecules

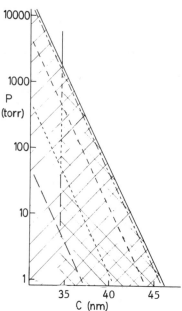

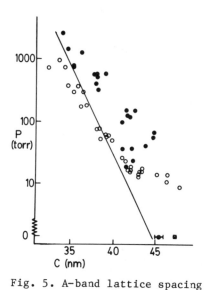

Fig. 4. Electrostatic pressure curves
calculated for conditions
appropriate to the relaxed
frog sartorius muscle at
short sarcomere lengths using
the linearized solution to
the Poisson-Boltzmann equation
but for different thick
filament charges. The
calculations assume a Debye
constant of 1.08 nm^{-1}, thick
and thin filament charge
diameters of 26 and 10 nm
respectively, a thin filament
filament charge of 15 e/nm and
a sarcomere length of 2.2 μm.
Thick filament charges are:
∞ ⎯⎯⎯, 100 e/nm ⎯ ⎯ ⎯ ,
10 e/nm ⎯ ⎯ , 1 e/nm ⎯ ⎯ ⎯ ⎯,
0.1 e/nm ⎯⎯ ⎯⎯. The left hand
vertical axis represents physical
contact between the thick and thin
filaments. The hatched region is
where lattice spacings are possible
using the above fixed parameters at
different thick filament charges
assuming no hydration layer. The
doubly-hatched region is as above
but assuming an hydration layer
1 nm thick around each filament.

Fig. 5. A-band lattice spacing
(C) as a function of
externally applied
osmotic pressure (P) in
chemically-skinned frog
sartorius muscle at
short sarcomere lengths
(1.9 - 2.5 μm).
(○) relaxed muscle;
(●) muscle in rigor.
Line is the electrostatic
pressure curve calculated
as in Fig 4 using a
thick filament charge of
32 e/nm.

link the actin and tropomyosin strands to form small bumps on the filament
surface every 40 nm along the filament length. The thick filaments contain
the protein myosin, which has a pair of enzymatic, globular "heads" attached
to a coiled-coil α-helical "tail". Most of the "tail" forms a backbone to
the filament about 15 nm in diameter, with the enzyme-containing "heads"

projecting out from the surface to a diameter about double that of the backbone.

In rigor muscle, the myosin "heads" are attached to actin, linking the filaments together, whereas in relaxed muscle the filaments are not connected. During contraction, it is generally believed that the projecting "heads" cycle through attached and detached states, swinging longitudinally during the attached phase to produce a sliding of the filaments past one another (Huxley and Brown, 1967). When stretched beyond a certain length, the filaments no longer overlap (and the muscle can no longer contract). In vertebrate muscle, this occurs at a sarcomere length of 3.6 μm. Each sarcomere (the repeating unit along the muscle fiber's length) contains one thick and two thin filament lengths of 1.6 and 1.0 μm respectively. Thus, depending on sarcomere length, each A-band will be a mixture of a "thick-thin filament" array and a "thick filament only" array. At short sarcomere lengths the former will be predominant; at sarcomere lengths greater than 3.6 μm only the thick filament array will be found.

In muscle from which the external cell membranes have been removed by chemical-skinning or glycerol-extraction, thus exposing the interior of the fibers to the external solution, one can osmotically compress the A-band in the same way as TMV gels. Sharp equatorial x-ray reflections characteristic of an hexagonal filament lattice are observed, from which the lattice spacing (and the electron density across the A-band cross-section) can be determined. Increasing the external osmotic pressure with dextran causes the lattice to shrink as shown in Fig. 5.

In relaxed muscles (i.e. in solutions containing ATP but no calcium, where the filaments are not linked by the projections or crossbridges) three regions of the pressure/spacing curve can be distinguished (Fig. 5, open circles). At pressures above 20 torr, there is a nearly linear relationship between the logarithm of pressure and the lattice spacing. The experimental data lie close to the electrostatic pressure curve calculated for a thick filament charge diameter of 26 nm, if filament charges given by Bartels and Elliott (1985) and other parameters appropriate to the experimental conditions are used. In this region of the curve, the magnitude of the slope of the electrostatic pressure curve is greater than that of the experimental data by a factor of about 1.6, similar to the difference in slope with simpler systems (Table 1). Stretching the muscle so that no thin filaments are left in the A-band (i.e. sarcomere length greater than 3.6 μm) causes a sharp drop in lattice spacing (by 7 nm), exactly equal to the change predicted in the electrostatic pressure curve on removing the thin filaments from the lattice. Thus in this high pressure regime, it appears that electrostatic forces are the dominant repulsive force in the lattice and that the filaments can be reasonably approximated by smooth cylinders.

At lower pressures, however, there is a much greater difference between the experimental and calculated curves. From 10 to 20 torr, there is a large (~5 nm) change in lattice spacing for very little change in pressure, whereas below 10 torr almost no change in spacing is observed (Fig. 5). In the latter case, there appears to be a balance between electrostatic and van der Waals forces: calculations of van der Waals forces using the procedure of Brenner and McQuarrie (1973) give an attractive force much the same as the estimated (repulsive) electrostatic force.

In the middle pressure range (10 to 20 torr), the data tell us that the lattice can be easily compressed. In relaxed muscle at these spacings, the only likely repulsive forces are electrostatic. But filaments with the diameters assumed above (i.e. thick filament charge diameter of 26 nm) do not give sufficiently large electrostatic forces whatever charge is assumed (cf. Fig. 4). Thus, the charge diameter must be changing. As the external pressure is decreased from 20 to 10 torr, the charge diameter increases by about 5 nm. Given the size of the charge diameter and the amount by which it changes, we believe that this change represents a collapse of the projections against the thick filament backbone. From other x-ray diffraction experiments (e.g. Huxley, 1968), it is known that the projections can move radially under certain conditions (e.g. change of state from relaxation to rigor). Thus here we seem to have another example of radial movement of the projections: i.e. in response to external pressure.

The A-band of muscle in rigor (i.e. in a solution without ATP and where thick and thin filaments are linked in the overlap region of the sarcomere) shows a very different relationship between pressure and lattice spacing (Fig. 5, filled circles). At higher pressure (above 30 torr), the spacing in rigor muscle is greater than in relaxed muscle. At pressures below about 20 torr, the reverse is true. Only at a pressure of about 25 torr are the spacings under the two conditions the same. Interestingly, this spacing is very close to that found in the muscle in vivo (Millman, 1981), where the intact cell membranes exert a small compressive pressure on the lattice (Matsubara and Elliott, 1972). But even in rigor, where the thick and thin filaments are linked, the dominant repulsive force in the lattice appears to be electrostatic. Changes in ionic strength of the external solution produce changes in the lattice spacings very similar to those predicted from electrostatic theory (Millman et al., 1983), and the slope of the experimental pressure/spacing curve at higher pressures (i.e. > 100 torr) is similar to that calculated for electrostatic pressure (Fig. 5).

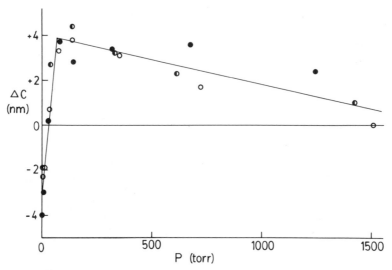

Fig. 6. Difference in A-band lattice spacing between muscles in relaxing and rigor solutions [ΔC = C(rigor − C(relaxed)] as a function of external pressure. Filled circles: rabbit psoas muscle at short sarcomere lengths. Half circles: frog sartorius muscle at short sarcomere lengths. Open circles: frog semitendinosus muscle at saarcomere lengths >3.6 μm. The line is the curve calculated for thick filament compressibilities of 5×10^{-13} and 10^{-14} m^3/N under relaxed and rigor conditions respectively.

Differences between relaxed and rigor conditions in the pressure/spacing relationship similar to those illustrated in Fig. 5 are found in frog muscle stretched beyond filament overlap and in rabbit psoas muscle at short sarcomere lengths. Thus the same differences between the relaxed and the rigor A-band are observed in different vertebrate striated muscles and are independent of the presence of thin filaments in the lattice. A graph of the difference between rigor and relaxed A-band spacings as a function of external pressure describes the same curve for all three muscle preparations (Fig. 6). This graph implies that the projections in rigor muscle have much greater radial stiffness than those in relaxed muscle, although the projections can move through a greater radial distance in relaxed as compared to rigor muscle. Quantitative analysis of the data of Fig. 6, assuming linear elasticities, gives extents of compression of 2.5 nm and 4.0 nm and compressibilities of 10^{-14} and 5×10^{-13} m^3/N for the rigor and relaxed muscle respectively. It should be noted that these compressibilities apply to the projections themselves, independent of whether or not they form links (or crossbridges) between the thick and thin filaments.

One can ask further what causes the change in radial compressibility of the projections between the relaxed and rigor states. It is still too early to answer this question definitely, but some clues have already come from experiments where the pH of the external solution was varied and the effect on A-band lattice spacing was studied. As the pH was increased or decreased (from 7.0) in either the relaxed or rigor states, the lattice spacing changed dramatically; decreasing as the pH was lowered to 5.5, increasing as the pH was raised to 8.5 (Fig.7). The total spacing change was more than 10 nm and occurred over the whole range of external pressures.

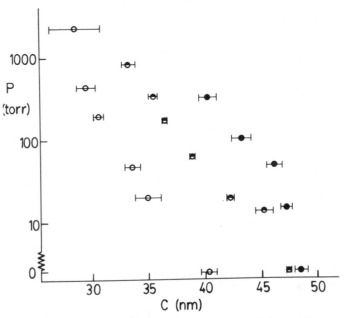

Fig. 7. A-band lattice spacing (C) (averages with standard errors) as a function of externally applied pressure (P) in relaxed frog sartorius muscle with conditions as in Fig. 5 (relaxed), except for the pH which was 5.5 – open circles; 7.0 – half circles (same data as Fig. 5); 8.5 – filled circles.

Such spacing changes cannot be accounted for by changes in filament charge alone (see Fig. 4), but imply substantial changes in charge diameter. Increasing the pH (which increases the filament charge) appears to shift the projections outwards from the backbone, thus increasing the effective thick filament charge diameter. Such an idea has already been suggested from biochemical studies (Ueno and Harrington, 1981). It seems plausible that under in vivo conditions, as well, the thick filament projections may be shifted away from the filament backbone towards the thin filaments where interaction can occur more readily. Such a shift in position is known to occur during the change from the relaxed state into either the rigor or contracting states (Huxley, 1968; Haselgrove and Huxley, 1973). Thus, electrostatic forces may have a major role to play in the contractile process itself.

In summary, it appears that electrostatic forces provide the major component of repulsive pressure which stabilizes the filament lattice in the A-band of striated muscle. Analysis of osmotic compression data in comparison with calculated electrostatic pressure curves gives strict limits on the position of the filament charge (i.e. the filament charge diameter) which in conjunction with knowledge of the mass density across the lattice, can be used to determine the affective position of charge in filament structure. As well, changes in charge position can be correlated with changes in lattice structure as the state of the muscle is changed. In particular, increase in charge on the thick filament backbone may be coupled to movement of parts of the thick filament – the projections – away from the backbone towards the thin filaments where interaction leading to contraction can take place.

ACKNOWLEDGEMENTS

I am grateful to Dr. Gerald Elliott and Tom Irving for helpful discussions and to Rosalind Blair for technical assistance. This work was done with grant support from the Natural Science and Engineering Council of Canada.

REFERENCES

April, E. W. , 1975, Liquid-crystalline characteristics of the thick filament lattice of striated muscle, Nature (Lond.), 257:139.

Bartels, E. M., and Elliott, G. F., 1985, Donnan potentials from the A and I bands of glycerinated and chemically skinned muscles, relaxed and in rigor, Biophys. J., in press.

Bernal, J. D., and Fankuchen, I., 1941, X-ray and crystallographic studies of plant virus preparations, J. Gen. Physiol., 25:111.

Brenner, S. L., and McQuarrie, D. A., 1973, Force balances in systems of cylindrical polyelectrolytes, Biophys. J., 13:301.

Egelman, E. H., and Padron, R., 1984, X-ray diffraction evidence that actin is a 100 A filament, Nature (Lond.), 307:56.

Elliott, G. F., 1968, Force-balances and stability in hexagonally-packed polyelectrolyte systems, J. Theor. Biol., 21:71.

Elliott, G. F., and Lowy, J., 1968, Organization of actin in a mammalian smooth muscle, Nature (Lond.), 219:156.

Gruen, D. W. R., and Marcelja, S., 1983, Spatially-varying polarization in H_2O: a model for the electric double layer and the hydration force, J. Chem. Soc. Faraday Trans., 2, 79(2):225.

Haselgrove, J. C., and Huxley, H. E., 1973, X-ray evidence for radial crossbridge movement and for the sliding filament model in actively contracting skeletal muscle, J. Mol. Biol., 77:549.

Huxley, H. E., 1968, Structural difference between resting and rigor muscle; evidence from intensity changes in the low-angle equatorial x-ray diagram, J. Mol. Biol., 37:507.

Huxley, H. E., and Brown, W., The low-angle x-ray diagram of vertebrate
    striated muscle and its behaviour during contraction and rigor,
    J. Mol. Biol., 30:383.
Irving, T. C., and Millman, B. M., 1983, Interfilament forces in
    vertebrate striated muscle at long sarcomere lengths, Biophys. J.,
    41:253a.
LeNeveu, D. M., Rand, R. P., and Parsegian, V. A., 1976, Measurement of
    forces between lecithin bilayers, Nature (Lond.), 259:601.
Lis, L. J., Parsegian, V. A., and Rand, R. P., 1981, Detection of the
    binding of divalent cations to dipalmitolyphosphatidylcholine
    bilayers by its effect on bilayer interaction, Biochem., 20:1761.
Loosley-Millman, M. E., Rand, R. P., and Parsegian, V. A., 1982, Effects of
    monovalent ion binding and screening on measured electrostatic
    forces between charged phospholipid bilayers. Biophys. J., 40:221.
Lowy, J., and Vibert, P. J., 1967, Structure and organization of actin in a
    molluscan smooth muscle, Nature (Lond.), 215:1254.
Matsubara, I., and Elliott, G. F., 1972, X-ray diffaction studies on
    skinned single fibres of frog skeletal muscle, J. Mol. Biol., 72:657.
Miller, A., and Woodhead-Galloway, J., 9171, Long range forces in muscle,
    Nature (Lond.), 229:470.
Millman, B. M., 1981, Filament lattice forces in vertebrate striated
    muscle: relaxed and in rigor, J. Physiol., 320:118P.
Millman, B. M., and Bell, R. M., 1983, Lateral forces between actin
    filaments in muscle, Biophys. J., 41:253a.
Millman, B. M., and Bennett, P. M., 1976, Structure of the cross-striated
    adductor muscle of the scallop, J. Mol. Biol., 103:439.
Millman, B. M., Irving, T. C., Nickel, B. G., and Loosley-Millman, M. E.,
    1984, Interrod forces in aqueous gels of tobacco mosaic virus,
    Biophys. J., 45:551.
Millman, B. M., and Nickel, B. G., 1980, Electrostatic forces in muscle and
    cylindrical gel systems, Biophys. J., 32:49.
Millman, B. M., Wakabayashi, K., and Racey, T. J., 1983, Lateral forces in
    the filament lattice of vertebrate striated muscle in the rigor
    state, Biophys. J., 41:259.
Morel, J. E., and Gingold, W. P., 1979, Does water play a role in the
    stability of the myofilament lattice and other filament arrays?, in:
    "Cell associated water," W. Drost-Hansen and J. Clegg, ed., Academic
    Press, London.
Parsegian, V. A., and Brenner, S. L., 1976, The role of long range forces
    in ordered arrays of tobacco mosiac virus, Nature (Lond.), 259:635.
Rand, R. P., and Parsegian, V. A., 1984, Physical force considerations in
    model and biological membranes, Can. J. Biochem. Cell Biol., 62:752.
Rome, E., 1967, Light and x-ray diffraction studies of the filament lattice
    of glycerol-extracted rabbit psoas muscle, J. Mol. Biol., 27:591.
Rome, E., 1968, X-ray diffraction studies of the filament lattice of
    striated muscle in various bathing media, J. Mol. Biol., 37:331.
Rome, E., 1972, Relaxation of glycerinated muscle: low-angle x-ray
    diffraction results, J. Mol. Biol., 65:331.
Shear, D. B., 1970, Electrostatic forces in muscle contraction,
    J. Theor. Biol., 28:531.
Squire, J., 1981, "The structural basis of muscular contraction," Plenum
    Press, New York.
Ueno, H., and Harrington, W. F., 1981, Cross-bridge movement and the
    conformational state of the myosin hinge in skeletal muscle,
    J. Mol. Biol., 149:619.

CONTRIBUTORS

Robert A. Aldoroty
The Mount Sinai Medical Center
New York, NY 10032

Ernest W. April
College of Physicians and Surgeons
Columbia University
New York, NY 10032

Laura S. Bakas
Instituto De Investigaciones
 Fisicoquimicas Teoricas y Aplicadas (INIFTA)
C.C. 16-Suc 4-1900
La Plata, Argentina

E. Bamberg
Max Planck Institute for Biophysics
Frankfurt
West Germany

E. M. Bartels
Open University
Oxford Research Unit
Foxcombe Hall
Berkeley Road
Boars Hill
Oxford, OX1 5HR
England

Martin Blank
Biological Sciences Division
Office of Naval Research
Arlington, VA 22217

John O'M. Bockris
Department of Chemistry
Texas A&M University
College Station, TX 77843

James D. Bond
Science Applications
 International Corporation
1710 Goodridge Drive
McLean, VA 22102

John A. DeSimone
Department of Physiology and Biophysics
Virginia Commonwealth University
Medical College of Virginia
Richmond, VA 23298

Shirley K. DeSimone
Department of Physiology and Biophysics
Virginia Commonwealth University
Medical College of Virginia
Richmond, VA 23298

E. Anibal Disalvo
Instituto De Investigaciones
 Fisicoquimicas Teoricas y Aplicadas (INIFTA)
C.C. 16-Suc 4-1900
La Plata, Argentina

G. F. Elliot
Open University
Oxford Research Unit
Foxcombe Hall
Berkeley Road
Boars Hill
Oxford, OX1 5HR
England

William K. Elwood
Departments of Anatomy and Statistics
University of Kentucky
A. B. Chandler Medical Center
Lexington, KY 40536

Eugene Findl
15730 Hartsook Street
Encino, CA 91436

Gunter Fuhr
Department of Biology
Humboldt-Universitat zu Berlin
Berlin, GDR

A. Georgallas
Department of Physics
Queen's University
Kingston, Ontario
Canada K7L 3N6

Roland Glaser
Department of Biology
Humboldt-Universitat zu Berlin
Berlin, GDR

I. S. Graham
Department of Physics
McGill University
Montreal, Quebec
Canada H3A 2T8

Howard Guthermann
BioResearch, Inc.
31 Rochester Road
Newton, MA 02158

M. Ahsan Habib
Department of Chemistry
Texas A&M University
College Station, TX 77843

Gerard L. Heck
Department of Physiology and Biophysics
Virginia Commonwealth University
Medical College of Virginia
Richmond, VA 23298

Felix T. Hong
Department of Physiology
Wayne State University
Detroit, MI 48201

R. A. Hughes
Open University
Oxford Research Unit
Foxcombe Hall
Berkeley Road
Boars Hill
Oxford, OX1 5HR
England

Kung-ming Jan
Departments of Physiology and Medicine
Columbia University
College of Physicians and Surgeons
New York, NY 10032

K. Janko
Physiology Institute
University of Saarlandes
L-6650 Homburg/Saar
West Germany

J. J. Kaufman
Department of Orthopaedics
Bioelectrochemistry Laboratory
Mount Sinai School of Medicine
One Gustave L. Levy Place
New York, NY 10029

Pawel Krysinski
Department of Physiology
Membrane Biophysics Lab
Michigan State University
East Lansing, MI 48824

Robert J. Kurtz
BioResearch, Inc.
315 Smith Street
Farmingdale, NY 11735

Jan Kutnik
Institute of Physics
Maria Curie-Sklodowska University
Lublin, Poland

Z. K. Lojewska
Department of Chemistry
University of Warsaw
Warsaw, Poland

P. C. Maccaro
Department of Orthopaedics
Bioelectrochemistry Laboratory
Mount Sinai School of Medicine
One Gustave L. Levy Place
New York, NY 10029

H. E. McKean
Departments of Anatomy and Statistics
University of Kentucky
A. B. Chandler Medical Center
Lexington, KY 40536

Israel R. Miller
Department of Membrane Research
The Weizmann Institute of Science
Rehovot, Israel

Barry M. Millman
Biophysics Interdepartmental Group
Physics Department
University of Guelph
Guelph, Ontario
Canada, N1G 2W1

Shinpei Ohki
Department of Biophysical Sciences
School of Medicine
State University of New York at Buffalo
Buffalo, NY 14214

Hiroyuki Ohshima
Department of Biophysical Sciences
School of Medicine
State University of New York at Buffalo
Buffalo, NY 14214

T. L. Okajima
Department of Physiology
Wayne State University School of Medicine
Detroit, MI 48201

A. A. Pilla
Department of Orthopaedics
Bioelectrochemistry Laboratory
Mount Sinai School of Medicine
One Gustave L. Levy Place
New York, NY 10029

R. Reinhardt
Physiology Institute
Universtiy of Saarlandes
L-6650 Homburg/Saar
West Germany

Agnes Rejou-Michel
Department of Chemistry
Texas A&M University
College Station, TX 77843

S. M. Ross
MRC Group in Periodontal Physiology
University of Toronto
Toronto, Ontario
Canada, M5S 1A8

J. T. Ryaby
Department of Orthopaedics
Bioelectrochemistry Laboratory
Mount Sinai School of Medicine
One Gustave L. Levy Place
New York, NY 10029

R. Schmukler
Department of Orthopaedics
Bioelectrochemistry Laboratory
Mount Sinai School of Medicine
One Gustave L. Levy Place
New York, NY 10029

Shlomoh Simchon
Department of Physiology
Columbia University
College of Physicians and Surgeons
New York, NY 10032

Stephen D. Smith
Departments of Anatomy and Statistics
University of Kentucky
A. B. Chandler Medical Center
Lexington, KY 40536

H. Ti Tien
Department of Physiology
Membrane Biophysics Lab
Michigan State University
East Lansing, MI 48824

V. S. Vaidhyanathan
Department of Biophysical Sciences
School of Medicine
State University of New York at Buffalo
Buffalo, NY 14221

N. Convers Wyeth
Science Applications International Corporation
1710 Goodridge Drive
McLean, VA 22102

M. J. Zuckermann
Department of Physics
McGill University
Montreal, Quebec
Canada H3A 2T8